Dictionary of Water Engineering

Dictionary of Water Engineering

Ken Nelson

With contributions from Charles Kerr
and Robert Legg

ITDG
PUBLISHING

Published by ITDG Publishing
Schumacher Centre for Technology and Development
Bourton Hall, Bourton-on-Dunsmore, Warwickshire CV23 9QZ, UK
www.itdgpublishing.org.uk

Originally published by Butterworths as the
Dictionary of Water and Water Engineering in 1973
This edition first published in 2005

ISBN 1 85339 490 4 hardback

A catalogue record for this book is available from the British Library.

ITDG Publishing is the publishing arm of the Intermediate Technology Development Group. Our mission is to build the skills and capacity of people in developing countries through the dissemination of information in all forms, enabling them to improve the quality of their lives and that of future generations.

Typeset by The Studio Publishing Services Ltd, Exeter, Devon EX4 8JN
Printed in Great Britain

Preface

Recent years have seen marked progress in the development of water science and technology, notably through greater emphasis on the needs of poorer communities and on environmental sustainability. Since the start of the Water Decade of the 1980s these changes have been most apparent in the proliferation of water and sanitation programmes; these must continue indefinitely if we are to conserve and improve water resources for all.

The advances in water engineering have required a review and updating of the earlier edition, resulting in the deletion of a few terms, the amendment of others, and the addition of new terms, notably terms appropriate for low-cost urban and community water supplies and wastewater systems. General engineering and management terms are outside the scope of this dictionary, as are trade names and manufacturing processes. This edition has been designed to meet the needs of engineers, technicians and students throughout the world. Above all, it is intended to provide a practical handbook for all those involved in sustainable programmes, from planners to field workers.

I am grateful to the Publishers for their valuable assistance with this new edition – in particular, for making available contributions in their specialist fields by Robby Legg (water resources and hydrology) and Charles Kerr (water supply and wastewater). I also wish to acknowledge the help given by the Australian National Committee on Large Dams and my many Australian colleagues.

Ken Nelson

Cross-references

These are indicated in the text by the use of small capital letters, for example *See* CLINOMETER. Sometimes a cross-reference given in the plural refers to an entry in the singular, and vice versa.

Acknowledgements

The publishers would like to thank the following organizations for their kind permission to use photographic images in this book.

ACWa Services Ltd. for Figure R.5.
AH Harris Ltd for Figures E.2. and H.3.
ITDG for Figure S.7.
Royal Meteorological Society for Figure R.2.
SANDEC for Figure S.3.
HR Wallingford for Figures F.3., P.3. and S.6.
WEDECO UV Systems plc for Figure U.1.

Authors

Ken Nelson ED, BSc, CEng, EurIng, FIEAust, FGS, MICE

Ken Nelson studied civil engineering at Cardiff University, where he was awarded the Colonel Page Prize in Engineering. He then enlisted in the RAF, but was later commissioned in the Army. Three years of active service in South East Asia included two years in Burma. He then returned to Cardiff University and completed a degree course. In 1950 he left for Australia, where he worked on irrigation, water supply and river improvement schemes, mostly in Victoria. He was employed as an Executive Engineer on various projects, and finally as Engineer-in-Charge, Farm Water Supplies. Since retiring from full-time work he has worked as a consulting engineer and technical writer.

Charles Kerr OBE, BSc(Eng), BSc(Econ), CEng, FICE, FCIWEM

Charles Kerr was a civil engineer and a specialist in water supply. After war service in the RAF he graduated BSc(Hons) in civil engineering from the University of Cape Town. Later he obtained a BSc in economics from the University of London. He spent his career working both for major firms of consulting engineers and in the water supply industry. For nearly 15 years he was the Engineer and Manager of the water supply authority serving a city in Central Africa. In recent years, and particularly following retirement from full-time consultancy work, he become well-known as an expert in low-cost water supply. During his long association with ITDG he made major contributions as a technical advisor, editor and author. For the last 15 years of his life he was technical editor of *Waterlines*, the journal for low-cost water supply and sanitation. Charles Kerr died in 2001 while this dictionary was in preparation.

Robert Legg MSc, DIC, BSc(Hons), EurIng, CEng, FICE, FCIWEM, FRMetS

Robert Legg graduated in Civil Engineering from the University of St Andrews, and, after serving in the British Army in the Far East, later did postgraduate work in London at Imperial College. Nearly all his career has been spent with consulting engineers, mainly on work related to water resources, hydraulic structures and applied hydrology. His work has taken him to many parts of the world, including Canada, Indonesia, USA, Central Africa and the Western Arctic. Robert Legg is a member of the British Dams Society and the British Hydrological Society.

Notes on units

Length

	mm	*m*	*yd*	*ft*	*in*	*km*	*miles*
1 mm	1	0.001	1.094×10^{-3}	3.281×10^{-3}	0.0394	1×10^{-6}	6.214×10^{-7}
1 m	1000	1	1.0936	3.281	39.37	1×10^{-3}	6.214×10^{-4}
1 yd	914	0.9144	1	3	36	9.144×10^{-4}	5.68×10^{-4}
1 ft	305	0.3048	0.3333	1	12	3.048×10^{-4}	1.894×10^{-4}
1 in	25.4	0.254	0.0278	0.0833	1	2.54×10^{-5}	1.578×10^{-5}
1 km	1.0×10^{6}	1000	1094	3281	3.937×10^{4}	1	0.6214
1 mile	1.609×10^{4}	1609	1760	5280	6.336×10^{4}	1.609	1

1 nautical mile (UK) = 1.853 km 1 degree of latitude is about 111 km

Area

	m²	*ft²*	*acre*	*hectare*
1 m^2	1	10.76	2.471×10^{-4}	1.0×10^{-4}
1 ft^2	9.29×10^{-2}	1	2.29×10^{-5}	9.29×10^{-6}
1 acre	4.047×10^{3}	4.356×10^{4}	1	4.047×10^{-1}
1 hectare	1.0×10^{4}	1.076×10^{5}	2.471	1

Volume

	m³	*L*	*UK gal*	*US gal*	*ft³*
1 m^3	1	1.0×10^{3}	2.200×10^{2}	2.642×10^{2}	35.32
1 L	1.0×10^{-3}	1	0.220	0.2642	3.532×10^{-2}
1 UK gal	4.546×10^{-3}	4.546	1	1.200	0.1605
1 US gal	3.785×10^{-3}	3.785	0.8326	1	0.1337
1 ft^3	2.832×10^{-2}	28.32	6.229	7.480	1

1 acre foot = 1.234 megalitres 1 cubic yard = 0.765 cubic metres

Discharge (Volume rate of flow)

	L/s	*m³/s*	*m³/d*	*UK gal/day*	*ft³/s*
1 L/s	1	1.0×10^{-3}	86.40	1.901×10^{4}	3.531×10^{-2}
1 m^3/s	1.0×10^{3}	1	8.640×10^{4}	1.901×10^{7}	35.313
1 m^3/d	1.157×10^{-2}	1.157×10^{-5}	1	2.200×10^{2}	4.087×10^{-4}
1 UK gal/day	5.262×10^{-5}	5.262×10^{-8}	4.546×10^{-3}	1	1.858×10^{-6}
1 ft^3/s	28.32	2.832×10^{-2}	2.447×10^{3}	5.382×10^{5}	1

Note: the cubic metre per second (m^3 /s) is also known as the 'cumec'
the cubic foot per second (ft^3 /s) is also known as the 'cusec'

Velocity

	m/s	*km/h*	*mile/h*	*ft/s*
1 m/s	1	3.6	2.237	3.281
1 km/h	0.2778	1	0.6214	0.9113
1 mile/hour	0.447	1.609	1	1.467
1 feet/second	0.3048	1.097	0.6818	1

Mass

	kg	*lb*	*t*	*UK ton*
1 kg	1	2.205	1×10^{-3}	9.842×10^{-4}
1 lb	0.454	1	4.536×10^{-4}	4.464×10^{-4}
1 metric tonne (t)	1000	2205	1	0.984
1 UK ton	1016	2240	1.016	1

1 US ton = 2000 lb

Density

	g/cm³	*kg/m³*	*lb/in³*	*lb/ft³*
1 g/cm^3	1	1000	0.0361	62.43
1 kg/m^3	1×10^{-3}	1	3.61×10^{-5}	0.0624
1 lb/in^3	27.68	27.68×10^3	1	1728
1 lb/ft^3	0.016	16.02	5.787×10^{-4}	1

Pressure

	kgf/cm²	*bar*	*kN/m²*	*lbf/in² (psi)*
1 kgf/cm^2	1	0.981	98.1	14.223
1 bar	1.02	1	100	14.504
1 kN/m^2	0.01	0.0098	1	0.145
1 lbf/in^2 (psi)	0.07	0.0689	6.89	1

1 Pa (pascal) = 1 N/m^2
1 N/mm^2 = 1 MN/m^2 = 1 MPa
101 325 Pa = 1 standard atmosphere (atm) = 1.01325 bar
100 kPa = 1 bar
10.33 m head of water = 1 atm
2989 Pa = 1 ft head of water = 22.42 mm of mercury (mmHg)
1 mmHg = 0.0394 inch of mercury (inHg)
1 MPa = 145 lbf/in^2 (psi)

Metric prefixes

Prefix	*Symbol*	*Factor by which the unit is multiplied*
giga	G	1 000 000 000
mega	M	1 000 000
kilo	k	1 000
hecto	h	100
deca	da	10
deci	d	0.1
centi	c	0.01
milli	m	0.001
micro	μ	0.000 001

A

abandoned shore line A seashore abandoned by depression of the sea level or by uplift of the land; a lake shore abandoned by lowering of the entire lake by evaporation or drainage, or by differential tilting of the land, resulting in the exposure of the lake floor at one end or the other. *See* COAST OF EMERGENCE.

ablation The waste of a glacier by melting; the removal of surface rocks by water action.

Abney level A hand instrument for taking angles or levels along steep slopes, frequently used for the preliminary rough levelling for small irrigation and water supply schemes. *See* CLINOMETER.

abrasion The action of natural agents, particularly flowing water, in wearing away soils and rocks. The rock fragments may be roughly rounded and abrade the beds of streams. *See* EROSION.

absolute density Mass per unit volume of particles (sand, soil or gravel), excluding VOIDS. *See* DENSITY.

absolute humidity *See* HUMIDITY.

absolute pressure The intensity of pressure above absolute zero. *See* GAUGE PRESSURE.

absorbed water Water held mechanically in a soil or rock and possessing physical properties which do not differ from ordinary water at the same pressure and temperature. *See* ADSORBED WATER.

absorbing well (or **well drain** or **drain well**) A well excavated or drilled mainly for drainage purposes. *See* DRAINAGE WELL, VERTICAL SAND DRAIN, WELLPOINT.

absorption A process in which a liquid or gas is taken into the body of another. *See* ADSORPTION, SORPTION.

absorption, flood *See* FLOOD ABSORPTION.

absorption loss The quantity of water absorbed by the rocks and soils in the wetted area during the initial filling of a dam, pond or canal. *See* SEEPAGE LOSS.

absorption (water) The sucking in or imbibing of water; the quantity of water which a rock or soil will incorporate, usually expressed in percentage terms of the original dry weight. A rock of low porosity can absorb or retain but little water, while a rock of high porosity may retain a large quantity. There may be wide variations even for a rock of one type. Absorption is affected by soil texture, surface slope, rate of precipitation and air temperature. Water may be absorbed from rainfall or indirectly from streams, although normally groundwater moves towards the streams.

abstraction The removal of water from any source, normally for the use or benefit of humankind.

abutments (of a dam) The zone of rock or soil adjacent to and supporting each end of a dam; in the case of a concrete or masonry dam the abutments transfer a portion of the horizontal thrust onto the rock sides of a valley. *See* ARCH DAM.

abutment wall *See* WING WALL.

Abyssinian tube well A pointed, perforated WELLPOINT which is driven into the water-bearing ground by ramming with a light pile hammer or by sledge hammer; water is extracted by pumping.

accelerated erosion The erosion of soil at a faster rate than that prevailing under normal or natural conditions; it may be caused by excessive rainfall, or by human activities such as removal of root-bound surface layers of soil. *See* SHEET EROSION.

access manhole A MANHOLE provided to give access for cleaning, internal inspection and maintenance. On a large pipeline a convenient position is at a hollow or summit and near main valves.

access road A track constructed to enable plant, equipment, materials and labour to reach a construction site. Frequently, major works such as dams are built in isolated regions and consequently the provision of an access road can be important in cost and time.

accidental lake A lake formed in a drainage area in which the outlet has been blocked by debris, such as a landslide. The lake may be large or small and is frequently of short duration. *See* LANDSLIDE LAKE.

accretion The building up of sediments from any cause as by stream action; the accumulation resulting from this process. *See* ALLUVIATION.

acidification (or **acidizing**) The use of acid, usually hydrochloric, to increase the water supply from a borehole which is failing owing to encrustations on screens and slotted pipes. It may also be used to improve the yield from bores in chalk or limestone by enlarging the fissures. *See* CALGON METHOD.

acidity The capacity of a water to react with hydroxyl ions (neutralize alkalinity) expressed quantitatively as mg/L of equivalent calcium carbonate. *See* ALKALINITY, PH VALUE.

acid rain Precipitation having a low pH, derived from atmospheric pollution, predominantly caused by emissions of sulphur dioxide or oxides of nitrogen from the burning of fossil fuels.

acid soil A 'sour' soil with a PH VALUE below 7.0.

acid water Water containing traces of free sulphuric acid, mainly by contact with iron pyrites. Some subsurface waters are corrosive and contain a high proportion of solids, particularly the sulphates of iron. *See* TOXIC WATER.

acre A US and British unit of area, frequently used to express the size of irrigation and catchment areas. Equal to 4840 square yards (or 0.405 hectares).

acre foot The volume of water which will cover 1 acre 1 foot deep, equal to 43 560 ft^3 (1233.5 m^3). It is a unit of measurement for reservoir and pond capacity, also irrigation storages (domestic or town water supply reservoirs are usually measured in millions of gallons).

acre inch The quantity of water which will cover 1 acre 1 inch deep, equal to 3630 ft^3 or 22 600 gal or 102.8 m^3.

activated carbon Particles of carbon from which hydrocarbons have been removed to give a large surface area enclosing minute pores. This provides a high absorptive capacity which is used to remove odours and residues of organic matter from potable waters and sewage effluents. It may be regenerated by heating.

activated sludge Sludge floc, produced when sewage is continuously aerated, comprising bacteria, protozoa and other micro-organisms with inert debris. In the activated sludge process, sewage is treated by mixing, agitation and aeration with activated sludge, which is then settled from the treated effluent and recycled as required.

active layer The surface soil or deposit which undergoes volume changes with temperature, i.e. swelling when freezing, and shrinking when thawing and drying. Also applied to the uppermost part of the PERMAFROST which thaws during milder weather and becomes waterlogged and forms MUDFLOWS on sloping ground. Usually structural foundations are taken down below the active layer.

active method A construction method used in PERMAFROST areas. The frozen deposits are thawed prior to construction and kept thawed or removed; the foundation materials used are not subject to frost heave and settlement. *See* PASSIVE METHOD.

actual velocity *See* EFFECTIVE VELOCITY.

adhesive water (or **pellicular water**) Water retained in a soil mass by molecular attraction; it forms a coating around the particles and may move from one particle to another. *See* INTERSTITIAL WATER.

adit Usually a rectangular heading or tunnel, horizontal or inclined, for tapping groundwater. Adits are often driven from a shaft and may be lined or unlined. Adits often form part of a well system in which wells and boreholes are sunk from the surface through the adits or from the floor of the adits. In some cases the adits provide temporary storage for the water.

adsorbed water Water retained in a soil mass by physicochemical forces. It has physical properties which differ from chemically combined water, or mechanically held water at the same pressure and temperature. *See* ABSORBED WATER.

adsorption A process in which materials adhere to the surface of a body, used in wastewater treatment to remove dissolved materials. *See* ABSORPTION, SORPTION.

advanced water treatment Any treatment process, physical, chemical or biological, intended to achieve a higher standard than secondary treatment. *See* PRIMARY TREATMENT, SECONDARY TREATMENT, TERTIARY TREATMENT.

aeration (water treatment) A continuous process transferring oxygen to a water by spraying the water in the air, agitating the water, bubbling air through the water or allowing the water to flow over a weir.

aeration, normal *See* NORMAL AERATION.

aeration zone *See* ZONE OF AERATION.

aerial photograph A photograph taken from aircraft. A sequence of such photographs, taken at a definite height and after making a series of parallel flights, can be used for map construction and geological interpretation. The photograph is usually vertical, i.e. taken by a camera pointing directly downward, although oblique photographs are used for some purposes. Aerial photographs are frequently used to determine catchment size. *See* GEOGRAPHICAL INFORMATION SYSTEM, INFRA-RED PHOTOGRAPHY, MOSAIC, PHOTOGEOLOGY, PHOTOGRAMMETRIST, REMOTE SENSING.

aerobic digestion A biological process by which the organic content of sludge is partly oxidized (digested) by aeration. *See* ANAEROBIC DIGESTION.

affluent A tributary stream or system; a stream flowing into a river or lake.

afflux The variation between high-flood levels downstream and upstream at a weir; the rise of water level above normal level on the upstream side of a contraction or obstruction in a channel.

afforestation The planting of trees, shrubs and cuttings of same on gathering grounds to retard runoff during and after heavy rains or the melting of snow. Plantations reduce erosion and the transport of trash and debris into rivers and reservoirs. *See* TREE PLANTING.

A-frame groynes A modified form of CRUCIFORM GROYNES in which two spars, with a third as crossbar, form 'A'-shaped frames instead of crosses. Green willow poles are planted against the groynes, on the upstream side, with their feet rooted in the river bed. These are small groynes used for protection along low eroding banks and also as short training walls, under conditions with only minor floods.

age (of **stream** or **river**) The stage reached in a stream's development or growth. *See* GRADED STREAM, MATURE RIVER, OLD RIVER, PROFILE OF EQUILIBRIUM, YOUTHFUL STREAM.

aggradation *See* ALLUVIATION.

aggraded floodplain An ALLUVIAL FLAT which is frequently covered by floodwater of a MATURE RIVER. When the waters subside, the plain is covered with a new layer of sediment. *See* GRADED RIVER FLOODPLAIN.

aggregate The crushed stone, sand, or gravel which is bound together to make concrete. *See* CHIPPINGS.

agricultural drain (or **field tile, land drain** or **field drain**) Earthenware or porous concrete pipes of about 75 mm (3 in) internal diameter, laid with open joints end to end, for subsoil drainage. *See* FRENCH DRAIN, SUBSOIL DRAINAGE.

A-horizon The uppermost deposit of a soil in which the surface layer is usually held together by plant roots. The decomposition products of various minerals, fine clay particles and soluble constituents are washed down into a lower layer known as the B-horizon of the soil. The B-horizon may be iron-rich containing grains of hydrated iron oxide, or lime-rich containing much calcium carbonate. The soil-forming decomposed rock, without humus, beneath the B-horizon is the C-horizon.

air, damp *See* DAMP AIR.

air, dry *See* DAMP AIR.

air-entrained concrete A concrete containing an agent to entrain about 3.5% air. It is sometimes used in mass concrete for dams and earth dam cutoffs, as it gives marked improvements in the workability and cohesion of the mix. In the USA and some other countries, air-entrained concrete is favoured on account of its superior resistance to freezing and thawing.

air face (dam) The downstream face or side of a dam (Figure A.1). *See* WATER FACE.

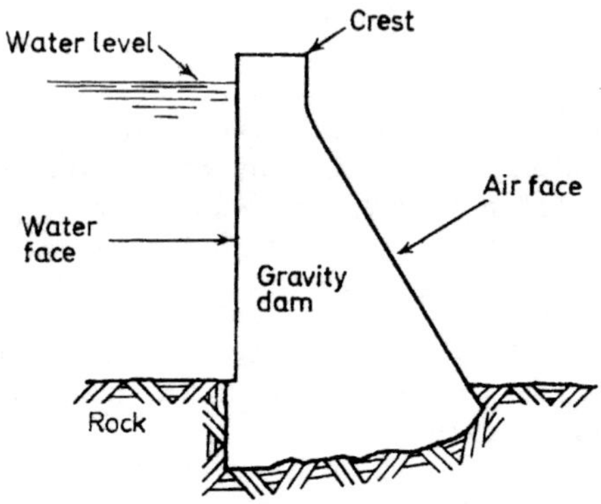

Figure A.1. Air face (dam)

air-lift pump An appliance (Figure A.2) for lifting water in a well or sump.

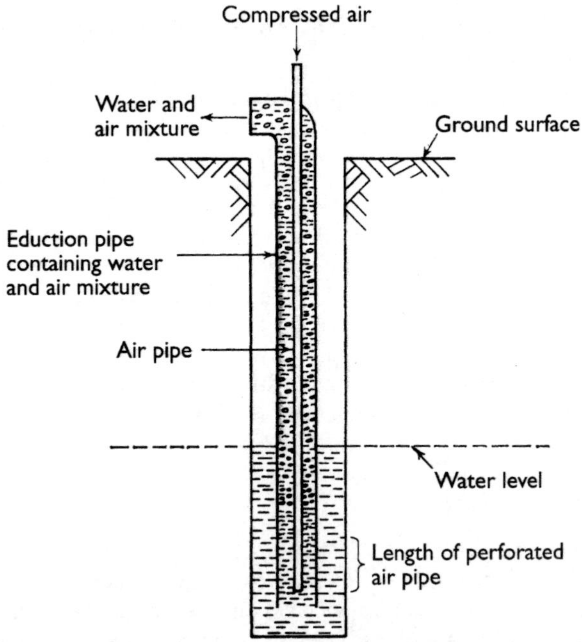

Figure A.2. Air-lift pump

It consists of an air compressor at the surface and two pipes hanging vertically, with one pipe usually within the other. The smaller pipe delivers compressed air to the depth at which water occurs. At this point a nozzle discharges the air into the free water and by aerating it causes its density to drop. This mixture of water and air is then forced upwards by the head of groundwater. Large volumes of dirty or gritty water can be removed in this way, but the consumption of compressed air is high. Air-lift pumps may be used to lift water into a reservoir or storage. *See* PULSOMETER PUMP, SUBMERSIBLE PUMP.

airline correction A correction required in measuring water depths to determine true depths. If large vertical angles are induced in log-lines by great depths, high velocities, inadequate sounding weight or any combination of these, two separate corrections are required: (1) airline correction for the section of line above the water surface and (2) wet line correction for the section under water.

air release valve A valve to release air or gas which tends to accumulate at high points on pipelines.

airspace ratio The ratio of (1) volume of water that can be drained, under the force of gravity, from a saturated soil sample to (2) total volume of voids.

air test A test used to check the condition of a pipeline or sewer by pressurizing sections internally with air; the rate of air loss provides an indication of watertightness. *See* INFILTRATION WATER.

air valve A valve, usually at the high points of a pipeline relative to the hydraulic gradient, to allow air to escape but not water; any valve to discharge air when pipe mains are being filled, or to admit air when they are being emptied.

air vessel A small air chamber fixed to the pipeline on the discharge side of a RECIPROCATING PUMP to act as a cushion to minimize the pulsations of the pump.

algal bloom The development of masses of algae (primitive aquatic plants) in a body of water, such as a lake or reservoir, where nutrient enrichment due to nitrogen and/or phosphorus has caused eutrophication. *See* ALGAL CONTROL, ALGICIDE, EUTROPHICATION.

algal control The treatment of water with chemicals to suppress algae. The most widely used algicide is copper sulphate (bluestone): the dose ranges from 0.05 to 10 mg/L, depending on the type of organism. *See* ALGAL BLOOM, ALGICIDE, COPPER SULPHATE.

algicide A chemical used to kill or control algal growths. *See* ALGAL BLOOM, ALGAL CONTROL.

alignment chart *See* NOMOGRAM.

alkaline soil A non-acid soil with a PH VALUE above 7.0.

alkalinity The capacity of a water to neutralize acids expressed in mg/L of equivalent calcium carbonate. *See* ACIDITY, PH VALUE.

Allen's method *See* SALT VELOCITY METHOD.

alluvial cone A deposit of sediment similar to an ALLUVIAL FAN but with steeper slopes; the material is usually finer than that in a DEBRIS CONE.

alluvial deposit Sediment deposited by streams; a deposit of sand or gravel formed along lengths of a river where the water velocity has diminished. *See* COMPETENCE.

alluvial fan A low fan-shaped heap of sediment deposited by a river when its gradient is suddenly diminished on entering a plain or open valley floor; deposits range from small heaps to large areas covering many square kilometres. Alluvial fans sometimes present engineering problems: culverts and bridges may be made useless, and roads or pipelines may be cut or deeply covered by debris.

alluvial flat A deposit of mud and silt in the lower course of a mature river; the sediment may be up to 9.1 m (30 ft) or more thick.

alluvial river A river that flows in a channel carved in its own deposits of alluvium. *See* REJUVENATION.

alluvial tract *See* PLAIN TRACT.

alluviation (or **aggradation**) The deposition or building up of sediments by stream or water action, particularly along rivers and estuaries where flow is retarded. *See* DEGRADATION.

alluvium A fine-grained deposit, composed mainly of mud and silt, deposited by a river. *See* FLOODPLAIN TERRACES.

altar A step in the wall of a dry dock. When the dock is empty, the altar holds the feet of the wooden shores which keep the vessel steady.

altitude The vertical height of a water surface, hydraulic point or ground measured from a reference plane or mean sea level. *See* BENCH MARK, STAGE.

alum Hydrated aluminium sulphate, $Al_2(SO_4)_3.18H_2O$, used as a coagulant. *See* COAGULATION.

alveus The channel or bed of a river occupied by normal flows of water.

Ambursen dam *See* FLAT-SLAB DECK DAM.

AMC Available moisture capacity. *See* AVAILABLE SOIL WATER.

American caisson *See* BOX CAISSON.

amplitude (1) The depth of a wave measured from the level of calm water; the extent of any oscillation or vibration. (2) The width of the meander of a river (Figure A.3).

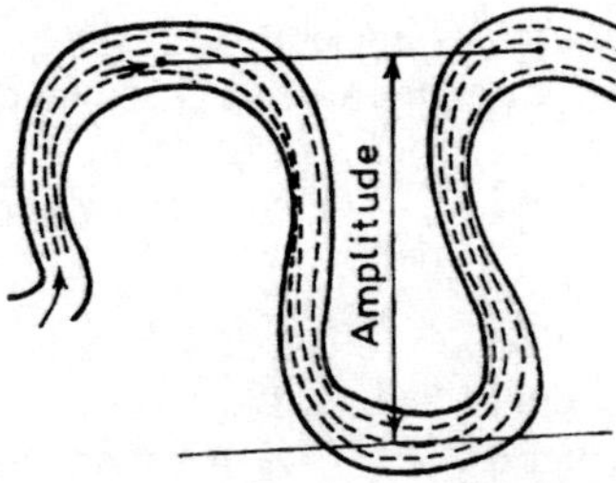

Figure A.3. Amplitude

anabranches Small interlacing streams (Figure A.4). *See* BRAIDED COURSE.

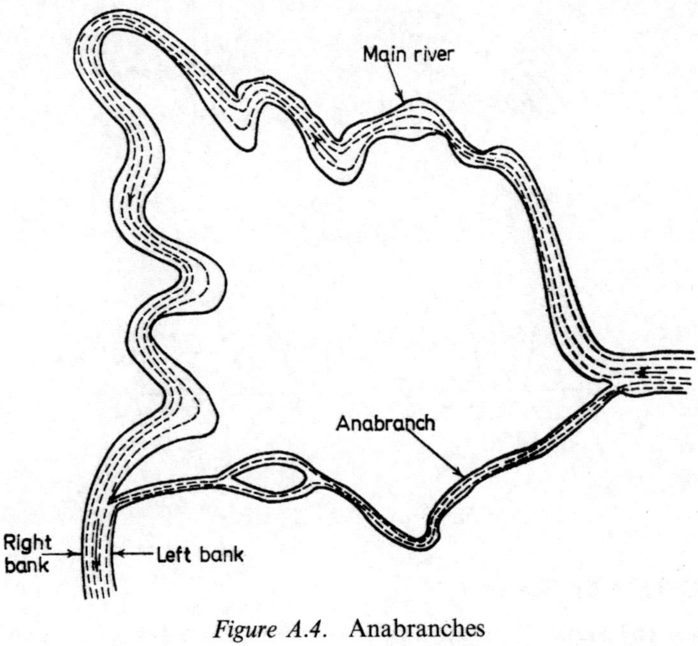

Figure A.4. Anabranches

anaerobic digestion The decomposition of sludge or strong organic waste through the action of micro-organisms, in the absence of oxygen, to reduce the concentration of organic matter and render the sludge innocuous. *See* AEROBIC DIGESTION.

analogy A scientific comparison between two phenomena: an electrical analogy has been applied to solve seepage problems, and alternating current is used as an analogy in the study of tides.

ancestral rivers A term for an ancient river system, usually of major dimensions. The limits of an ancestral river may be delineated by aerial photographs. The meander pattern of existing streams may change suddenly, indicating that they are flowing over an ancestral river course. *See* PRIOR RIVERS.

anchorages (pipes) Thrust or anchor blocks, usually of concrete, placed along a pipeline to prevent or reduce the transmission of longitudinal stress from pipe to pipe (Figure A.5). They are often necessary on bends or when a main is on a gradient steeper than about 1 in 3. In firm ground any concrete required along the pipeline is keyed into the bottom and sides of the trench.

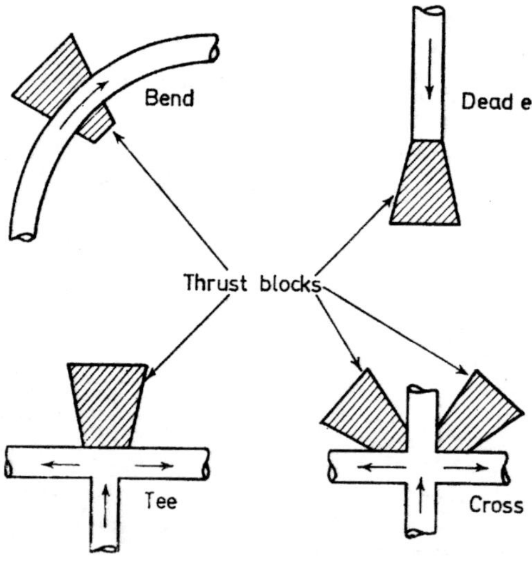

Figure A.5. Anchorages

anchorage spud *See* SPUD.

anchor and collar A strong metal hinge for lock gates; it is built into the masonry or concrete of the lock, and a hole in a projecting plate receives the pintle of the gate.

anchor buoy A buoy fixed in position by an anchor. *See* BELL BUOY.

anchored trees (bank protection) The use of bushy growing trees, well anchored together, to protect an eroding river bank (Figure A.6); it is rapid and inexpensive where suitable trees are available locally. The trees are laid along the bank, butts upstream, and anchored by wire ropes or cables to posts or buried logs set well back from the bank. To prevent any loose trees from floating or moving at high water, they may be weighed down by concrete blocks or STONEMESH sausages. Willows and vegetation are often planted behind the trees. The method has been used widely in south-eastern Australia and in New Zealand. *See* CRIBWORK, UNDERWATER APRON.

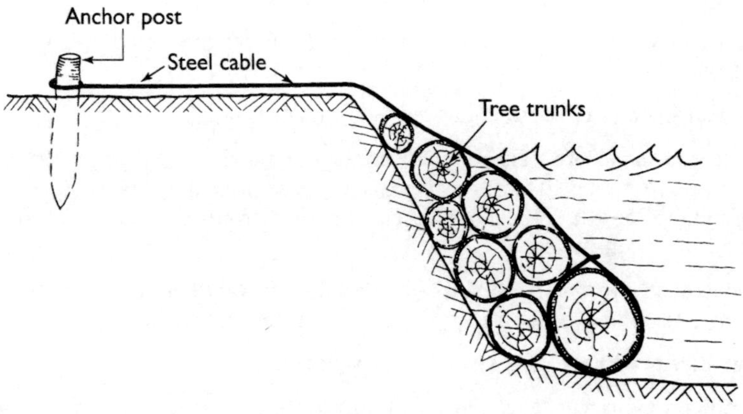

Figure A.6. Anchored trees (bank protection)

anchor-ice A layer of ice up to about 50 mm (2 in) thick on the bed of a river, sometimes formed during clear cold nights; it usually becomes free and floats to the surface after sunrise.

anemometer An instrument for measuring total wind travel in a given time, or for continuously recording the velocity and direction of wind movements. *See* BEAUFORT SCALE.

angle of repose The maximum slope, measured from the horizontal, at which rocks, soils and loose material on the banks of canals, rivers, dams or embankments remain stable. *See* BATTERS.

angle of slide The angle (measured in degrees from the horizontal) of the sloping slides of a canal, reservoir or other cutting, at which a LANDSLIDE will start. The angle varies considerably, and the water content of the bank material has a major influence. *See* STABILITY NUMBER.

angularity correction The correction made to the observed velocity when the direction of water flow is not exactly at right angles to the DISCHARGE SECTION LINE. *See* MODIFIED VELOCITY.

angular stone Broken stone obtained from the quarry, as distinct from gravel, boulders or trimmed stone; it may be large and used for groynes etc., or small as used for concrete aggregate. *See* STONEMESH, STONE RIPRAP.

anhydrous Without water, especially WATER OF CRYSTALLIZATION.

annual depletion rate The average rate at which withdrawals deplete storage in a reservoir of groundwater during a period of one year.

annual flood The maximum momentary discharge measured in a WATER YEAR.

annual flood peak series *See* PARTIAL DURATION SERIES.

antecedent moisture The moisture content of a soil at the commencement of a runoff period. It is often given as an index determined by summation of weighed daily rainfalls for a specified period (from 10 to 20 days) before the runoff considered.

antecedent stream A stream that deepens its valley and maintains its course in spite of slow tilting of the ground. *See* DEFEATED STREAM.

antecedent wetness *See* ANTECEDENT MOISTURE.

antiflood valve A CLACK VALVE or NON-RETURN VALVE so positioned in drainage pipes or channels that water discharges at low levels but flooding by backflow is prevented.

anti-seep diaphragm *See* CUTOFF COLLAR.

antitranspirants Monomolecular films of high alcohols which are introduced onto the open water surface of storages to reduce EVAPORATION losses.

apparent velocity (groundwater) The apparent rate of motion of groundwater in the ZONE OF SATURATION; it may be expressed by

$$V = Q/A$$

where Q is the volume of water passing through a cross-section of area A per unit of time. *See* EFFECTIVE VELOCITY.

approach channel *See* CHANNEL OF APPROACH.

approach velocity The mean velocity of water measured in the CHANNEL OF APPROACH.

appropriated rights (water) An individual's rights to the exclusive use of water, based strictly on priority of appropriation and the beneficial use of the water, and without limitation of the place of use to RIPARIAN LAND. *See* WATER RIGHTS.

apron A hard floor or surface on the bed or banks of a river or canal, or below chutes, spillways, groynes or the toes of dams, to minimize scour or seepage. It may consist of bagwork, mattress, timber, mass concrete, riprap, or preferably reinforced concrete. *See* STILLING POOL.

apron, falling *See* UNDERWATER APRON.

aquaprivy Similar to a SEPTIC TANK, an aquaprivy provides primary treatment for wastewater. Both consist of sealed tanks buried in the ground and receive household wastewater, which is partially treated within the tanks. Whereas a septic tank is separated from the house and also receives sullage, the aquaprivy is located directly below the latrine structure.

aqueduct A generic term for works including pipes, tunnels, open or covered channels, canals and elevated structures for the bulk conveyance of water from source or storage to an area of consumption. *See* TRUNK MAIN.

aquiclude A geological formation of low porosity which absorbs water slowly but does not transmit it freely enough to form useful supplies for a well; any saturated deposit, containing an interconnected system of interstices, but which does not yield useful supplies of water.

aquifer A relatively permeable or fissured deposit which yields useful supplies of water when tapped by a well. *See* ARTESIAN AQUIFER, ARTESIAN BASIN.

aquifer, recharge of *See* RECHARGE OF AQUIFER.

aquifuge A relatively impervious rock which has no interconnected fissures, voids or openings and hence cannot absorb or transmit water. *See* IMPERVIOUS ROCK.

aquitard A semi-porous rock bed which slows the percolation of water but still allows some seepage to pass.

Arany's number A number indicating the moisture content of a soil at which the material just starts to flow and to behave as a fluid. It is determined by pulling a thread of the soil out with a pestle as water is mixed into it. The moisture is at Arany's number when the thread starts to bend.

arch dam A dam which is curved in a horizontal plane, convex upstream. The water load is transmitted horizontally by arch action to the valley or canyon walls, and the dam must therefore be built on solid rock, as yielding ground could cause instability and failure. The construction material is commonly concrete, although brick or stone is sometimes used. *See* MULTIPLE-ARCH-TYPE DAM.

Archimedean screw A lifting device for water comprising a spiral screw rotating in an inclined cylindrical casing (Figure A.7). It is used mainly for low-head irrigation.

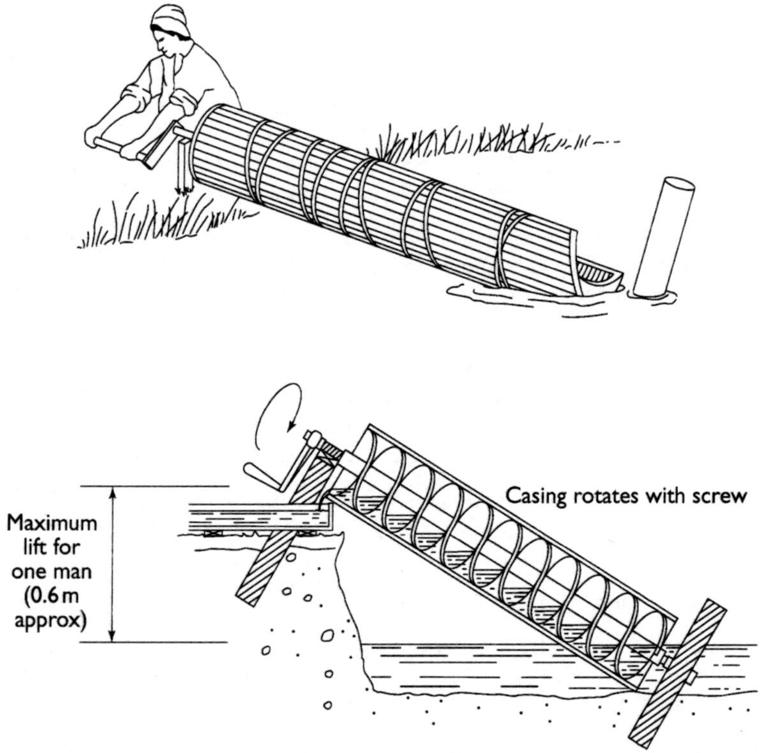

Figure A.7. Archimedean screw

Archimedes' principle A principle credited to Archimedes (287–212 BC), the celebrated Greek mathematician and engineer. It states that when a body is wholly or partly immersed in a fluid, the total uplift on the body is equal to the weight of the fluid displaced.

area, catchment *See* CATCHMENT AREA.

area curve A curve showing an area of cross-section, or other area of a stream, channel or structure, as a function of elevations.

area, discharge *See* DISCHARGE AREA.

area, drainage *See* CATCHMENT AREA.

areal rainfall The average fall of rain over an area. In regions where a rainfall radar system is in operation it can be measured directly; otherwise it must be estimated from one or more observations of point rainfall taken at rain-gauges. Many techniques are available for estimating areal rainfall from such data, including a simple average, the THIESSEN POLYGON or more complex computation methods.

areal reduction factor The correction applied to a point rainfall observation, taken to be representative of an area, to estimate the AREAL RAINFALL. It is always less than 1.0 and is very largely a function of rainfall duration and the area's size.

area of aquifer, intake *See* INTAKE AREA OF AQUIFER.

area of artesian flow The area overlying an aquifer in which the water is under sufficient pressure to rise above the surface. *See* ARTESIAN PROVINCE.

area of groundwater discharge The surface area embracing all points at which groundwater is discharged as springs or as evaporation from soils and plants.

area of influence (well) The surface area around a well or borehole with the same horizontal extent as that part of the underlying water-table that is lowered by pumping from the well at a given rate of discharge. *See* CONE OF DEPRESSION.

areic area An area with no flowing streams. *See* DRY VALLEY.

arête *See* COMB RIDGE.

arid Dry, parched; relating to climate or land where agriculture requires irrigation because precipitation is inadequate. *See* SEMI-ARID.

arid cycle *See* CYCLE OF EROSION.

arroyo A narrow valley or gully, usually with steep banks, in an arid region; it is located above the water-table and often dry owing to infrequent rainfall or cloudbursts.

arterial drain A MAIN DRAIN which has secondary drains leading into it.

artesian aquifer (or **confined aquifer**) An aquifer in which the water is under pressure and confined beneath impermeable deposits (Figure A.8).

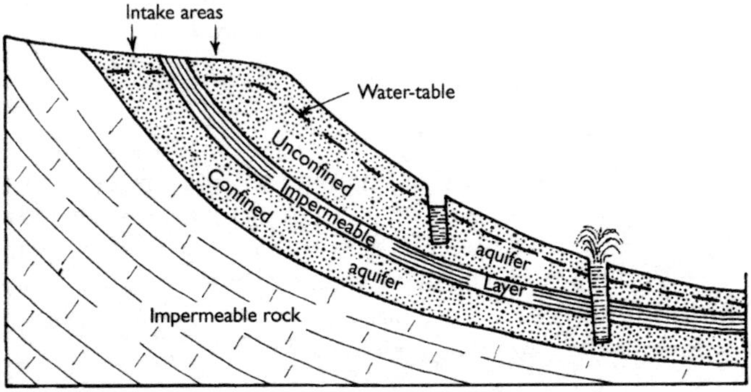

Figure A.8. Artesian aquifer (or confined aquifer)

artesian basin An ARTESIAN AQUIFER in the form of a broad basin (Figure A.9). The regional artesian basin of Queensland, Australia, covers some 1 550 000 km^2 (600 000 square miles) and supplies water to many deep wells in arid country. The chalk of the London Basin is another example. Sands and sandstones form the principal source of artesian and other water. Limestone deposits may yield water under favourable conditions.

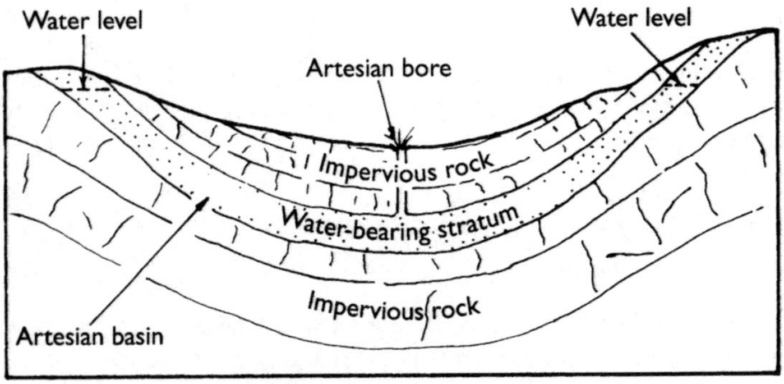

Figure A.9. Artesian basin

artesian discharge The flow of water from an ARTESIAN WELL or ARTESIAN SPRING; the volume of water thus discharged.

artesian flow area *See* AREA OF ARTESIAN FLOW.

artesian head The PRESSURE HEAD of ARTESIAN WATER.

artesian head, negative A term sometimes applied to a well in which the HYDROSTATIC PRESSURE is negative and the free water level is below the existing water-table level.

artesian head, positive A term applied to an aquifer or well in which the static pressure is positive and the free water level is above the existing water-table level. *See* POSITIVE CONFINING BED.

artesian pressure The pressure of the water in an ARTESIAN WELL.

artesian province A region within which ARTESIAN AQUIFERS occur and where geological conditions are in general similar. Although a general similarity exists, the water yield and its quality may vary from point to point.

artesian slope An ARTESIAN AQUIFER which slopes or dips beneath impermeable strata. Water seeping down from the outcrop into the aquifer at depth is stored there under pressure beneath the impervious cover.

artesian spring (or **blow-well**) An uncommon type of SPRING in which the ARTESIAN WATER emerges along a fracture or opening in the impervious deposit overlying the aquifer.

artesian water Groundwater which is under pressure and confined beneath relatively impermeable deposits. The water pressure causes it to rise above the level at which it is encountered in drilling.

artesian well A well which has tapped artesian water; a well in which the

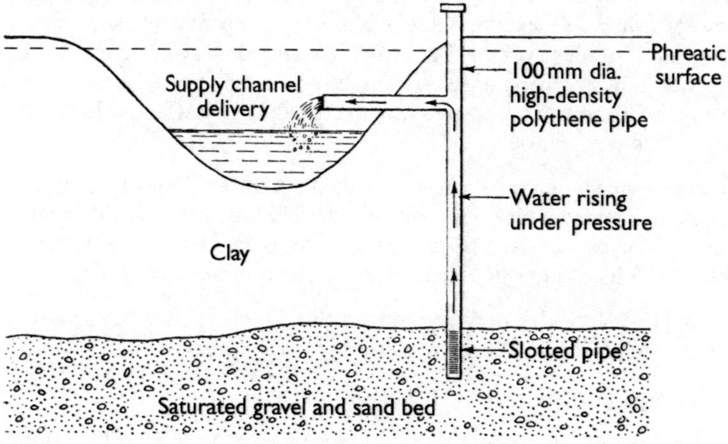

Figure A.10. Artesian well

hydrostatic pressure of the water is sufficient to cause it to rise in the well and sometimes reach the land surface (Figure A.10).

artesian well (or **free-flowing bore**) **drainage** A well put down for tapping shallow water, under artesian or near-artesian conditions, for drainage and not for usage. A typical installation is a high-density polythene pipe of 100 mm (4 in) diameter, the lower end being slotted to allow the entry of water which at the surface is discharged via a 'T' piece to a deep drain. In favourable localities, batteries of such wells are used to lower the water-table over considerable areas. *See* WELLPOINT.

artificial catchment A small catchment which has been treated, wholly or in part, to increase the runoff. In the iron-clad catchment one section of the area is covered with flat iron sheets. Some form of bituminous preparation is another method of covering the ground. The roaded catchment (see Figure R.8) involves the compaction of the area by rollers after it has been subdivided into a series of roads. The roads are about 7.3 m (24 ft) wide with a cross-fall from crest to drain of about 1 in 20. The table drains have a slope of approximately 1 in 200.

artificial harbour A sheltered area of sea formed by building BREAKWATERS. *See* NATURAL HARBOUR.

artificial recharge A means of augmenting the natural infiltration of surface water or rainfall into underground rocks or channels. The recharge may be by spreading of water, by changing the natural conditions, by some form of construction or by putting down wells. *See* INVERTED WELL.

artificial storages (streams) Measures to reduce the irregularity of stream flow by which the excess water is temporarily stored by the erection of dams and similar works. The water so stored may be used for town supplies, power or irrigation. A dam may be used wholly or partly for the storage of floodwater and so reduce the flooding of lowlands along a river. *See* NATURAL STORAGES.

asbestos–cement pipes Pipes made of a mixture of asbestos fibre and Portland cement which can withstand pressure and also internal and external corrosion. Asbestos is now known to give rise to serious risks to health and its use has now been discontinued in many countries.

ashlar Freestones as they come from the quarry; stones cut square and dressed. Ashlar facing is often used in masonry dams.

Ashley type groynes *See* IMPERMEABLE GROYNE.

astyllen A low barrier or dam placed across an ADIT to retard the outflow of mine water.

Atlantic-type coastline A coastline in which faulting, and not folding, plays a dominant role, and its line is not parallel to the trend of the mountain chains. *See* PACIFIC-TYPE COASTLINE.

atmidometer (or **evaporimeter** or **atmometer**) An instrument for measuring EVAPORATION.

atmosphere (1) The gaseous matter surrounding the earth. Composition varies very slightly according to altitude and region; it is mainly nitrogen and oxygen, with traces of carbon dioxide, argon, helium, etc. In addition, air contains water vapour and dust particles in very variable amounts. (2) A unit of pressure: that pressure which will support a column of mercury 760 mm (29.92 in) high at 0 °C (32 °F), sea level, and latitude 45°; 1 normal atmosphere = 101.325 kN/m^2 (14.72 lb/in^2). *See* HUMIDITY.

atmospheric water The water present in the atmosphere; it may be gaseous, liquid or solid. *See* CLIMATE, RAINFALL, WEATHER.

atoll A roughly circular coral reef enclosing a LAGOON (Figure A.11).

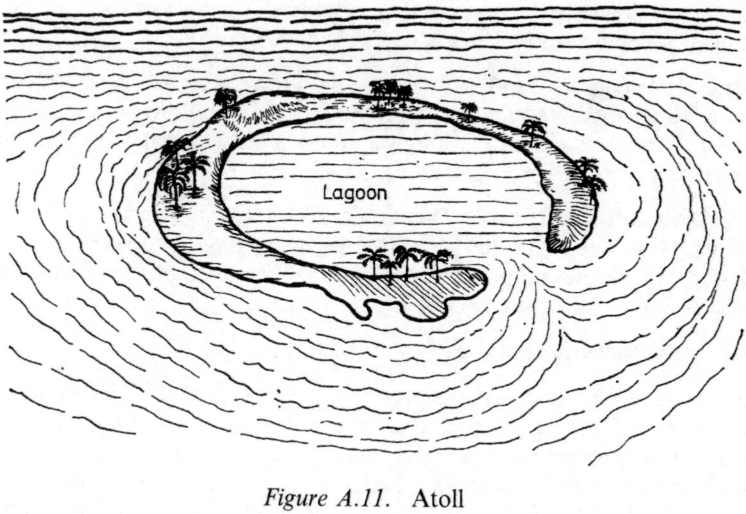

Figure A.11. Atoll

atrazine A nitrogenous herbicide which is not effectively removed by conventional water treatment processes but can be adsorbed by activated carbon.

attached groundwater That portion of groundwater retained on the surfaces of particles, against the force of gravity, during drainage or pumping. *See* SPECIFIC RETENTION.

attenuation The reduction of a physical quantity by either natural or artificial processes. Typical examples are the reduction of peak flood flow rates in rivers by RETARDING RESERVOIRS or through reservoirs by FLOOD ROUTING.

auger A power- or hand-operated boring tool, varying from about 38 mm to 0.6 m (1.5 to 24 in) in diameter. Augers are often used to investigate the foundation conditions of earth dams and the availability of suitable construction material in the borrow pit area. *See* EARTH AUGER.

autographic rain-gauge An instrument that maintains a continuous – or semi-continuous – record of rainfall. Two types are in common use: those that record the total fall of rain within a few minutes, and those that make a continuous digital or graphical record of rainfall intensity. Records from these gauges are more valuable that those from daily or weekly storage gauges, as in addition to total falls of rain they provide data on storm profiles and rainfall intensities.

automatic irrigation An irrigation system which is automatically controlled by electrical, pneumatic or mechanical devices.

automatic pump A pump provided with a FLOAT SWITCH which starts and stops the pump automatically depending on the water level in the suction sump or storage. Pumping stations are now often automatic in operation, with indicators and recorders designed for remote reading.

automatic self-closing valves Valves fitted to large aqueducts or pipelines which close automatically when the discharge exceeds a predetermined figure, and so prevent possible damage to the pipeline or to turbines, etc. downstream. *See* FLOAT VALVES.

automatic siphon spillway *See* SIPHON SPILLWAY.

automatic water treatment plants Water treatment plants equipped with automatic cleansing filters and also possibly with automatic chemical equipment. All chemicals are delivered in bulk as liquids and stored in tanks which feed service tanks. The latter, in turn, are automatically kept full and the chemicals are also automatically mixed in the proper proportions inside the tanks.

auto-regressive time series A TIME SERIES in which at least part of each successive term is related to one or more of the preceding terms, usually by CORRELATION COEFFICIENTS, such as in a Markov chain (*see* MARKOV

PROCESS). The NUMERICAL SIMULATION MODELS of many parameters in water sciences are based on auto-regressive time series.

auxiliary protection (groynes) Works and practices to protect GROYNES from damage and premature failure, and also to reduce maintenance costs. Various types of MATTRESSES and APRONS are used to reduce scour, and may be placed at the root and head of the groyne. At the root, aprons may be carried up the bank above flood level. At the head, protection is provided on the stream bed round the bend where flow is swift. These aprons and mattresses are made flexible to maintain contact with the ground. *See* PILOT CHANNELS, SINKER GROYNE, STONEMESH MATTRESS.

available chlorine The total oxidizing power of products such as chlorinated lime and hypochlorites, expressed as a percentage of chlorine available.

available moisture capacity (AMC) *See* AVAILABLE SOIL WATER.

available soil water (or **available moisture capacity**) The soil water that is available or can be taken up by plants at any moisture content between FIELD CAPACITY and PERMANENT WILTING POINT. Sandy soils contain less available water than clay soils, and loams form an intermediate group.

available water-holding capacity (soils) The difference between the percentage of water at FIELD CAPACITY and the percentage of water at wilting point. *See* WILTING COEFFICIENT.

avalanche A large slide of ice, snow and debris down a mountain slope. *See* LANDSLIDE.

average annual rainfall For a specified place, the mean of the annual amounts over the period for which rainfall statistics are available.

average velocity (groundwater) The mean distance covered by mass of groundwater per unit of time; it equals total volume of groundwater passing through unit cross-sectional area per unit of time, divided by porosity of material passed through. *See* APPARENT VELOCITY.

avulsion (1) The sudden removal of land by flood; the accelerated erosion of shore by storm waves. (2) The severance or division of an area of land by a stream cutting across the narrow neck of a horseshoe bend. *See* OXBOW LAKE.

axial-flow pump A propeller pump in which the water flow is axial; the principle is similar to that of a ship's propeller. The pump is used for lifting large quantities of water over small heads. *See* PUMP.

axis of dam *See* DAM, AXIS OF.

B

Bacillus coli A rod-like germ sometimes found in drinking water and often derived from sewage pollution.

back acter An EXCAVATOR fitted with front-end equipment comprising a jib with an arm and bucket. Although designed primarily for vertically sided trenching, it is also useful for bulk digging below track level and for many excavations for drainage, ditching and other hydraulic works.

backblowing A method of improving the water yield from boreholes, particularly in fissured rocks. It is not unlike SWABBING in action and is sometimes more effective. It employs compressed air (1) for pumping in the bore by air lift, or (2) by pumping in air (when the top of the hole is sealed) until the maximum pressure is obtained and then releasing it suddenly. *See* SHOT-FIRING.

back cutting The extra excavation carried out to make up a canal or other hydraulic embankment when the material available from other sections of the work is insufficient. *See* BALANCED EARTHWORKS.

backfilling (or **backfill**) Material excavated at a site or construction and reused for filling, as in groynes and trenches for pipelines, etc.

back-inlet gully A cast-iron or glazed stoneware branch entry to a drain, water-sealed and covered by a grating, but open to the air. It receives rainwater or discharge from waste pipes under the grating but above the water seal.

backshore The area of shingle or sand beach between normal and abnormal high-water marks. *See* FORESHORE.

backwash The back flow of water which follows the breaking of a wave: it scours and carries the finer material towards deeper water, where it is temporarily deposited. *See* UNDERTOW.

backwashing The reversal of flow through a rapid gravity sand filter to remove flocculent matter trapped both on the surface and within the media, which not only causes loss of filtering head but could also lead to the build-up of mud-ball accretions within the sand. It is often preceded by air scouring through the bed to break up the flocculent blanket.

backwater A branch or side channel off the main course of a river; water held back in a channel or stream by a dam, regulator or obstruction; water stored during high tide and discharged during low tide. *See* AFFLUX.

backwater curve A curve representing the BACKWATER surface upstream along an open channel resulting from a dam or weir or obstruction (Figure B.1); it may extend upstream for several kilometres or miles. *See* STREAM PROFILE.

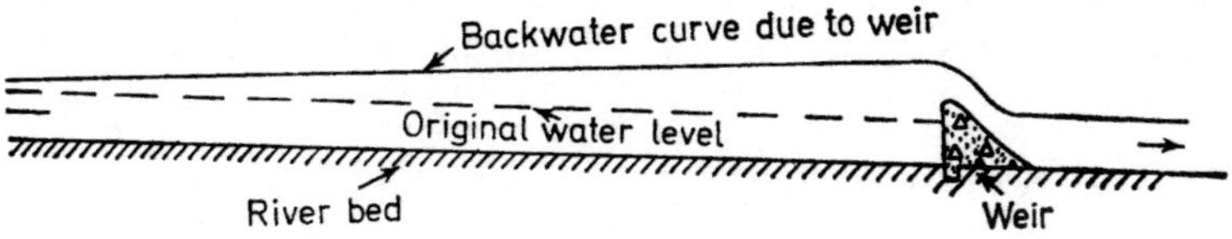

Figure B.1. Backwater curve

bacteria Primitive micro-organisms lacking chlorophyll, which are usually simple unicellular rods or spheroid cells, but are occasionally filamentous. The three principal types are aerobic bacteria, anaerobic bacteria, and coliforms found in sewage and decaying vegetation.

bacteria bed (or **biological filter** or **contact bed** or **continuous filter** or **percolating filter**) A layer of FILTER MATERIAL, such as clinker or rock, which will expose sewage effluent to the oxidizing action of micro-organisms in the air. *See* SLOW SAND FILTER.

baffle One of a series of plates, vanes or guides set in a channel or conduit to secure more uniform flow conditions by damping eddies.

baffle-pier A wall, mass or obstruction erected in the path of high-velocity water to dissipate energy and reduce scour; a pier on the apron of an overflow dam. *See* GROYNES.

bagwork A form of bank protection used on rivers and sea walls. Small bags are filled with dry concrete and tamped by hand against the section to be protected; gravel is sometimes used instead of concrete. Bagwork along sea walls is often held together by steel dowel rods driven through the bags. *See* REVETMENT.

bailer (or **sand pump** or **sludger**) A pump for removing sludge, or sand and water or mud, from a well or borehole. It consists of a tube or pipe with a valve at the lower end which opens upwards. An opening at the side at the upper end of the pipe enables the sediment water to be discharged.

Bailey bridge A military bridge of welded lattice construction developed in Britain during World War II by Sir Donald Bailey, and since used extensively for water crossings in many countries.

bai-u The season with abundant rains in southern Japan and in parts of China (where it is termed MEIYU).

balance bridge *See* BASCULE BRIDGE.

balanced earthworks Earthworks in which the 'cuts' equal the 'fills'. The ideal is not easily achieved but is a target when planning the layout of trenches, water supply channels and other hydraulic works. *See* MASS HAUL DIAGRAM.

balance storage *See* PONDAGE.

balancing tank *See* EQUALIZING RESERVOIR.

band screen An endless moving band of wire mesh or perforated steel plates used at INTAKE WORKS to retain and remove solid matter from the water. The water passes through the screens and the solids are washed off by water jets. The washings are collected in a launder and returned to the river or source well below the intake. *See* SCREENS, DRUM SCREEN.

bank (or **embankment**) The elevated ground along the sides of a stream or lake or channel; an elevation in the sea; a shoal; any ridge or earth slope formed or trimmed to a definite shape. *See* RIGHT BANK.

bankful discharge The DISCHARGE at the BANKFUL STAGE.

bankful stage The STAGE when a stream fills its channel and just overflows its natural banks.

bank materials The soils and other superficial deposits used in the construction of embankments or earth dams. The material must be sufficiently impervious to reduce seepage to a safe rate and must also be sufficiently stable to permit economic BATTERS. Material containing about 25% clay and the remainder sand and silt makes good impervious banks. Too much clay weakens the bank and also makes it liable to volume changes with changes in moisture content, while insufficient clay can cause seepage. Bank materials are usually tested to determine their suitability before construction. *See* CLAY, SAND, TOPSOIL COVER.

bank protection Measures to combat erosion along river and canal banks and along coastlines. They may include costly works of a permanent or semi-permanent nature, such as SHEET PILES, stone gabions, etc., or more temporary works which might last only a short period until vegetation has been established along the eroded banks. In some cases, bank protection to prevent erosion and to check MEANDER MIGRATION has helped to stabilize a river for 4 or 5 kilometres (3 or 4 miles) downstream. *See*

GROYNES, MATTRESS, PERMANENT BANK PROTECTION, REVETMENT, RIPRAP, SEA WALLS, SLOPING OF BANKS.

bank, raw *See* RAW BANK.

bank revetment *See* REVETMENT.

bank sloping *See* SLOPING OF BANKS.

bank storage (or **lateral storage**) The water absorbed by the banks of a stream, reservoir or lake at high water levels and returned to it, wholly or in part, as the water level falls. *See* CHANNEL STORAGE.

bar A barrier; a hard ridge of rock cutting across a river bed; a bank of gravel or silt at the mouth of a river or harbour and often unstable. Bars at the mouth of rivers may be formed in several ways: (1) across the mouth of a tidal estuary they may be due to littoral or shore currents which drift material across the mouth of the estuary, or to eddies and still water produced by currents at the entrance; (2) wind and waves may drive sediment across the mouth of a river entering a lagoon or bay of a tidal sea; (3) the checking of the current when sediment-bearing rivers enter inland seas or lakes often causes the load to be deposited as a bar.

bar (pressure) A unit of pressure approximately equal to a pressure of 1 atmosphere. (10 bars = 1.0 N/mm^2 = 145 psi)

barge A flat-bottomed boat used to carry heavy loads by canal or river. In the UK it has a beam of 4.3 m (14 ft) or more and the loads range from 70 to 150 t. On the continent of Europe loads up to 500 t may be transported.

barge bed A place where barges can moor and rest on a mud bottom at low tide near the bank of a river. A wall is usually built to retain the river bank. *See* CAMP SHEATHING.

barminutor A wastewater bar screen fitted with an automatic shredding device designed to travel up and down the screen, so as to reduce the size of the retained material sufficiently to be carried on through the screen with the flow. *See* BAR SCREEN.

Barnes's formula This formula, which is based on MANNING'S FORMULA, was used to calculate the velocity of flow in sewers:

$$v = 72\ R^{0.7}\ S^{0.5}$$

where v is the velocity in m/s, R is the HYDRAULIC RADIUS in m, and S is the slope of the sewer in m/m.

barrage A low barrier or dam, incorporating a series of gates, built across a river to regulate its upstream level, or for irrigation. A barrage may be erected to prevent ingress of salty sea-water or to impound in the loop of a river tidal water used for scouring out the lower channel and thus maintaining its capacity.

barrier basin A basin formed by landslide debris in a narrow valley. A lake may be formed upstream from the obstruction. Other types of barrier basins include hollows behind morainic barriers, lagoons and bays which have been closed in by BAY-MOUTH BARS, and depressions made by a sheet of lava which chokes a valley. *See* ICE GORGE.

barrier beach (or **sand reef**) A type of beach ridge formed a short distance offshore; it sometimes occurs where the coast has a gentle slope and the waves break some distance out, causing beach material to be deposited there. Storm waves may build up the barrier above water level with the formation of a lagoon between the mainland and the barrier. In time the lagoon may become filled with sediment. *See* BEACHES.

bar screen A wastewater screen with parallel bars, usually vertical, fixed in a channel to remove floating debris. The screenings are removed manually or automatically. *See* BARMINUTOR.

bascule bridge (or **balance bridge** or **counterpoise bridge**) A river bridge hinged at the bank to enable ships to pass through, by raising the portion over the navigable channel and lowering the portion over the bank behind the hinge. The construction material is usually aluminium alloy to minimize the weight to be moved by the machinery.

base course A layer of specified material and thickness placed on the subgrade (the prepared and compacted ground below a structure or road) to spread the load, minimize frost action and provide drainage.

base-exchange method A reversible chemical process. Some substances acting on solutions containing ions can replace some of the ions with others. A clay that has absorbed hydrogen ions can be changed to a sodium clay by infiltration of sodium solutions. The process decreases the permeability of the clay. *See* ZEOLITE PROCESS.

base flow The sustained flow of streams from snow and glaciers or from underground storage and not related to direct runoff; sustained flow resulting from drainage of large lakes or springs or other outflow of groundwater, as opposed to surface runoff. *See* PERENNIAL STREAM, HYDROGRAPH TIME BASE.

base level The level of the body of water into which a river flows, usually the sea or a large lake. *See* TEMPORARY BASE LEVEL.

base map (or **working map**) The map used during GEOLOGICAL MAPPING; a topographic contour map or aerial photographic map for plotting the geologic information obtained in the field. If used to investigate water resources, all porous and impervious beds are marked, and also old or existing wells, springs, streams, etc.

base period The period when the BASE FLOW of a stream is exceeded as a result of heavy rainfall or a storm.

basic properties, of soils Soil properties such as consolidation (or compaction), shear strength and permeability. These properties affect the behaviour of a soil at an engineering site. The results of consolidation tests are useful when the deposit is used as 'fill' material in earth dams or as foundation material. The shear strength and permeability of a soil are important in connection with dam foundations and earth dams, and also when excavations are made in sands. *See* DAM, PERMEABILITY.

basin Topographically, either an area drained by a river and its tributaries or low-lying land encircled by hills. Geologically, an area in which the stratified rocks dip towards a central spot, and the rocks are said to possess a CENTROCLINAL DIP. *See* BARRIER BASIN.

basin irrigation A form of irrigation in which an earth ridge is erected around a flat area of land and water is introduced to form a pond.

basin recharge The portion of precipitation that remains in a basin as groundwater, surface storage, soil moisture, etc.; the difference between precipitation and runoff and other losses for a given period or storm.

basin, river *See* CATCHMENT AREA, RIVER BASIN.

bathotonic reagent A substance which diminishes the SURFACE TENSION of a fluid.

battered-bank system A heavy timber method of bank protection, similar to the RAW-BANK SYSTEM except that the bank is sloped to a convenient batter. It is used in Australia and New Zealand. *See also* ANCHORED TREES.

battering The artificial sloping back of a river or other bank to a flatter grade to reduce caving or collapse of the earth. The procedure is often followed by BEACHING with stone or by grassing or other vegetation to bind the soil layers.

batter level A CLINOMETER for measuring the slope of BATTERS.

batters (or **side slopes**) The artificial, uniform slopes of the sides of earth dams, canals, embankments, etc. Earth embankments or farm dams should not be steeper than 1 in 3 on the water or upstream side, and 1 in 2 on the downstream side (Figure B.2).

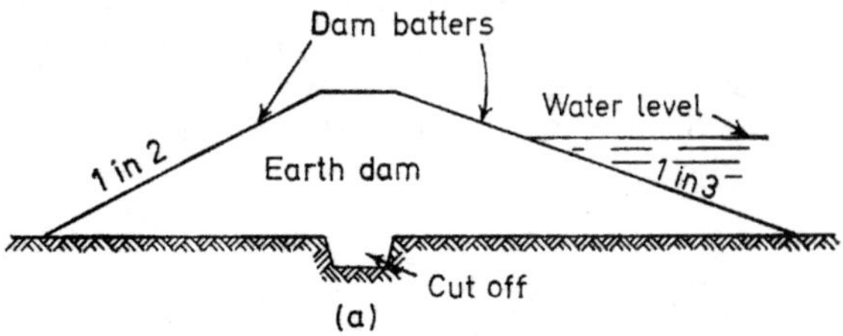

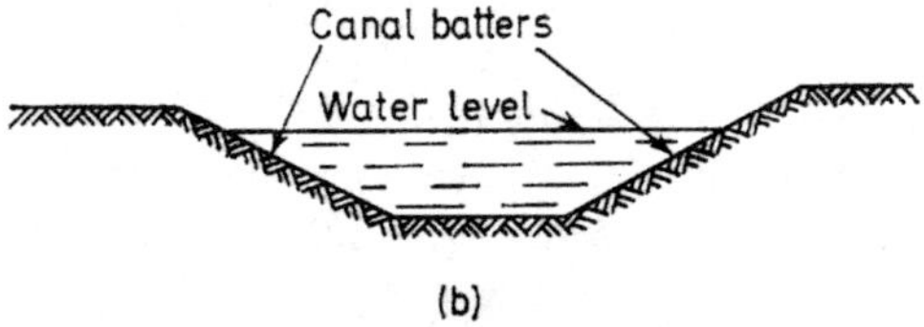

Figure B.2. Batters

battery of wells A number of wells within a convenient radius and connected to a main pump or other lifting appliance for withdrawal of water. *See* WELL FIELD.

bay A portion of the sea which extends into an open-mouthed land formation.

bay (irrigation) The area of ground between adjacent CONTOUR DITCHES, CONTOUR BANKS, BORDER CHECKS or BORDER DITCHES.

bay-head bars Banks of silt or gravel formed at the heads of bays. *See* MID-BAY BARS.

bay-head beaches A feature associated with irregular coasts, consisting of the accumulation of water-washed sediments at the heads of bays; the material was worn and transported from the neighbouring headlands.

bay-mouth bars SPITS that have continued to accumulate and finally closed the entrance to a bay.

bay outlet or **stop** (irrigation) *See* SLUICE BOX.

Bazin's formula An empirical formula, proposed by Bazin in 1897, giving the value of the coefficient *C* in CHÉZY'S FORMULA in terms of HYDRAULIC RADIUS, *R*, and a coefficient of roughness, *M*, a numerical constant depending on degree or kind of roughness. Thus:

Metric units	*Imperial units*
$C = \dfrac{87}{1 + M/\sqrt{R}}$	$C = \dfrac{157.6}{1 + M/\sqrt{R}}$

B. D. Richards method A procedure for predicting the PEAK RATE OF RUNOFF and a FLOOD HYDROGRAPH from an ungauged CATCHMENT AREA following a STORM of any selected RETURN PERIOD. This method uses only rainfall data and catchment characteristics, most of which can be measured from contoured maps. It was developed from the RATIONAL METHOD in the UK by B. D. Richards (1944), but can be used anywhere in the world where rainfall data is available. In view of the relatively small amount of data required, the accuracy of the flood predictions obtained is not great. However, and particularly for steeper areas extending over between about 10 to 1000 km^2, they are often acceptable for engineering design. This method needs a considerable amount of numerical calculation, but computer programs are available.

beached bank The part of a river bank covered with stone above and below the normal water level. *See* BEACHING.

beaches The shore areas of the sea or lake washed by the tide; all shoreline deposits consisting of sand, pebbles or boulders, and formed principally by the action of waves and longshore currents. In general, beach sediments are coarser at the top than they are near the water, and along sea beaches pebbles often accumulate above high-tide level, grading downward into sand and finally mud below low-tide level. Mud flats are usually exposed at low tide along very low shelving shores.

beaching A layer of stones for revetting below the level of stone pitching a reservoir or embankment (Figure B.3): the layer is from 0.3 to 0.6 m (1 to

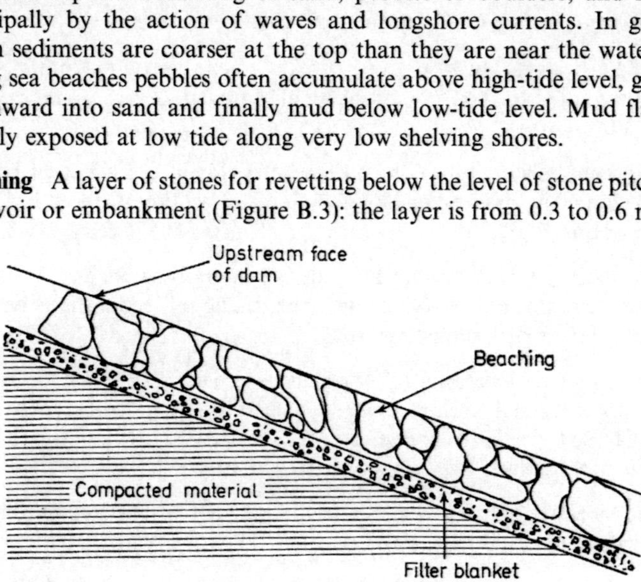

Figure B.3. Beaching

2 ft) thick and stone size from 75 to 200 mm (3 to 8 in). The term is also applied to a layer of stones of about 150 mm (6 in) diameter to prevent scour due to wave action. *See* STONE RIPRAP.

beach ridge A deposit of beach sediment formed by wave action. Frequently formed along a relatively flat beach where the UNDERTOW is weaker than the inward movement of the waves, with the result that silt is shifted and cast up near the water's edge to form a beach ridge. *See* BARRIER BEACH.

beam engine A type of steam engine with a vertical cylinder, introduced as early as 1700. Initially used widely in mining to drive the CORNISH PUMP, but soon adapted for water pumping, marine and general industrial use.

bear-trap dam A type of dam or obstruction to river flow consisting of hinged leaves. The pressure of water admitted to the inside causes the leaves to be raised and held up; the obstruction is lowered by draining the interior.

Beaufort scale An arbitrary scale, first introduced for sailing ships, with a range from 1 to 10, for indicating wind velocities and based on observed effects, such as bending of trees.

bed-building discharge *See* DOMINANT FORMATIVE DISCHARGE.

bed-building stage The STAGE of a stream when there is maximum silting or building up of the bed by deposition of sediment. *See* ALLUVIATION, OLD RIVER, SILTING.

bedding, land Elevation of the surface of fields into a series of parallel beds or 'lands' by grading or ploughing and separating them with shallow surface drains.

bedding plane The surface marking the boundary between a rock stratum and the stratum above or below; the plane of separation between individual layers of sedimentary rock.

bed erosion The deepening of a stream by erosion of its bed. It may be caused by artificial means, such as bend cutting, CANALIZATION or GROYNES. Bed erosion increases channel capacity but may create other problems if excessive. *See* TRAINING WALL, CHECKDAM.

bed load The weight or volume of boulders, pebbles and gravel rolled or moved by a stream along its bed per unit time. *See* LOAD.

bed load function The relationship between bed load and discharge in a given cross-section of stream.

bed load sampler An appliance for measuring the rate of movement of material along the bed of a stream; the thickness of material differs at different periods and for different streams.

bed profile A curve indicating the elevation and shape of the bed of a stream. It may be a longitudinal curve or a transverse curve at a cross-section. *See* STREAM PROFILE.

bedrock (or **rockhead**) The plane or division between the surface UNCONSOLIDATED MATERIAL and the solid rocks; the first hard, solid rock encountered in a well, borehole or excavation.

bed scour The wearing away of the rock, soil or deposited material from the bed of a river or sea by flowing water or waves. *See* LOAD, SCOUR, SCOUR PROTECTION.

bed slope The inclination of the bed of a stream along its course, given as difference in elevation per unit horizontal distance.

beheaded stream A stream which has been severed from its original drainage area by another and more powerful stream; the area now draining into the second stream. *See* RIVER CAPTURE, OXBOW LAKE.

bela A term commonly used in India for a sandbank or island in a river. *See* BAR, SHOAL.

Belanger's critical velocity *See* CRITICAL VELOCITY.

bell buoy A BUOY which gives audible warning by an attached bell. *See* CAN BUOY.

bell dolphin A large bell-shaped steel or concrete fender for mooring ships at sea. The bell is mounted on a cluster of driven piles of timber, steel or concrete. *See* DOLPHIN.

bellmouth overflow (or **spillway**) A structure built on a reservoir to pass overflow discharges, consisting of a vertical tower with a bellmouth inlet which passes the water downwards to an outlet conduit. Two bellmouth overflow shafts may be constructed, one near each end of the dam.

belt of phreatic fluctuation The mass of rocks in the LITHOSPHERE in which FLUCTUATION OF THE WATER-TABLE occurs.

belt of soil water The upper near-surface part of the ZONE OF AERATION containing SOIL MOISTURE.

belt of weathering *See* ZONE OF WEATHERING.

bench flume An open conduit which rests on a bench cut from the hillside and usually constructed of timber or reinforced concrete, frequently adopted when the soil materials are unsuitable for earth channels (Figure B.4).

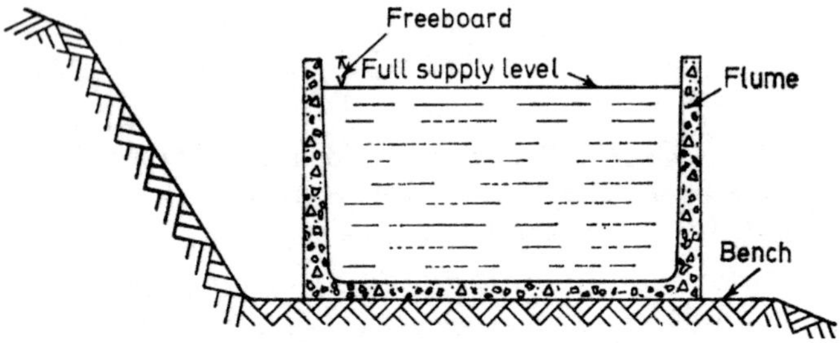

Figure B.4. Bench flume

bench mark (or **BM**) A fixed point or mark the level of which relative to a given datum is known, used by engineers and surveyors to determine the level of other points or surfaces, such as reservoirs, dams, canals, streams and harbours. *See* CONTOUR LINE.

bench terrace A terrace or small embankment of earth formed along the contour of sloping ground to regulate runoff and reduce erosion. A bench terrace has a level or nearly level top and a steep or vertical downhill face. *See* RIDGE TERRACE.

benthal deposit The material accumulated on the bed of a stream, river, lake or sea. Organisms living within the benthal deposit are known as benthic organisms or benthos.

bentonite A naturally occurring, very fine clay substance with a maximum particle size in the 2 micron range, having a very high swelling property when moistened. It is used in reservoirs, ponds, irrigation and water supply channels, and in small farm dams to reduce seepage. The typical application is about 5 kg/m^2 (1 lb/ft^2) of surface. The bentonite is covered by a SAND BLANKET to prevent cracking during the period when the dam or channel is empty. To support a large head of water, bentonite must be used together with sand or soil to increase its structural strength. *See* LINING.

berm (or **berme**) A narrow ledge; a flat shelf which breaks or modifies the uniformity of an earth slope. It intercepts earth or stones rolling down the slope and it also strengthens the embankment, and is often used in large earth irrigation channels and dams (Figure B.5).

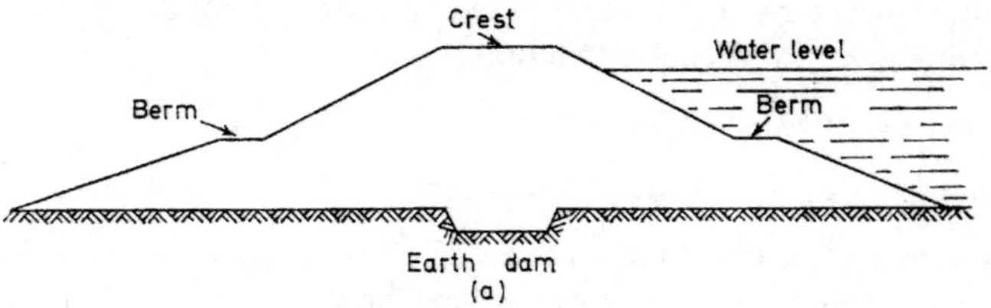

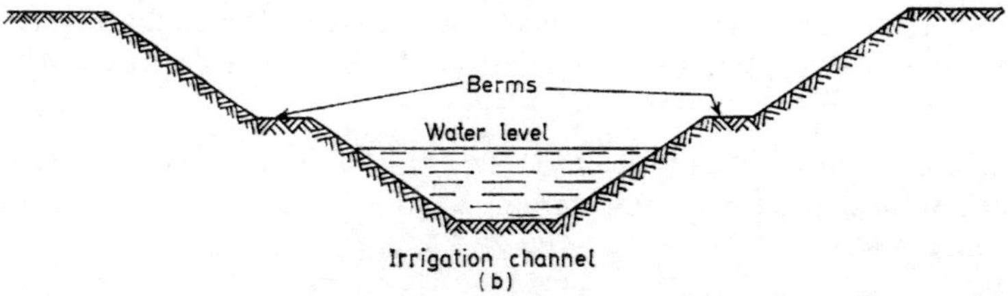

Figure B.5. Berm

Bernoulli's theorem A basic theorem that relates to the flow of fluids. When applied to flows in conduits, it states that when a perfect incompressible fluid flows in a steady stream, then, if friction and eddy current losses are neglected, the total energy is constant. The equation is:

$$p/w + z + v^2/2g = \text{constant}$$

where p is the pressure of fluid, w is the specific weight of the fluid, z is the height of the fluid above an arbitrary datum, v is the velocity of the fluid, and g is acceleration due to gravity. When applied to open channels the term p/w is essentially constant, so simplifying the equation.

berth The place where a ship lies at anchor, or at a wharf for loading or unloading.

berthing impact The forces to which structures, such as piers and jetties, are subjected during the berthing of vessels; it is generally estimated from the kinetic energy of a large vessel and an assumed berthing velocity of about 150 mm/s (6 in/s).

betrunked Relating to the partial submergence of a river system; said of streams that were formerly connected flowing into the sea by separate mouths. *See* BEHEADED STREAM.

B-horizon The soil layer between A-horizon and C-horizon. *See* A-HORIZON.

bifurcation structure A structure or gate placed in a conduit to divide the water flow into two separate conduits.

bilharzia *See* SCHISTOSOMIASIS.

billabong (Australia) *See* OXBOW LAKE.

biodegradable Organic materials that can be destroyed or mineralized by micro-organisms populating soils, natural bodies of water or wastewater treatment systems.

biodiversity The variety of all life on earth, from mammals to micro-organisms. Unpolluted water, together with oxygen, is a vital necessity to maintain all the interdependent diversity of species and their ecosystems. *See* ECOSYSTEM, SUSTAINABILITY.

biological film (or **microbial film**) The film of micro-organisms on the surfaces of the inert materials forming the medium in a biological filter or BACTERIA BED.

biological filter *See* BACTERIA BED.

biological (or **biochemical**) **oxygen demand (BOD)** The quantity of dissolved oxygen used in the biochemical oxidation of organic matter, usually when a sample is incubated for five days at 20 °C (68 °F) under specified conditions. BOD gives a rough measure of organic matter and is used as an indicator of wastewater strength. *See* CHEMICAL OXYGEN DEMAND.

biota Animal or plant life of an ecosystem such as a stream.

biotic indicator An indicator of the biota to assess water quality of a river or stream, which can be expressed as a biotic index, ranging from the most polluted to the highest quality.

black water Wastewater from water closets and urinals. *See* GREY WATER.

Blaney–Criddle method An empirical method of estimating crop water requirements that has been developed in the western United States. It assumes that crop water requirements can be estimated from a function of temperature, hours of daylight and empirical coefficients. It gives good results, but only if used in the arid regions for which it was developed.

blanket grouting A form of grout treatment used under concrete dams and to a lesser extent under earth dams. It is used to consolidate the foundations of a dam where the foundation rock is of good quality but jointed. Usually the entire foundation area to a depth between 6 and 15 m (20 and 50 ft) is treated with low-pressure grouting through holes drilled on a 3 or 6 m (10 or 20 ft) grid. The grout pressures are controlled to prevent the upheaval of the foundation rock. *See* GROUTING.

bleaching powder Calcium hydroxide (slaked lime) saturated with chlorine gas to give an available chlorine of about 35%; the powder is rather unstable and the strength is unreliable.

blind drain *See* RUBBLE DRAIN.

blizzard A violent snowstorm with intense cold and high wind.

block-in-course Blocks of hard stone, of variable length but uniform depth not exceeding 0.3 m (12 in) laid in courses in dock walls.

blockwork Stone or pre-cast concrete blocks weighing from 10 to 50 t used in the construction of BREAKWATERS and similar marine structures to resist wave action. *See* RUBBLE-MOUND BREAKWATER, TETRAPOD, TRIBAR.

bloom *See* ALGAL BLOOM.

blowing well A water well from which a current of air is periodically blown out. *See* BREATHING WELL.

blow-off An outlet fitted to a pipeline or sewer at a low point for emptying or for discharging water or sediment. *See* AIR VALVE.

blow-out (1) A failure due to TUNNELLING or PIPING in earth dams or channels. (2) The sudden loss of compressed air from a tunnel or caisson, very dangerous in certain conditions.

blow-well *See* ARTESIAN SPRING.

bluffs High banks presenting steep fronts, as at the extremes of river meanders.

BM *See* BENCH MARK.

BOD *See* BIOLOGICAL (OR BIOCHEMICAL) OXYGEN DEMAND.

body current (lakes) The general movement of the water from inlet to outlet of a lake; the velocity is low and may be noticeable only at the head or foot of the lake. The flow is not necessarily direct from inlet to outlet. *See* SEICHES.

boil (1) The seepage or inflow of water and fine sand or silt at the bottom of an excavation due to a high external water pressure. The remedies include reducing the pressure difference or drainage or lowering the water-table in the vicinity of the excavation. *See* PIPING, WELLPOINT. (2) A vertical eddy developed in a river during a flood. It produces an upthrow of water to the surface, causing it to 'boil'. *See* SAND BOIL, TIDAL CURRENT.

boison A topographic basin where the streams converge to a central area or point. *See* RADIAL DRAINAGE.

bollard A strong cast-iron post fixed in a quay structure for mooring vessels. *See* BELL DOLPHIN.

bonded gravel screen A WELL SCREEN of an improved type and made from particles of gravel bonded together with a resin adhesive. Its yield performance is at least equal to the ordinary metal well screen but it is much cheaper to construct.

bone oil (or **Dippel's oil**) A polluting substance sometimes used by military forces to make water supplies undrinkable. When added to water it gives a nauseating smell and a taste so objectionable that it cannot be drunk in sufficient quantities to poison a person, hence it does not contravene the Hague Convention.

boning A field method for obtaining levels or of ensuring a correct gradient of earthworks, pipelines, etc. with the aid of BONING RODS.

boning rod A T-shaped staff of timber about 1.2 m (4 ft) long and rods used to set individual pipes on the correct gradient, or to extend given points at the same level. *See* SIGHT RAIL.

boom An obstruction or a floating barrier of timber extending across a harbour or river mouth.

booster pump An automatically controlled variable-speed impeller-type pump often with electric drive; it is usually installed on the track of the main to increase the pressure. A booster pump is sometimes used to feed small areas situated above the level of the source in a gravity system.

border check A ridge or mound of earth formed down the slope of an irrigation area to control the spread of water.

border check irrigation The use of low earth banks to confine the water to runs or bays, often adopted in pasture irrigation. *See* CONTOUR FURROWS.

border dikes (USA) Earth ridges erected around a field or area to impound irrigation water. *See* DIKE.

border ditch A shallow channel or ditch formed down the slope of irrigated land. The ditch facilitates drainage and the ditch banks control the spread of water.

border irrigation Flood irrigation of land areas between BORDER DIKES.

bore (1) *See* TIDAL CURRENT. (2) The internal diameter of a pipe.

borehole A hole drilled into the ground, from the surface or from subsurface excavations, to secure geological information or for the drainage or abstraction of water, or for access to hydraulic works, etc.

In the UK the trend is towards larger boreholes, from 0.6 to 1.0 m (24 to 40 in) and depths for tapping water from 120 to 180 m (400 to 600 ft); they rarely exceed 300 m (1000 ft). *See* WELL.

borehole casing A plain or perforated pipe of steel or other material inserted in a borehole, often used in weak or loose ground. The pipe sections may be coupled or flush jointed, being smooth externally and internally. In recent years bronze, stainless steel, asbestos cement and plastic casings have been used. *See* OPEN-END WELL, PERFORATED-CASING WELL, SCREENED WELL.

borehole log A record, mainly of the rocks penetrated during the drilling of a borehole, and prepared by the master driller or field geologist. The information recorded varies with the objectives in putting down the hole. *See* WELL RECORD, WELL SECTION.

borehole pattern The disposition or spacings of a group of boreholes put down for water supplies or for geological information. Where the deposits are relatively uniform, the bores may be positioned at the corners of squares or equilateral triangles. The pattern depends on the objectives and the economic aspects.

borehole pump In general a HANDPUMP or electrically driven CENTRIFUGAL PUMP. It may be used to provide water for farming, for irrigation, or for dewatering areas. *See* SHAFT-SINKING PUMP, SUBMERSIBLE PUMP.

borehole samples The soil or rock chippings, sludge or core extracted during drilling and used to ascertain the nature of the ground penetrated in the borehole. The diamond drill and shot-drill yield rock cores, while percussive drills yield chippings and sludge. Samples can also be obtained by earth augers. *See* CORE, SLUDGE SAMPLE.

borehole surveying The lowering of instruments into a borehole to determine whether the hole has deviated from the vertical plane or the planned direction of drilling. Horizontal or angle diamond-drill holes over about 90 m (300 ft) deep are liable to deviate badly. Vertical holes may also become crooked, but not usually to the same extent. Deviations may be caused by (1) lack of care or drilling defects, (2) shear zones or fault planes or (3) inclined rocks of varying hardness.

bore, hydraulic *See* HYDRAULIC BORE.

bore, tidal *See* TIDAL BORE.

borrow pit (or **borrow area**) An excavation or area from which soil or rock is taken to build an earth dam, embankment or other construction (Figure F.1). *See* CONSTRUCTION MATERIALS.

bottle silt sampler A device for taking samples of silt and consisting of a bottle held in a frame and suspended from a pipe. A lever at the top of the pipe operates a rubber stopper fitted to the bottle. *See* BOTTOM SAMPLER.

bottom contraction A reduction in cross-sectional area of water overflowing a weir crest caused by contraction of a NAPPE. *See* END CONTRACTION.

bottom ice *See* SUBSURFACE ICE.

bottomland (USA) A term for FLOODPLAIN.

bottom sampler An appliance for picking up sample material from the bed of the sea or other body of water, consisting of a sounding lead with adhesive on the underside. *See* BOTTLE SILT SAMPLER.

bottomset beds *See* DELTA.

boulder A large block of rock which has been transported some distance from its parent bed by ice or water and become rounded.

boulder clay A clayey deposit containing some rocks up to the size of boulders, which often show striations indicating glacier movement; the deposit is typically unstratified. *See* GLACIAL DRIFT.

boulder well (or **cavity well**) A water well put down into a thick aquifer composed of sand, gravel and boulders.

bound gravel Hard rounded or lenticular masses of gravel and sand sometimes occurring along the zone of FLUCTUATION OF THE WATER-TABLE, formed by the deposition of cementing substances around the constituent grains. Bound gravel may cause difficulties when putting down water wells; also, the masses may be mistaken for bedrock.

Bourdon pressure gauge An instrument which can be used to measure water pressure and PORE WATER PRESSURE. It consists essentially of a tube, of oval cross-section, which tends to straighten as the internal pressure is increased.

bournes VALLEY SPRINGS in which the flow is usually confined to winter rainfall, during which period they give rise to streams known as LAVANTS. Bournes or nail-bournes are terms also applied to temporary streams formed in otherwise DRY VALLEYS after heavy rainfall or very wet seasons.

box caisson (or **American caisson** or **stranded caisson**) A large box of steel, or reinforced concrete, with an open top, constructed onshore and floated out into a river or seaway and sunk at the site selected for a foundation to form part of the permanent structure. The box, which enables excavation to be done in relatively dry conditions, is used for bridge piers and similar works. *See* CAISSON.

box culvert A box-shaped conduit for carrying a small flood flow under roads or embankments. *See* CULVERT.

box dam A COFFERDAM which entirely encloses a site or area.

box drain A small box-shaped brick or concrete drain.

box gauge A tide GAUGE operated by a float inside a vertical box with openings in the base for the entry of tidal water.

Boyle's law A relationship between the pressure (p) and volume (v) of a perfect gas when the temperature is constant, i.e. pv = constant. It was discovered by Robert Boyle (1627–91).

brackish A descriptive term applied to water which contains salts, usually at a concentration of 1000 to 10 000 mg/L; this is significantly less than sea water. *See* BRINE, FRESH.

Bradford gauge A large-capacity RAIN-GAUGE used for weekly or monthly readings at less accessible stations, in which the amount of water collected is measured by means of a graduated rod immersed in the liquid. *See* RAIN RECORDER.

braided course A very wide river which has separated into a number of entwined channels called ANABRANCHES. The waters usually carry a large amount of detritus.

Bransby-Williams formula A formula to calculate the TIME OF CONCENTRATION:

$$t_c = 14.6 L\, A^{-0.1}\, S^{-0.2}$$

where t_c is the time of concentration in minutes, L is the length of the channel from divide to outlet in km, A is the catchment area in km^2, S, the slope, is that of a linear profile having the same area under it as the actual profile of the stream, in m/m.

breadth (stream) *See* WIDTH.

breakaway A term sometimes applied to a new channel cut by a river during flood periods. *See* CUTOFF.

breakers (waves) Waves which collapse or whose crests fall over. Waves approaching a shelving shore change in form and become both higher and shorter; the crests become steeper and sharper, and finally they collapse. Waves of a given height break in the same depth of water, and the zone or LINE OF BREAKERS is that along which the incoming waves collapse. *See* FORCED WAVES.

breakpoint chlorination Chlorination of water or wastewater until the chlorine demand has been satisfied, when the free available residual chlorine increases proportionately to the incremental dose.

break-pressure tank A small open tank sited on a gravity pipeline to reduce the maximum pressure of water in the pipe column. A series of such tanks is used and placed at the level of the HYDRAULIC GRADIENT of a gravity water main in undulating country. Each tank in turn takes the discharge from the main, and a new section of the main takes off and enters the next tank in the series. Since the water main operates under a much lower pressure than that in a continuous main, considerable savings in costs are often possible.

breakwater (or **mole**) Any structure to break the force of waves, applied especially to a wall constructed out into the sea to safeguard a natural or artificial harbour. *See* BLOCKWORK, FLOATING HARBOUR, RUBBLE-MOUND BREAKWATER, TETRAPOD.

breathing (river) The rise and fall in water level of a river. *See* TIDAL RIVER.

breathing well A water well in which air is alternately blown out and sucked in; the phenomenon appears to be related to barometric pressure. The blow-out of air is usually strong and noticeable, while the intake of air is not so apparent. *See* PERIODIC SPRING.

breeze concrete A concrete with poor fire-resisting qualities; however, it is cheap and nails can be driven into it. Made by mixing three parts of coke breeze (the finer coke products from coke ovens), one of sand, and one of PORTLAND CEMENT. *See* STANDARD MIX.

bridge abutment The portion at each end of a bridge works where the roadway joins to the bridge structure proper. Because of the contraction of this point and consequently the increased velocity of flow, the abutments are protected by suitable anti-erosion works. *See* ABUTMENTS, WING WALL.

brine Any water in which the content of dissolved salts is greater than sea water, such as the Dead Sea. *See* FRESH.

broad-base terrace A ridge-like terrace with gently sloping sides and a rounded top, with a dish-shaped channel to control erosion by diverting runoff along the contour of sloping ground at a non-scouring velocity. The terrace is from 4.5 to 9 m (15 to 30 ft) wide and from 0.25 to 0.5 m (10 to 20 in) high, and may be level or have a grade towards one or both ends. *See* CONTOUR FURROWS.

broad-crested weir A WEIR with a flat or gently sloping crest, with a length along the line of flow considerably greater than the height of nappe over it (Figure B.6). *See* BARRAGE.

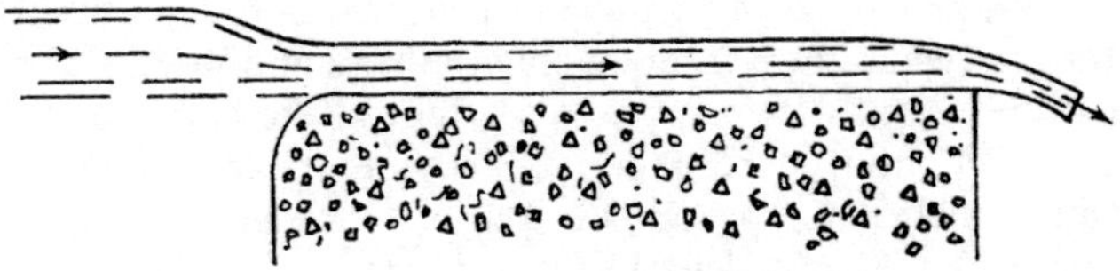

Figure B.6. Broad-crested weir

broad irrigation A method of sewage disposal, without pipe drains, sometimes used on agricultural or waste land. The sewage flows over and soaks into the ground, which has been carefully levelled to avoid local accumulations of sewage.

brook A general term for a small stream flowing in a natural surface channel; the flow is usually turbulent, and the channel rugged and stony with many irregularities; it is usually restricted to watercourses with a discharge below about 0.14 cumecs (5 cusecs).

brush matting A matting or cover consisting of branches placed on land to reduce erosion and conserve moisture while some form of vegetal cover is being established; a matting consisting of mesh wire and brush placed along a stream bank to reduce erosion. *See* RIPRAP.

brushwood Small tree branches, saplings or shrubs often used in river improvement work. They are placed close together and parallel. The material should be strong enough to resist breaking, thin enough to be flexible, and reasonably straight. Green willow branches and the straight-growing stems of the tea-tree make good brushwood. *See* FASCINES.

brushwork *See* BRUSH MATTING, BRUSHWOOD, FASCINES.

BS British Standard(s). *See* BSI.

BSI British Standards Institution. An independent organization that is approved by the British government. It publishes standards and specifications for a large number of materials, products and manufacturing processes. These standards are primarily intended for use in the UK, but most products exported from the UK are also in accordance with BSI standards. Major projects in overseas countries that have not yet established local standards are often specified in BSI standards.

bucket (1) A cup on the perimeter of a PELTON WATER WHEEL; the container of a bucket elevator or DREDGER. (2) The reversed curve in a spillway profile which deflects the water horizontally at its base from the steep overflow face onto the apron.

bucket-ladder dredger (or **elevator dredger**) A dredger that has a rigid, ladder-like frame around which moves a continuous chain of buckets. The ladder frame extends down from the ship to the floor of the river or ocean. The buckets are sent down empty to the foot of the ladder and, as they turn round the bottom of the ladder, they cut into the waste material. They return with a full load on the top side. The load is discharged into a hold in the vessel, or into a chute that leads to a barge.

bucket pump A reciprocating pump sometimes used in wells or boreholes or for dewatering purposes. *See* LIFT PUMP.

buffer solution A control solution prepared with a combination of chemicals which enable it to resist any marked change in pH after adding moderate amounts of either a strong acid or alkali.

buffer strips Strips of erosion-resisting vegetation or grass formed along the contour below or between cultivated fields.

built terrace Material deposited on the sea floor beyond the shore platform or in sheltered areas in bays; a beach deposit consisting of material derived from the weathering of a headland and shore platform.

bulk density The mass per unit volume of any material including moisture, generally expressed in kg/m^3 (lb/ft^3). *See* DRY DENSITY.

bulkhead A partition or structure in a conduit or tunnel to stop the flow of water; an upright partition in a ship to form separate compartments.

bulking The increase in volume or swelling of a dry sand or shale when it absorbs water. The increase in volume when rock or soil is excavated ranges up to 50%. *See* SWELLING PRESSURE.

bulldozer A high-powered caterpillar-mounted tractor to the front of which is fitted a concave buffer-blade. As the machine moves forward the blade levels the ground. Used on earth dams and channels. *See* TILT DOZER.

bund A continuous vertical or near-vertical wall or structure acting as a bank protection along the front of the margin of a river or channel; it gives strength and stability in the space available. In India the term is applied to a LEVEE, but in other parts of the Far East it refers to almost vertical structures to retain earth along quays and riverside drives.

buoy An anchored float to show navigable courses, reefs or natural obstructions such as bars. *See* ANCHOR BUOY.

buoyancy The upward thrust exerted by a fluid upon a body immersed in the fluid, equal to the weight of the fluid displaced. *See* ARCHIMEDES' PRINCIPLE.

buoyant foundation (or **buoyant raft**) A foundation consisting of a continuous slab of reinforced concrete which extends under the entire structure or works; it is used in areas such as river estuaries where the supporting material consists of nearly fluid silt or mud. It is so designed that the weight of the concrete raft plus the load it carries is about equal to the weight of the saturated silt it displaces. *See* RAFT FOUNDATION.

buried channels Usually preglacial depressions or valleys filled with glacial drift. Many of the present streams occupy the wholly or partly filled preglacial valleys. Tunnels and other works in these channels often present engineering problems, extra costs and perhaps modifications in design. On the other hand, the channels often yield useful supplies of shallow water.

buried erosion surface A surface of UNCONFORMITY below ground level.

bush logs Rough, unshaped tree trunks, of various lengths and sizes, for use in river and bank protection works. They are usually cut at, or near, the site where they are required for use.

butterfly gate A GATE used to control flow on hydraulic structures; similar to the BUTTERFLY VALVE, it is easily operated because it is perfectly balanced.

butterfly valve A circular disc fitted inside a pipe and hinged at two pivots on a diameter, often used in large pipes for controlling the flow, especially in HYDROELECTRIC POWER schemes. The valve opens and closes very easily because it is well balanced.

buttress A concrete or masonry projection or pier built out at right angles to a wall to reinforce its resistance to earth thrust or water pressure.

buttress dam A modified form of GRAVITY DAM in which BUTTRESSES are incorporated to direct thrusts to the foundation. A buttress dam, as compared with a gravity dam, requires less concrete but a considerably greater quantity of SHUTTERING. In the massive buttress dam, the upstream face may be flat and nearly vertical, or each buttress may have a steeply inclined round-head or diamond-head. *See* PRESTRESSED GRAVITY DAM, ROUND-HEAD BUTTRESS DAM.

butts The thickest ends of trees cut down and removed. When used for bank protection, the butts point upstream and are well anchored. *See* RAW-BANK SYSTEM.

bye channel *See* BY-WASH.

bypass A conduit or pipe arrangement for directing water flow around instead of through another conduit, pipe or valve.

bypass channel An artificial channel for directing flood water in excess of a river's capacity; the channel is often lined with LEVEES on each side to increase its capacity.

by-wash (or **bye channel** or **natural escape**) A type of spillway or escape channel for a small reservoir. It passes flood waters around a dam, as distinct from spillways passing flood water through or over a dam. It is termed a masonry or paved escape when such protective works are required, or a NATURAL ESCAPE when the ground is not paved. Some engineers consider that a by-wash for large FARM DAMS should be capable of handling flows for a 1-in-100-year flood. The failure of many farm dams is probably due to a faulty, or lacking, by-wash (Figure C.1). *See* DROP-INLET SPILLWAY, OVERSHOT SPILLWAY.

C

cable tool A sharp chisel-edged bit used in drilling a deep well; it is lifted and dropped to break the rock by impact. The rock fragments or cuttings are removed from the hole by a bailer, which is a section of pipe with a foot-valve through which the cuttings enter. *See* ROTARY DRILL.

caisson A rectangular or cylindrical watertight structure for holding back water or waterlogged sediments from a site while digging proceeds to solid ground upon which a foundation can be constructed. The caisson finally forms part of the foundation structure. *See* BOX CAISSON, OPEN CAISSON, PNEUMATIC CAISSON, SHIP CAISSON.

caisson disease (or **decompression sickness**) A disease suffered by workmen who spend long hours in the high pressure environment of a pneumatic caisson or under similar conditions, such as those of tunnel construction or deep-sea diving.

caldera lakes Lakes that occupy enlarged volcanic craters.

Calgon method The use of Calgon (sodium hexametaphosphate) to increase the supply of water from boreholes in chalk, clays and silts. *See* SWABBING.

calm belt *See* DOLDRUMS.

camber (1) The additional bulk or height given to a dam embankment to allow for settlement of the material (Figure C.1). (2) A small tidal basin or dock.

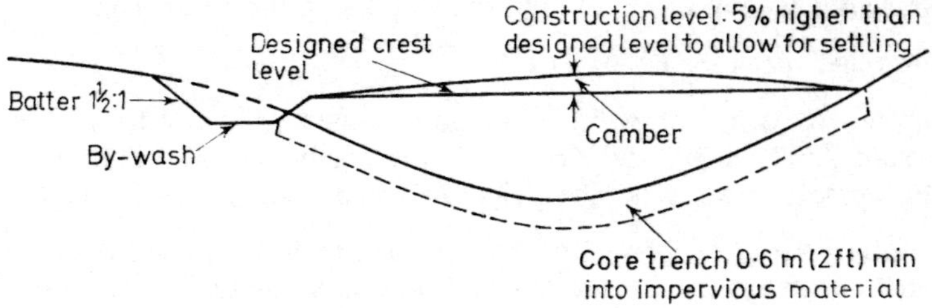

Figure C.1. Camber

camel A large hollow steel float secured to a ship to elevate and float it across a shallow area. *See* SAUCER.

camp sheathing (or **camp shedding** or **camp sheeting**) A wall to hold back a river bank; construction often consists of two connected rows of timber piles 1.8 to 3.0 m (6 to 10 ft) apart, with timbers or earth packed into the space between. *See* BARGE BED.

canal An artificial waterway or channel, usually for the passage of vessels or boats. It may also be used for connecting two or more bodies of water, for irrigation, water supply or water power, or for other applications. *See* LATERAL CANAL.

canal, inundation *See* INUNDATION CANAL.

canalization (1) A defined channel cut through a swamp area with levee banks on both sides which are constructed from material excavated from the channel. The canalized stream takes the floodwater from the upper reaches through the swampy ground, while the rain falling on the swamp itself may be removed by independent drains. The canal capacity is often designed to cope with small to medium floods. Drop structures may be necessary to reduce the grade and to give a non-scouring velocity. (2) The division of a river into reaches separated by dams, weirs and control structures. Canalization assists navigation, prevents or reduces flooding, and provides water for irrigation and sometimes HYDROELECTRIC SCHEMES.

canal lift A mobile tank to allow barges to change levels in a canal, other than via a lock. Vessels enter the tank at one end and the tank is hauled on wheels up or down an incline, or lifted and lowered vertically. *See* LIFT.

canal, main irrigation (or **feed canal** or **main canal** or **storage canal**) A canal conveying water from a stream to an off-channel storage reservoir; a conduit or channel conveying water from a storage reservoir or point of diversion to lateral or branch canals or to a group of irrigated farms.

can buoy A buoy with a flat top showing above the water level.

canoe fold *See* SYNCLINAL CLOSURE.

canyon A deep narrow valley or gorge, often with a swiftly flowing stream. Arid climate and firm rocks are conditions favourable to the development of canyons. The Colorado canyon is a good example.

capacity Receiving power or holding power; ability (as wells) to yield water or streams to transport sediment; volume of artificial or natural containers, storage or basins, etc. *See* CAPACITY OF STREAM, CAPACITY OF WELL, GROSS CAPACITY, INFILTRATION CAPACITY, INVERTED CAPACITY, MAXIMUM CAPACITY, OUTLET CAPACITY, SPECIFIC CAPACITY, TESTED CAPACITY, TOTAL CAPACITY, WATER CAPACITY, WATER-HOLDING CAPACITY.

capacity curve (or **storage curve** or **volume curve**) (1) A graph showing the relation between the volume of water in a tank, reservoir, or storage and the level of the water surface (Figure C.2). (2) A graph of the capacity rate of flow in a conduit, or pipe.

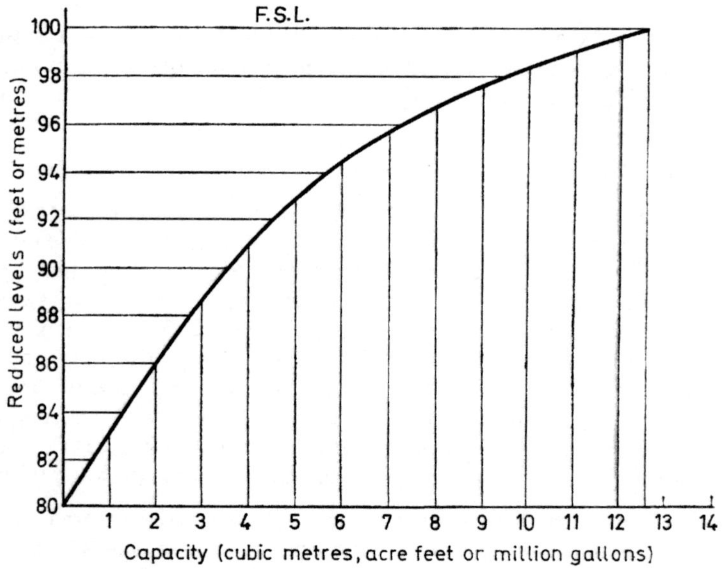

Figure C.2. Capacity curve

capacity factor In the case of channels or canals for irrigation, the ratio of average supply to authorized full supply or capacity.

capacity formula A hydraulic formula to determine the discharge or capacity of a channel:

$$Q = AV$$

where Q is the discharge or capacity (cumecs or cusecs), A is the cross-sectional area of channel (m^2 or ft^2) and V is the mean velocity of water (m/s or ft/s).

capacity of stream (1) The ability of a stream to transport material in terms of total weight. *See* LOAD, TRANSPORT COMPETENCY. (2) The largest flow of water a stream can carry without overflowing its banks.

capacity of well The ability or competence of a well to yield water under certain specified conditions. *See* ECONOMIC YIELD, MAXIMUM CAPACITY, TESTED CAPACITY, TOTAL CAPACITY.

capacity of well, maximum *See* MAXIMUM CAPACITY.

capacity of well, tested *See* TESTED CAPACITY.

capillary action A generic term for the movement of liquids due to capillary forces, or forces due to unbalanced intermolecular attraction at the liquid boundary; the rise or movement of water in the interstices of a soil or rock specifically in the CAPILLARY FRINGE.

capillary fringe The zone or belt of ground immediately above the water-table and containing FRINGE WATER or CAPILLARY WATER.

capillary porosity *See* TOTAL POROSITY.

capillary pressure (or **seepage force**) If the ground is being drained from inside an excavation, the capillary pressure tends to cause instability and the earth faces may collapse. If the ground is being drained at points outside an excavation (*see* WELLPOINT) the capillary pressure may help to stabilize the earth. The capillary pressure varies with the nature of the ground and ranges from 0.0063 to 0.063 N/mm^2 (0.9 to 9 lbf/in^2) in silt and much lower in sand.

capillary rise (or **height of capillary rise**) The height to which water rises above the free water level owing to CAPILLARY ACTION.

capillary water Water retained in rocks or soils above the water-table; water in the CAPILLARY FRINGE.

captive float (or **reefing float**) A canvas skirt, with stiffening rings spaced at regular intervals, supported by a float resting on the surface of water. By reefing, or clamping, the required number of stiffening rings to the float to shorten the canvas skirt, its length is adjusted to the depth of the channel.

carbonate hardness A term which has replaced the earlier 'temporary hardness'. *See* HARDNESS OF WATER.

carcinogen A substance that can cause cancer.

carry-over storage (1) The reserve of water retained in a reservoir after wet or normal years to ensure that the minimum outflow demands are met during drought years. (2) A reservoir which meets these requirements.

cascade (1) A stream irregularity intermediate between a WATERFALL and RAPIDS; a stretch of stream where the slope is considerable but not sufficient to cause the water to drop vertically. (2) A series of vertical steps in a sewer or drain, separating lengths where the slope is normal. A cascade avoids blockages with small or sluggish flows or scouring of pipe by high flows on steep slopes.

cataract A rush of water; a downpour of rain, or, especially, the flow of a large body of water over a precipice. The most famous of these are the Victoria Falls on the Zambezi in Africa, the Niagara in North America, the Orinoco in South America, the Falls of the Rhine at Schaffhausen and the Cascade of Gavanni in the Pyrenees. *See* WATERFALL.

catastrophic floods *See* RETURN PERIOD.

catch basin (USA) A CATCH PIT.

catch drain *See* CATCHWATER DRAIN.

catch feeder A DITCH used for irrigation.

catchment area (or **drainage area** or **drainage basin** or **gathering ground**) The area of land from which rainwater or snow melt drains into a reservoir, pond, lake or stream, usually expressed in hectares or square kilometres (acres or square miles). Catchment areas are often managed to slow down runoff, absorb precipitation and avoid erosion.

catchment area of aquifer *See* INTAKE AREA OF AQUIFER.

catchment management Management of catchments to improve the quantity and quality of runoff water. The measures include safety against fire and damage, prevention of erosion, and maintenance of ground cover. In general, the forested catchment gives the best results. *See* FIRE CONTROL.

catchment yield The total volume of water which flows from a CATCHMENT AREA over a given period, often expressed as the equivalent depth of water (mm or in) over the catchment area.

catch pit A pit or small chamber located at an accessible point in a drain, ditch or sewer to collect solid matter and thus prevent blockage of the inaccessible parts of the drains. The pit is periodically cleaned.

catchwater drain (or **catch drain** or **catchwater** or **drain** or **interception channel**) A surface ditch or drain on the rise side of an excavation or earthworks to intercept and divert surface water clear of the working site or urban water storage. *See* GRIP.

catchwaters Open channels or pipes to intercept the runoffs from adjoining catchment areas and conduct the water into the IMPOUNDING RESERVOIR. The channels are formed at steady slopes gradually crossing the contours. Pipes are often preferred, although both methods have their merits and demerits.

caterpillar gate (USA) A type of heavy steel hydraulic gate in a spillway used to control flow. It is mounted on crawler tracks with steel rollers and operates on inclined rails along each side of the opening.

causeway A raised road or way over water or wet marshy ground, consisting of an earth bank or fill material and a paved surface.

caverns Solution channels and cavities formed in rocks, usually limestone, by water percolating along joints, bedding planes and other paths of weakness. *See* SINKS, SOLUTION CAVITIES, SWALLOW HOLES.

cavern water Water occupying caverns and solution cavities. *See* SUBTERRANEAN STREAM.

caving The wearing away, undercutting and collapse of soils and rocks forming the banks of a river by the action of flowing water. *See* UNDERCUT SLOPES, UNDERCUTTING.

cavitation The formation of cavities during pumping at high speeds, resulting in corrosion of metal parts due to liberation of oxygen from the water. It may occur with centrifugal pumps or in water turbines near the DRAFT TUBE.

cavity well *See* BOULDER WELL.

celerity of wave The velocity of propagation of a wave through a liquid, compared with the undisturbed velocity of liquid through which the disturbance is propagated. The celerity of a gravity wave of relatively large wavelengths in an open channel is equal to $\sqrt{(gd)}$, where d is depth of water and g is acceleration due to gravity – 9.8 m/s^2 (32.2 ft/s^2).

cellular cofferdam A COFFERDAM with a double wall consisting of a succession of cells in contact. Each cell may consist of a steel SHEET PILE ring, of diameter 18.3 m (60 ft) and filled with sand. This is an American DOUBLE-WALL COFFERDAM, sometimes adopted in major schemes.

cementation Processes involving the injection of cement grout into fissured rocks to render them relatively watertight during excavation work, tunnelling and shaft sinking; it is also used extensively in earth and concrete dams and cutoff works for the prevention of leakage. *See* GROUTING.

cement ditches *See* LINED DITCHES.

central drainage tunnel *See* DRAINAGE TUNNEL.

centre of pressure The point on a plane immersed surface through which the resultant of all the pressure on the surface acts. The principle is applied in SELF-ACTING MOVABLE FLOOD DAMS.

centrifugal pump A pump in which the water enters near the centre of a high-speed rotating IMPELLER and through which it flows radially under centrifugal force. Its kinetic energy is then converted into pressure energy in the casing. It is a compact pump and suitable for direct electric drive, but the water must be reasonably clean. Centrifugal pumps of various types are widely used for irrigation work and water engineering generally. *See* PUMP.

centrifuge A machine rotating at a very high speed and used widely for cleaning, sizing and dewatering minerals; it is also used in soil mechanics and in sewage for dewatering sludge.

centroclinal dip A condition where the rocks or ground dip towards a common point or area. *See* BASIN.

cess A drainpipe, often porous, laid along the foot of channel excavations. It may be 230 mm (9 in) or more in diameter.

cesspit (or **cesspool**) A brick- or concrete-lined tank placed underground for collecting sewage in districts where no sewage treatment exists. The pit is pumped out periodically. The cesspit is becoming obsolete and the SEPTIC TANK is preferred. *See* AQUAPRIVY.

chain gauge A gauge consisting of tagged or indexed chain tape or other line, with a weight at the end which is lowered to touch the surface of water. It is often used for water level measurements at bridges. *See* ELECTRICAL TAPE GAUGE, WIRE WEIGHT GAUGE.

chain of buckets An arrangement used in the early days of mining and engineering for raising water, dewatering shaft sinkings and excavations. It consisted of a series of buckets attached to a chain or belt which was revolved around a wheel by men or animals. It was a primitive form of the bucket-ladder dredger for removing water.

chain of locks A series of connected LOCKS in which each CHAMBER is followed immediately by another; the HEADGATE of each lower lock forms the tailgate for the lock above it.

chalk A soft white or greyish type of LIMESTONE, which can be up to 99% pure calcium carbonate. The rock itself is relatively impermeable, but water storage and movement readily occurs in cracks and fissures. This makes chalk an excellent AQUIFER and WELLS and ADITS driven into it normally give high yields, but of hard water.

chamber (or **lock bay**) The space enclosed between the upper and lower gates with reference to a canal lock.

channel A natural or artificial groove, ditch or open conduit for the flow of water; a WATERCOURSE or CANAL; a narrow sea; that stretch of a stream where water normally flows. *See* BYPASS CHANNEL, BY-WASH.

channel axis A line joining the midpoints in successive cross-sections of a channel. *See* STREAM PROFILE.

channel capacity In artificial channels, the maximum flow carried when the channel is running at FULL SUPPLY LEVEL; in natural channels, the safe maximum flow of water that a stream can carry without overflowing its banks.

channel check A small structure, with or without gates, built across a farm supply ditch or channel to regulate and control the flow of water, usually made of steel or concrete. *See* CHECK.

channel, diversion *See* DIVERSION.

channel improvement The improvement of flow characteristics of a natural or artificial channel, thus increasing its carrying capacity. Methods include clearing, cleaning, and excavation or dredging to enlarge the area. The term is sometimes applied to measures to prevent channel erosion. *See* RIVER TRAINING.

channel line The path or line of maximum flow along a stream, usually coinciding with the STREAM PROFILE.

channel of approach A channel which leads up to a measuring weir, gate, tube or orifice. *See* APPROACH VELOCITY.

channel precipitation That part of precipitation that falls directly on the water surface of a canal, stream or any open channel.

channel revetment A lining or facing to protect a channel bank from erosion; it may consist of stone, concrete, piling, etc.

channel roughness The unevenness, irregularity, or texture of the channel surface in contact with the flowing water. *See* CHÉZY'S FORMULA, KUTTER'S FORMULA, MANNING'S FORMULA, ROUGHNESS COEFFICIENT.

channel storage The water stored in a channel during periods when the flow exceeds the discharge capacity of the channel. *See* BANK STORAGE.

check A structure built to regulate or raise the water level in a supply channel; an area of land irrigated by water confined by earth ridges.

check and drop A structure which combines the functions of both a CHECK and a DROP; the water level may be raised upstream of the structure and dropped on the downstream side.

checkdam A low, temporary type of dam or barrier erected across a narrow watercourse to retard flow, minimize erosion and promote the deposition of silt. It may be of stonemesh, with an apron of similar construction or of timber. The establishment of a thick growth of trees or bushes along the banks and bed of the channel also retards velocity and scour.

check irrigation An irrigation method whereby earth ridges are built around an area, partly or wholly to retain water. *See* BASIN IRRIGATION, BORDER DIKES, SQUARE CHECK IRRIGATION.

check valve *See* CLACK VALVE.

chemical gauging (or **chemi-hydrometry**) A method of measuring quantity or flow of water by introducing a chemical solution of known concentration at a cross-section upstream, and then determining the degree of dilution of the solution at another cross-section downstream. The method is sometimes used along mountain streams where current meters cannot be used, and where measuring structures would be costly or difficult to erect. *See* DILUTION METHODS, SALT VELOCITY METHOD.

chemical oxygen demand (COD) The amount of oxygen used in the chemical oxidation of both organic and inorganic matter present in a sample of water or wastewater under standard conditions. *See* BIOLOGICAL OXYGEN DEMAND.

chemical precipitation The settlement of sewage aided by FLOCCULATION.

chemical water treatment The addition to hard water of chemical substances that break down impurities. The residue is either precipitated for subsequent retention in a filter or driven off as a gas, or passed on in solution in harmless or less harmful form. Lime or soda or a combination of both is generally used with or without the addition of COLLOIDS or ZEOLITES. *See* COLLOIDAL WATER TREATMENT, HARDNESS OF WATER.

chemical weathering The action of rainwater containing weak acids on stone structures, rocks and soils. The acids are derived from atmospheric gases such as carbon dioxide and sulphur dioxide. In industrial areas, chemical weathering may be serious on account of the higher content of corrosive impurities in the air and rainwater. Limestone or calcium carbonate is dissolved by rainwater containing carbon dioxide, and is held in solution as calcium bicarbonate, thus:

$$CaCO_3 + H_2O + CO_2 \rightarrow Ca(HCO_3)_2$$

chemi-hydrometry *See* CHEMICAL GAUGING.

chemise A wall built as a protective lining to an earth bank.

chevron drain (or **herringbone drain**) A system of rubble-filled trenches or open jointed pipes set out in herringbone pattern; the water flows into main drains laid out in the direction of maximum slope (Figure C.3).

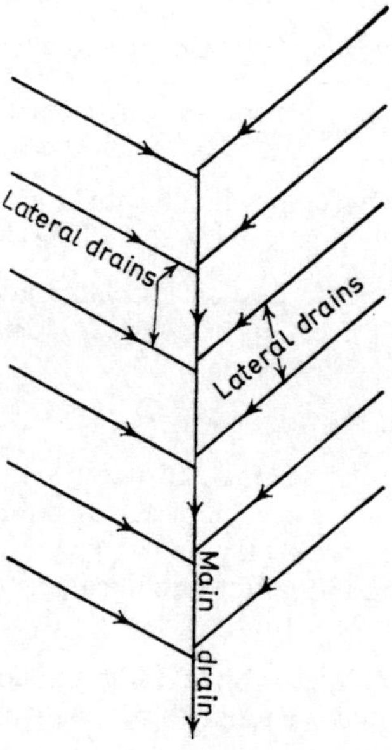

Figure C.3. Chevron drain

Chézy's formula An empirical formula, put forward by Antoine Chézy in 1775:

$$V = C(RS)^{0.5}$$

where V is the mean velocity of flow, R is the hydraulic radius, S is the slope of the channel, and C is the Chézy roughness coefficient. Estimates of the coefficient C can be made by BAZIN'S FORMULA and KUTTER'S FORMULA.

chimney A conspicuous pillar-like projection of rock on a coast which has been separated from the sea cliff by wave action around it. *See* STACK.

chimney drain A vertical or near-vertical drain in an earthen embankment, downstream of the CORE, to relieve the build up of PORE WATER PRESSURE in the downstream earth shell. The chimney drain continues as a blanket drain on the foundation to the downstream toe. The drain material may be constructed of sand, gravel or rock, with a thickness up to 3 m (Figure C.4).

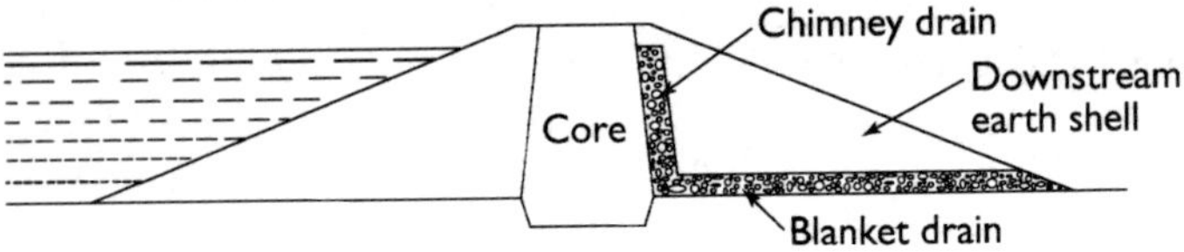

Figure C.4. Chimney drain

chippings Crushed stone fragments ranging from 3 mm to 25 mm (1/8 in to 1 in). *See* AGGREGATE.

chironomid A non-biting midge. Some species are indicators of organic pollution in rivers, while others graze on biological filters.

chloramines A compound formed during chlorination from reaction with ammoniacal nitrogen present in a water or wastewater, which affects chlorine demand.

chlorinated lime *See* BLEACHING POWDER.

chlorination Treatment of sewage and water with chlorine to aid purification; DISINFECTION of water with bleaching powder or other substance containing chlorine. Chlorine may impart or generate obnoxious tastes and odours and as a result various modifications of the treatment have been developed. *See* OZONIZATION.

chlorine demand The difference between the amount of chlorine added to a water or wastewater and the residual chlorine remaining at the end of a specified contact time.

cholera An infectious and often fatal disease of the small intestine caused by the bacterium *Vibrio cholerae*, resulting in severe vomiting and diarrhoea. The disease is transmitted by the faecal oral route, as in the case of polio and typhoid. It was a major cause of death in cities until the end of the 19th century, and continues to cause worldwide epidemics. There is no effective vaccine: the best control measures continue to be interventions in drinking water, sanitation and hygiene education.

C-horizon The soil-forming decomposed rock below the soil layers. *See* A-HORIZON.

chute A steep conduit or channel for conveying water to a water wheel or to a lower level without causing erosion. *See* HEAD RACE.

Cipolletti weir A form of MEASURING WEIR with a trapezoidal opening, with maximum width at the top and side slopes of 1 horizontal to 4 vertical (Figure C.5). *See* WEIR.

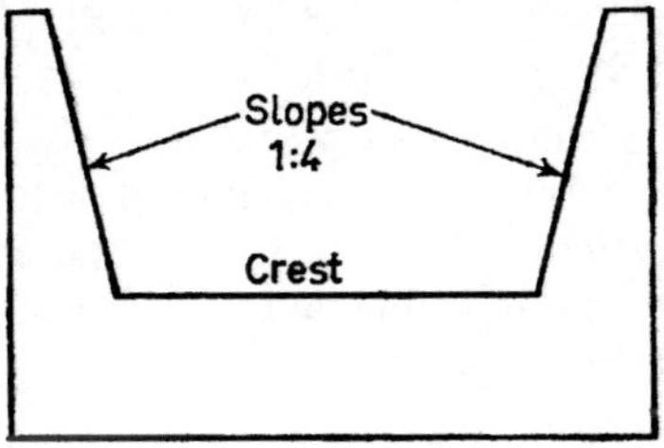

Figure C.5. Cipolletti weir

circle of influence The circular area of ground at the surface corresponding to the circular base of the inverted CONE OF DEPRESSION.

circular-arc method The use of the SLIP SURFACE OF FAILURE principle when investigating the possibility of slides in dam embankments, foundation excavations and clay banks.

circulating pump Normally a centrifugal pump used to maintain a continuous flow of water through a treatment or cleaning plant, or through a condenser of a steam plant.

cirque glaciers (or **hanging glaciers** or **glacierets**) Small glaciers which are perched well above valley glaciers, and occupy depressions in mountain slopes.

clack valve (or **check valve**) A simple type of NON-RETURN VALVE or REFLUX VALVE, most suitable for smaller diameter pipes (Figure C.6).

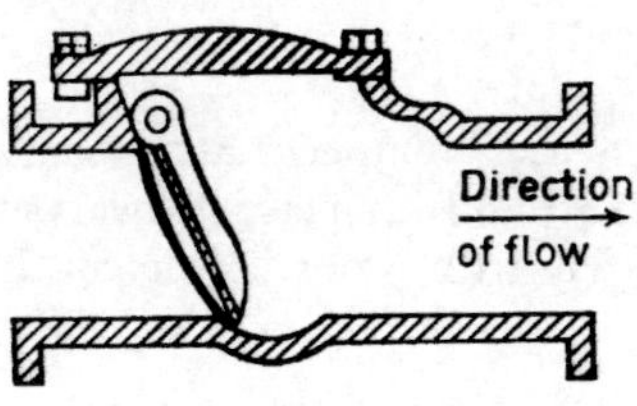

Figure C.6. Clack valve

clapotis The lapping of waves reflected from a wall, breakwater or bund that is above the water level.

clap sill *See* LOCK SILL.

clarification To clean or remove suspended or colloidal matter from liquids. *See* CHEMICAL WATER TREATMENT, SUSPENSION.

clarifier A sedimentation tank used to remove settleable solids from water or wastewater. *See* SETTLING TANK.

clay A fine-grained earthy deposit with a grain size less than 0.002 mm, and composed mainly of hydrous aluminium silicates with small amounts of other minerals. In water engineering, clay is important as it is practically an impermeable mass and will not permit the passage of water, or only at a very slow rate. *See* SAND, SILT.

clay blanket A construction to reduce the seepage of water through the bed or banks of earthen supply dams or channels (Figure C.7). It usually consists of a clay layer, about 0.6 m (2 ft) thick, placed on the slope of the banks or on the bottom of the bed. *See* HEELING.

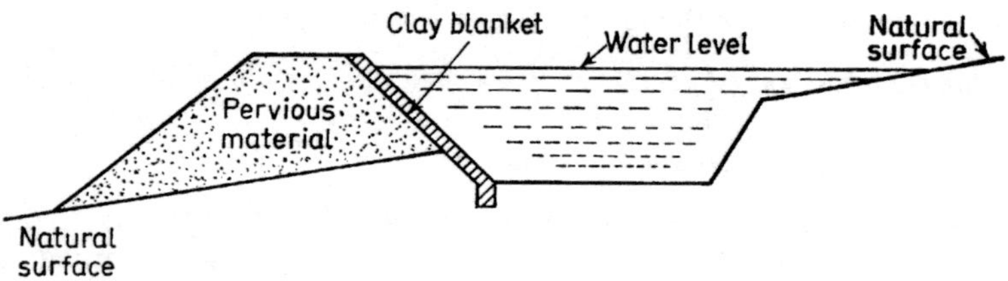

Figure C.7. Clay blanket

clay core A watertight core or barrier, consisting of puddled clay, in dam construction and embankments. *See* CUTOFF TRENCH, PUDDLE.

clay cutter The shaft-driven bit fitted at the lower end of the suction pipe in a SUCTION-CUTTER DREDGER. It extends the effective range of the dredger considerably.

clay lining *See* LINING.

claypan A bed of dense, compact, and relatively impervious clay occurring near the surface, stiff or plastic when wet and hard when dry. Its presence may interfere with water movement during the drainage or irrigation of land.

clay puddle (or **puddle** or **puddle clay**) A plastic substance formed by mixing clay with about one fifth of its weight with water. It is used as a waterproofing material in hydraulic engineering, as under PUDDLE. The coefficient of permeability of puddle clay usually ranges from 3 to 9 mm/year (0.01 to 0.03 ft/year).

clay sealing (earthen channels) The injection of clay into earthen channels to seal the voids in permeable sections and thus reduce seepage losses. The treatment has effected seepage reductions of up to 70%, and the treatment costs are low. The plant consists of a small centrifugal pump and a drum in which the water is mixed with pulverized dry clay, and then spilled down a chute into the channel. *See* GROUTING.

clay slip The downward movement of a substantial portion of a clayey bank or mountain slope, caused mainly by the presence of water along slippage planes. A clay slip has two manifest characteristics: (1) a pronounced heave at the base or toe, and (2) tension cracks along the top of the bank. *See* LANDSLIDE.

cleading A timber lock gate, or the boarding of a cofferdam.

clear-water reservoir *See* SERVICE RESERVOIR.

climate The average condition of a land area with respect to all atmospheric and meteorological influences, mainly of temperature, dryness, pressure, wind, evaporation, light, etc. *See* ARID, SEMI-ARID, WEATHER.

climate, continental A climate typical of land areas separated from the moderating influences of the sea by mountain barriers or by distance, and marked by relatively large seasonal and daily temperature changes.

climate, oceanic (or **marine climate**) A climate typical of land masses near the oceans, which affect the humidity and modify the temperature and temperature variations.

climatic year A hydrologic and meteorological record extending over a continuous period of twelve months. *See* WATER YEAR.

climatology The science of climate; the systematic measurement and study of climatic elements, including statistical relations, frequencies, mean values, normals, variations and distribution. *See* MICROCLIMATE.

clinometer A hand-held instrument for sighting along inclined planes and for measuring the angle of slope of the sides of earth dams, canals and embankments. *See* ABNEY LEVEL, BATTER LEVEL.

closed drainage (or **non-contributing area**) A land area where surface runoff has no outlet to the sea and flows into lakes, ponds, or other depressions or sinks.

closer A sheet pile specially cut or fabricated to close a COFFERDAM in cases when a standard pile will not fit the space.

closing dike A structure erected across a branch channel of a river to regulate, reduce or stop the flow of water. A similar structure is sometimes used in conjunction with other works to control the flow in a breakaway.

Clostridium A genus of spore-forming anaerobic bacteria. Some species form resistant spores which can survive in water much longer than other faecal indicators.

closure *See* STRUCTURE CLOSURE.

cloud A mass of visible condensed watery vapour suspended in the atmosphere high above general ground level; at low levels or in contact with the land surface it is called mist or FOG.

cloudburst A violent rainstorm, in which extremely heavy precipitation falls over a short period. In treeless areas the runoff cuts deep gullies and often does severe damage to land and property. Streams often become overwhelmed and have their courses changed.

cloud seeding Under certain conditions, the use of chemical substances, such as silver iodide, to cause clouds to produce rain. A specially equipped aircraft is used.

cloud velocity gauging A method of measuring velocity in a channel. A known quantity of dye solution is introduced at a station, and velocity is measured at the moment a concentration of the solution passes two other given stations located downstream and separated from the injection point by a short turbulent reach. *See* DILUTION METHODS.

clough A sluice gate in a CULVERT.

coagulation The coalescence of fine suspended matter by the addition of floc-forming substances, such as ALUM, FERROUS SULPHATE, *MORINGA OLEIFERA* seeds and POLYELECTROLYTES.

coarse-grained soil A soil in which sand and gravel predominate, and the grains range from 0.5 to 2 mm and over. It is a soil type least affected by moisture-content changes, and most surface rain becomes GRAVITATIONAL WATER. *See* CLAY.

coarse strainers Coarse drum and bar screens or band strainers placed at water supply intakes to keep out trash and debris, and so prevent obstruction or damage to pumps, pipelines and valves.

coast The zone between the inland limit of manifest marine influences and the outer limit of breaking waves. *See* BACKSHORE, FORESHORE, SHORE.

coastal engineering A branch of civil engineering concerned mainly with works to arrest the erosive action of water and waves. *See* COAST PROTECTION.

coastal lagoon A LAGOON running between the mainland and a sand spit, and along which a river flows before gaining an outlet to the sea. Usually it breaks open a new direct outlet only in flood time. In some areas the sand spit is breached periodically to protect the lands around the lagoon from inundation.

coastal plain The area exposed along a coast where the sea retreats from the land mass; the sea floor exposed when the land suffers relative uplift.

coastal plain swamps Uplifted portions of the sea floor which contain an excess of moisture, primarily because of their flatness and consequent poor drainage; the water is usually fresh. *See* DELTA-PLAIN SWAMPS.

coastline The margin of the land along the coast, which may comprise a line of cliffs.

coast of emergence The coast where the sea retreats from the land mass as a result of earth movements or EUSTATIC MOVEMENTS.

coast of submergence A coast where the sea advances over the land because of earth movements. The sea enters the valleys, and the coast and headlands are more strongly attacked by the waves. *See* DROWNED VALLEYS.

coast protection The local protection of the coast by arresting the erosive action of water and waves with SEA WALLS or by GROYNES. The erosion of dunes may be arrested by the planting of shrubs and grasses to bind the surface layer and encourage the growth of a vegetal cover.

COD *See* CHEMICAL OXYGEN DEMAND.

coefficient of conductivity *See* COEFFICIENT OF PERMEABILITY.

coefficient of contraction (C_c) A value related to the VENA CONTRACTA. When a jet of water flows through an orifice under pressure, the area of the jet (A_j) is less than the area of the orifice (A_o). Thus:

$$A_j = C_c A_o$$

where C_c is the coefficient of contraction (Figure C.8).

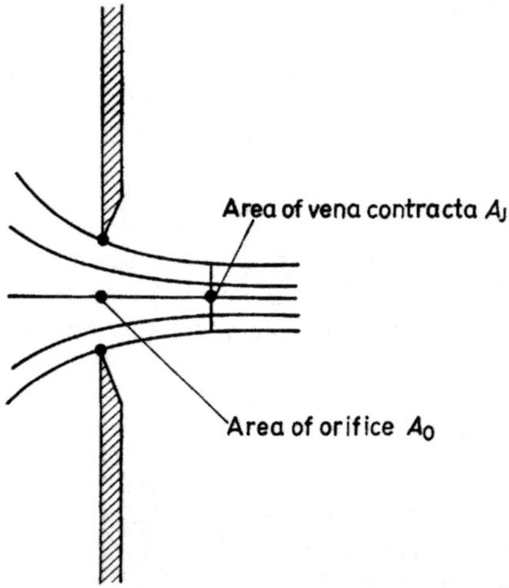

Figure C.8. Coefficient of contraction

coefficient of discharge (C_d) The ratio of the actual discharge of water through a weir, pipe or orifice to the theoretical discharge. The average value for an orifice is about 0.62. *See* EFFECTIVE AREA OF AN ORIFICE.

coefficient of imperviousness (USA) The IMPERMEABILITY FACTOR.

coefficient of permeability The rate of flow of water, usually in litres or gallons per day through unit cross-section of soil under a HYDRAULIC GRADIENT of unity and at a specified temperature. It is used to calculate seepage in earth channels and dams, and also drawdown in wells. Also called coefficient of conductivity, coefficient of transmission, hydraulic conductivity, transmission constant, unit of permeability.

coefficient of regime The ratio of the maximum annual flow to the minimum annual flow.

coefficient of rugosity *See* ROUGHNESS COEFFICIENT.

coefficient of transmissibility *See* TRANSMISSIBILITY COEFFICIENT.

coefficient of transmission *See* COEFFICIENT OF PERMEABILITY.

coefficient of uniformity *See* UNIFORMITY COEFFICIENT.

coefficient of velocity The ratio between the actual discharge velocity and the theoretical discharge velocity of water flowing through an orifice. The average value for a sharp-edged orifice is 0.98.

coefficient of viscosity A quantitative expression of the friction between the molecules of a fluid when in motion. The ability of a rock or soil to transmit water through the void spaces varies inversely as the coefficient of viscosity of the water.

cofferdam A temporary watertight case, barrier or dam, forming an enclosure in submerged or waterlogged ground, to exclude water and permit free access to the area within; sheet piling may be used, or a dam built above the ground to exclude water. Cofferdams may be formed to about 9 m (30 ft) below water level, and are sometimes used in deep wet excavations, foundations, and bridge building. *See* DOUBLE-WALL COFFERDAM.

cohesive soil A CLAY or clayey SILT, as opposed to SAND, which is a cohesionless or frictional soil.

col A depression or pass in a mountain range; a strip of land connecting a range of hills with an outlying feature.

cold front *See* FRONTAL SURFACE.

cold spring A spring whose water has a temperature below the mean annual temperature of the atmosphere in the area; or, in localities where there are also warm THERMAL SPRINGS, the term refers to springs whose water temperature is more normal and fairly constant at the annual average air temperature.

Colebrook–White equation This largely empirical expression is the most precise expression in general use for estimating the head loss to friction when water flows through pipes or open channels. For pipe flow it relates the velocity of flow V to a complex function of pipe diameter D, a measure of surface roughness k_s, kinematic viscosity ν and slope i, where g is acceleration due to gravity.

$$V = -2\sqrt{(2gDi)}\,\log_{10}\left[\frac{k_s}{3.7D} + \frac{2.51\nu}{D\sqrt{(2gDi)}}\right]$$

Modifications to the equation are needed for conduits of other shapes and channels. The Colebrook–White equation is too complex to be easily used directly, but tables of results computed using it are readily available.

coliform bacteria A group of bacteria found in the intestine of most animals, but also present in soil and vegetation. In general they are harmless, but tests for their presence are commonly used to indicate the presence of sewage pollution.

collecting area The surface area from which an aquifer receives its supply of water. The area may be large or small and near to, or at some distance from, a well supplied by the aquifer.

collecting system All the drains or sewers in a sewerage system between the buildings from which the sewage originates and the sewage disposal works. *See* COMBINED SYSTEM.

colloidal water treatment The treatment of hard water by the addition of COLLOIDS – often proprietary substances. The colloid forms a film around the solids in the water and prevents their adherence to pipes and metal work. *See* CHEMICAL WATER TREATMENT, HARDNESS OF WATER.

colloids Substances in a very fine state of subdivision, which are suspended rather than dissolved in water; they diffuse through membrane very slowly or not at all.

colony count *See* PLATE COUNT.

colour velocity gauging Determining water velocity by introducing a suitable dye and recording the time it takes to float along a known distance downstream. *See* CLOUD VELOCITY GAUGING.

columnar section *See* VERTICAL SECTION.

combination well A type of OPEN WELL which is connected to an INFILTRATION TUNNEL or to one or more other wells. *See* OPEN PIT.

combined chlorine *See* RESIDUAL CHLORINE.

combined system A drainage system in which the surface and soil waters are carried in the same drains and sewers. *See* SEPARATE SYSTEM.

comb ridge (or **arête**) A rocky spur or sharp-crested ridge between adjacent valley glaciers.

commanded land Generally, land capable of being irrigated by gravity from the outlets of water supply channels.

comminutor A device that shreds or grinds the larger suspended materials in wastewater prior to sedimentation.

common excavation A contract term used in earth dam specifications. It includes all material other than ROCK EXCAVATION but includes weathered or soft rock and cemented materials, which are usually ripped for loading purposes. It includes boulders up to 0.76 m^3 (1 yd^3) in volume.

communication pipe A pipe connecting the mains to the boundary of the consumer's property.

compacted yards (or **fill yards**) A contractor's term used in the pre-metric era to describe the quantity of soils placed and compacted in earth dams. Materials excavated were called CUT YARDS. In the contractor's jargon a yard meant one cubic yard. One cubic yard is equal to 0.765 cubic metres.

compaction *See* SUPERFICIAL COMPACTION.

compaction plant Machines fitted with rollers to increase the density of soils in earth dams and embankments. The number of passes required varies with the soil type, moisture content and kind of roller used. *See* PADFOOT ROLLER, SMOOTH-DRUM ROLLER, VIBRATING ROLLER.

compaction, water of *See* WATER OF COMPACTION.

compensation water The water which must legally be released from a reservoir to meet the requirements of the downstream users that received a water supply before the dam was constructed.

competence (stream) The ability of a stream to carry material; the largest pebble or boulder which a stream can transport. The transporting power of a stream increases at the rate of the fifth or sixth power of its velocity; thus, after heavy rains even boulders are moved along the bed. *See* TRANSPORT COMPETENCY.

complete flood control *See* FLOOD CONTROL (COMPLETE).

composite unit graph A graph or tabular presentation covering several unit hydrographs for important sub-areas of a basin or drainage area. Allowances are made for time of flow from sub-area outlets to the main gauging station, and runoff for each sub-area is computed independently. The graph gives the estimated total flow at the outlet of the basin.

composting latrine A communal dry toilet (from which urine is drained away) with sealed vaults to receive the solids for composting. Vaults, usually in duplicate for alternate use, are often designed to be cleaned out and reused every six months.

compound coast A coast affected by emergence during one period and by submergence at a later period, both movements being of roughly equal magnitude.

compound cross-section A CROSS-SECTION of a stream or channel in which the width abruptly increases above a certain level of the sides (Figure C.9).

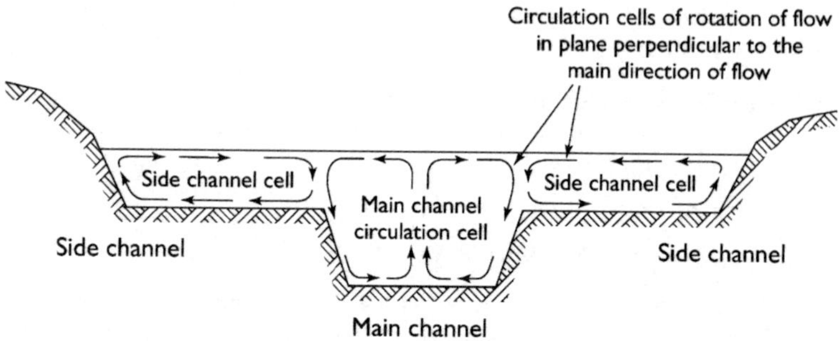

Figure C.9. Compound cross-section channel flow

compound dredger A DREDGER equipped with a bucket-ladder for removing mud from the bed of a river, or other water, and also a CLAY CUTTER. *See* SUCTION-CUTTER DREDGER.

compound hydrograph A HYDROGRAPH showing flow during an intermittent storm when the effects of one sub-storm continue into the next.

compound weir A flow-measuring structure consisting of a combination of a TRIANGULAR NOTCH and one or more RECTANGULAR WEIRS. It is used to determine stream flows over a wide range of dry weather and flood conditions (Figure C.10).

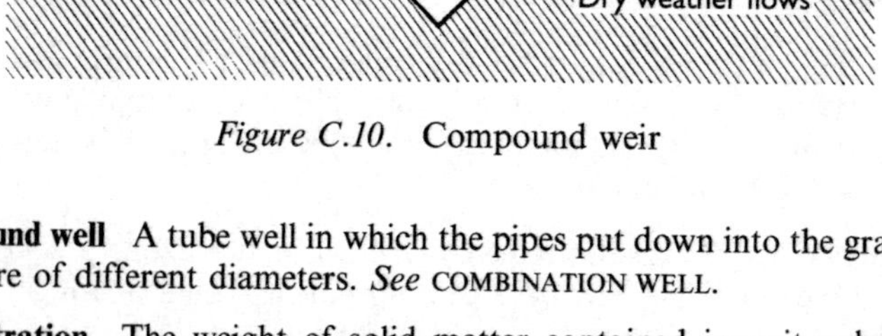

Figure C.10. Compound weir

compound well A tube well in which the pipes put down into the gravel or sand are of different diameters. *See* COMBINATION WELL.

concentration The weight of solid matter contained in unit volume of water or other fluid. *See* SEDIMENT WATER, SUSPENSION.

conceptual model An idealized and usually simplified representation of a structure or process in such a form that it becomes most suitable for analysis. Nearly all engineering calculations are based directly or indirectly on some type of conceptual model, as without simplification the analysis needed for design would be excessively and unnecessarily complicated. Analysis of conceptual models is usually done numerically, but physical or analogue methods can be used. Conceptual methods range from the very simple to the very complex, intended only for analysis with powerful computers.

concrete A construction material consisting of an intimate mixture of cement (which acts as a binder), stone, sand and water. It hardens to a stone-like mass, and is used widely for hydraulic structures. *See* BREEZE CONCRETE, PRESTRESSED CONCRETE, REINFORCED CONCRETE, CYCLOPEAN CONCRETE.

concrete bagwork *See* BAGWORK.

concrete dam Any dam mainly constructed from concrete, usually including ARCH DAMS, GRAVITY DAMS, and BUTTRESS DAMS. In view of the cost of concrete, the design is prepared as accurately as the data available allow. Also, the foundations for a concrete dam must possess considerably greater bearing strength than those considered adequate for an earth dam. *See* PRESTRESSED GRAVITY DAM.

concrete piling Driven concrete piles, used as in TIMBER SHEET piling, as a protection against bank erosion. It lasts longer than timber and is not liable to attack by MARINE BORERS. The addition of a stone apron improves the construction but is rather costly. *See* SHEET PILES.

concrete pipes Plain or reinforced concrete pipes from about 75 mm to 2 m (3 in to 6 ft) in bore, and porous when used for SUBSOIL DRAINAGE. They are not commonly used for conveying water under pressure, though they can be made to withstand moderate pressures of about 52 m (180 ft). *See* PRESTRESSED CONCRETE.

condensation (of vapour) The change of vapour into the liquid or solid state; water vapour may change to cloud, rain, dew, sleet, or snow, according to prevailing atmospheric temperature. *See* DEW POINT.

condensation nucleus The central particle around which condensation of water vapour starts and gathers in the free air.

conditioning (sludge) The treatment of sludge to facilitate dewatering by physical or chemical methods.

conduit A general term for any artificial channel, flume, ditch, canal, pipe or other appliance for conveying water.

cone of depression The inverted cone-shaped depression in the water-table around a well, shaft or borehole in which pumping is in operation

(Figure C.11). The rate of pumping and recharge of groundwater will affect the shape and volume of the cone. *See* CIRCLE OF INFLUENCE, DRAWDOWN, WELL INTERFERENCE.

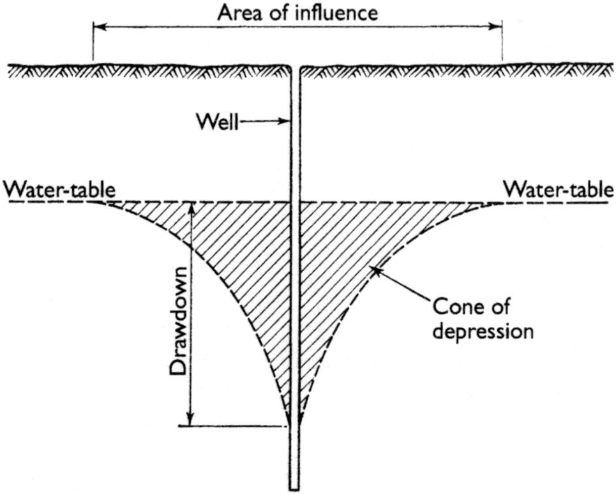

Figure C.11. Cone of depression

cone of exhaustion A term for CONE OF DEPRESSION.

cone of influence A term for CONE OF DEPRESSION.

confined aquifer *See* ARTESIAN AQUIFER.

confined eddy *See* EDDY (CONFINED).

confined water (USA) A term for ARTESIAN WATER.

confining bed A relatively impervious deposit directly above or sometimes below a water-bearing rock; an impervious rock which confines water under pressure in an underlying aquifer.

confluence The flowing together or uniting of streams, such as the point where a tributary joins the main stream. *See* BRAIDED COURSE.

conjugate depths Fluids flowing in open channels and at a velocity above the CRITICAL VELOCITY can satisfy BERNOULLI'S THEOREM for the SPECIFIC ENERGY of flow at two distinct depths. These two depths are known as CONJUGATE DEPTHS; at the deeper the flow will be SUB-CRITICAL and at the shallower SUPER-CRITICAL.

conjunctive use The linking and joint development of a number of different sources to enhance supply, provide operational flexibility and improve reliability. Typically a surface reservoir is linked to groundwater sources so that during periods of low rainfall supplies can be supplemented by ABSTRACTIONS pumped from wells.

connate water (or **'fossil' water**) Native or original water in the interstices of a sedimentary rock which was not expelled during consolidation. *See* JUVENILE WATER, METEORIC WATER.

consequent streams Streams whose channels were determined by the initial topography or slopes of a newly formed land mass. *See* DRAINAGE PATTERN, SUBSEQUENT STREAM.

conservation storage Water stored in channels, ponds, and reservoirs during heavy precipitation and later released for power, irrigation or municipal use.

conservation survey *See* SOIL CONSERVATION SURVEY, WATER PROSPECT MAP.

consolidation The slow volume reduction of a cohesive soil (clayey soil) under applied loads, as in foundations, earth dams and embankments. During this volume reduction some of the pore water is expelled. *See* SUPERFICIAL COMPACTION.

consolidation settlement The gradual settlement of a loaded clayey deposit; the process may be accelerated by VERTICAL SAND DRAINS. *See* SETTLEMENT.

constant rate of injection method A measurement of discharge in an open channel by injecting a constant flow of a solution of known concentration at a station upstream and followed by a dilution measurement at another station downstream where the mixing is complete. *See* SUDDEN INJECTION METHOD.

construction materials Materials obtained from borrow pits, quarries and excavations, such as sand, gravel, building stone, road metal, ballast, clay, lime or limestone, etc., and prepared for use in engineering works, such as retaining walls, dams, canals, groynes, docks, harbours and breakwaters. *See* BORROW PIT, CLAY, MATERIALS SURVEYS, SAND.

consumptive use The amount of water used by living plants plus that transpired and evaporated. *See* EVAPO-TRANSPIRATION.

contact aerator A tank in which sewage is treated or aerated by injecting compressed air.

contact bed *See* BACTERIA BED.

contact erosion valley A valley carved by erosion along outcropping lines of weaker rocks at conformable bedding contacts, at igneous contacts, or at upturned unconformities and faults. Valleys formed along fault lines are termed FAULT-LINE VALLEYS, where the block of softer beds becomes the site of the valley.

contact load The material transported by a stream along its bed by sliding and rolling. *See* COMPETENCE, LOAD.

contact spring (or **stratum spring**) An outflow of groundwater at or along the base of a permeable bed resting on an impervious bed, often seen at the base of an escarpment (Figure C.12). Springs of this type usually give an increased flow after a rainy period. *See* ARTESIAN SPRING, FAULT SPRING.

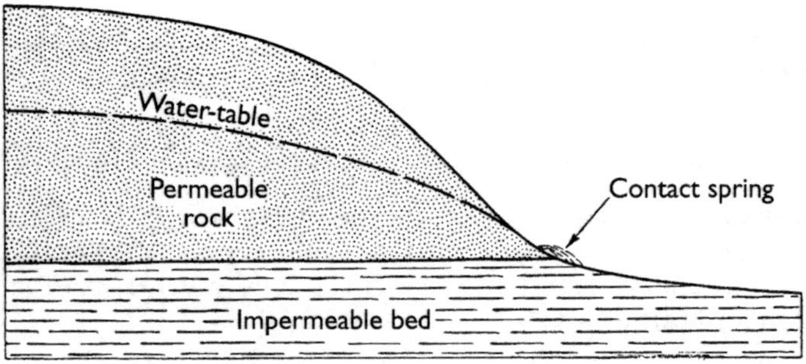

Figure C.12. Contact spring (or stratum spring)

contact tank A tank used in the chemical treatment of a water or wastewater to ensure adequate contact time, such as for chlorination.

continental ice sheets Thick and widespread layers of ice or ICE CAPS in regions where the SNOW LINE is low, such as that covering the Antarctic continent.

continental shelf The submarine continuation of a continent; a submerged platform sloping gently seaward to the CONTINENTAL SLOPE.

continental slope The true boundary of a continental mass with a steep slope downwards to the deep ocean floor.

continuity, law of *See* LAW OF CONTINUITY.

continuous delivery (irrigation) A scheme by which each irrigator's allotted quantity of water is delivered at a continuous rate. *See* ROTATION IRRIGATION.

continuous filter *See* BACTERIA BED.

continuous stream A stream which flows without interruption or contains water along its entire course between any two points. *See* INTERRUPTED STREAM.

contour *See* CONTOUR LINE.

contour bank A ridge of earth built along a contour line for irrigation, or as a soil conservation measure.

contour ditch A ditch or furrow cut along a contour line.

contour furrows Furrows ploughed along the contour lines of pasture or other land: they allow water to penetrate the soil but prevent serious erosion, as the water follows the rows at a low velocity. *See* CONTOUR LINE, CONTOUR PLOUGHING.

contour interval The difference in height or elevation between adjacent contour lines. *See* BENCH MARK, WATER-TABLE CONTOUR PLAN.

contour line (or **contour**) A line drawn on a plan or map to represent an imaginary line on the surface or underground of a definite level at all points. *See* STRUCTURE CONTOURS.

contour plan A plan showing principally the measured or estimated surface contours of the bedrock, water-table, upper surface of an aquifer, or any other deposit of importance during the construction of a reservoir, canal or other project. At a reservoir site the necessary detailed contouring may be carried out by ground or aerial surveys. The latter are speedy if large-scale maps are available, and the cost is often lower than with ground surveys. *See* DAM SITE INVESTIGATION.

contour ploughing (or **terracing**) The ploughing of furrows at very low slopes along which scour by water is very slight. It is adopted as a soil conservation measure on slopes adjoining reservoirs, canals and other works in valleys. *See* CONTOUR FURROWS.

contracted weir A MEASURING WEIR with a rectangular notch, which is narrower than the channel. If one side of a measuring weir is flush with the channel side it is said to have one END CONTRACTION, or if both sides of the weir are some distance away from the channel sides then it is a weir with two end contractions. *See* BOTTOM CONTRACTION.

contraction (weir) The combined effect of end contractions and bottom contraction. *See* CONTRACTED WEIR.

contraction, bottom *See* BOTTOM CONTRACTION.

contraction, end *See* END CONTRACTION.

control (1) The structural features of an OPEN CHANNEL that determine the

STAGE of the open channel at a given point for a specific discharge. (2) A section or area which has received standard treatment or no treatment, and is used in research investigations as standard for comparison with an experimental area; a comparative research often used in irrigation investigations. *See* CONTROL REACH.

control flume Part of an open channel suitable for measuring the flow of water; a short contraction of flume where CRITICAL DEPTH occurs. *See* RATING FLUME, VENTURI FLUME.

control meter A structure in the channel of a stream where CRITICAL DEPTH is obtained by decreasing the width or raising the bottom, or both combined.

control of flood *See* FLOOD CONTROL.

control reach A length of channel or stream, upstream from a natural or artificial barrier, along which the water level is relatively stable at some or all stages of discharge. Control may be partial or complete. Partial control exists if downstream fluctuations have some effect on the upstream water level; control is complete if the upstream water level is entirely independent of downstream fluctuations. *See* CONTROL.

control section *See* CONTROL REACH.

control valve A valve which regulates the flow of water in a pipeline and gives a constant discharge irrespective of heads or pressures; a term sometimes applied to a DISCHARGE VALVE.

convectional rain Rain resulting from the upward movement of air heated above the temperature of its surroundings. Thundery rain in a temperate region is commonly typical of convectional rain.

convergence zone, intertropic *See* INTERTROPIC FRONT.

conveyance loss The loss of water, largely unavoidable, in irrigation and water supply channels. It includes all losses (evaporation, seepage, structure leakage, absorption, etc.) between HEADWORKS and farms or consumers.

co-operative drainage The joint action of two or more mine owners, landowners or farmers to drain their properties on the most efficient and economic lines. The cost is shared between the owners on an agreed basis. In mines, pumping stations may be installed at suitable centres, or drain tunnels so positioned as to intersect mine workings below the lowest ore or coal deposit, or at some other intermediate level, depending on local geology and topography. The cost and length of the drain tunnel and the dewatering benefits are closely investigated. *See* DRAINAGE TUNNEL, LAND ACCRETION.

copper sulphate A chemical prepared from copper and sulphuric acid sometimes used to control algal growths. It should be used with great care under controlled conditions as it is poisonous in high doses. A product, copper sulphate pentahydrate, is proving to be a non-toxic effective disinfectant of water-borne bacteria, and could be a most useful disinfectant of water in the future. *See* DISINFECTION.

coppicing A VEGETATIVE PRACTICE for the protection of river and other banks. It involves the cutting back of willows or other vegetation to produce a thick, low growth of branches and suckers, which form a dense root growth to bind the soil layers together.

core (1) The cylinder of rock extracted during drilling with the diamond drill or shot-drill. The core is taken from the core barrel and may be about 25 mm (1 in) or more in diameter and up to 6 m (20 ft) or more long. The total length of core recovered compared with the total length of the hole may be from 80 to 95%, depending on the nature of the rocks and the care taken in drilling. The water content of the core is often examined for quality. (2) The above-ground watertight barrier of an earth dam; where it is below ground it is known as the cutoff. *See* CUTOFF TRENCH.

core trench A term sometimes applied to the CUTOFF TRENCH.

core wall *See* CORE (2).

Cornish pump A pump introduced early in the 19th century and used widely, especially in mining, for practically 100 years. A single-acting steam engine transmitted the power for pumping through the action of a cumbersome beam.

corrasion The mechanical work performed by wind and particularly rivers in eroding soils and rocks. The corrasive work of a river is performed mainly by the coarse, angular rock fragments as they are carried and rolled along the bed. *See* CORROSION.

correlation coefficient A statistical term which estimates the likelihood that two apparently independent variables are related to each other. This coefficient is commonly termed *r*, and can take values between 0.0 and 1.0. If *n* pairs of observations (X, Y) are available an estimate of *r* is given by

$$r = \frac{\Sigma\left\{(X_i - \overline{X})(Y_i - \overline{Y})\right\}}{\sqrt{\left\{\left[\Sigma(X_i - \overline{X})^2\right]\left[\Sigma(Y_i - \overline{Y})^2\right]\right\}}}$$

where X_i and Y_i are a pair of samples in the data set and $\overline{X}$ and $\overline{Y}$ are the mean values of all the *n* values of X_i and Y_i. The significance of this estimate depends on both the values of n and r; $r = 0.0$ suggesting there is no connection and $r = 1.0$ that there is a direct relationship.

correction, airline or **wet line** *See* AIRLINE CORRECTION.

corrosion (1) The chemical wearing away of soils and rocks, particularly by rivers; it is of minor importance compared with that of corrasion. (2) The wearing away of pipes: external corrosion may be caused by the nature of the ground in which they are placed, and internal corrosion by the aggressive quality of the water carried. The remedies include lining and coating the pipes, using non-metallic pipes, chemical treatment of the water, electrolytic countermeasures and avoiding corrosive soils. *See* ENCRUSTATION.

coulée A deep gully or ravine that flows only in the wet season.

counter-arched revetment A REVETMENT in brickwork with arches between counterforts, similar to a MULTIPLE-ARCH-TYPE DAM, chiefly used in a cutting.

counter drain A drain constructed along the toe of a dam, canal or embankment to collect and remove seepage and so stabilize the bank.

counterforts BUTTRESSES erected to support and strengthen structures and RETAINING WALLS. They are placed at intervals along the structure and at right angles to it.

counterpoise bridge *See* BASCULE BRIDGE.

coursed blockwork Blockwork for breakwater construction in which 10–50 t pre-cast concrete blocks are laid like masonry in bonded horizontal courses. *See* SLICED BLOCKWORK.

cove A sheltered inlet or small bay.

coverage (irrigation) Generally, the maximum depth of water in farm ditches in relation to the land to be irrigated. A coverage of less than 75 mm (3 in) is not considered adequate for efficient farm irrigation.

covered open channel *See* OPEN CHANNEL.

covered reservoir A SERVICE RESERVOIR enclosed by reinforced concrete roofing or sometimes by prestressed concrete. The roof is often covered with soil.

cover, soil The protective mat of vegetation over the soil surface. It retards erosion and runoff, and may consist of low-growing shrubs and herbaceous plants.

crack An opening which may develop in an earth dam or embankment due to tensile stresses. *See* TRANSVERSE CRACK, LONGITUDINAL CRACK.

cradle (1) A structure which supports and is shaped to fit an open or closed conduit. (2) A low steel framework running on rails on a slipway, on which a ship rests and moves during construction, repairs or launching.

crag Marine and estuarine gravels and sands, referred to the Pliocene epoch, and forming sandbanks or estuarine deposits near the mouth of a large river; it is well developed in the east of East Anglia (UK).

crag and tail A surface feature in which a tail of relatively soft rock is left behind a protecting crag of hard rock, sometimes seen in glaciated country where a resistant mass of igneous rock lies in the path of the moving ice.

crater lakes Permanent lakes that occupy depressions of volcanic origin. *See* CALDERA LAKES.

creek A stream or small river; a short arm of a stream: a small, narrow inlet of sea coast, a bay or cove; any turn or winding.

creep (water) The movement of water and the resulting erosion around and under a structure. *See* PIPING.

crest (1) The top of a dam, weir, spillway or bank, often restricted to the overflow portion (Figure A, B.5(a), C.13, R.1, R.4). (2) The peak of a flood or wave. (3) The summit of a hill or mountain, or the line connecting the highest points along an anticline.

crest gate (or **spillway gate**) A gate located on the spillway crest of a dam to regulate the flood discharge and thus maintain or lower the water level. *See* FLASHBOARD, ROLLER GATE, SECTOR GATE, SLIDING GATE.

crest level, design *See* DESIGN CREST LEVEL.

crevasse (1) A gap in the NATURAL LEVEES of a river caused by high water flows (USA). (2) A crack of considerable depth and width which has developed in a GLACIER.

crib dam (or **crib wall**) A wall, dam or barrier consisting of pre-cast concrete members or rectangular interlocking timber, forming bays or cells, laid one over the other to a batter of from 1 in 6, to 1 in 8. The cells are filled with stone or other heavy material. *See* GRAVITY DAM.

crib groynes GROYNES made of logs; they may be triangular, rectangular or any other shape. They are filled with stone and often used in the upper reaches of rivers where durable hardwoods and stones are commonly available. *See* RECTANGULAR CRIB GROYNE, TRIANGULAR CRIB GROYNE.

cribwork A means of revetment along eroding banks or for small groynes. It consists of boxes made of logs fixed *in situ*, and may be rectangular, triangular or an other shape. The boxes are filled with stone or boulders gathered from the river bed. The logs are spiked or wired together, or sometimes a length of reinforcing steel is passed through bored holes at the intersection points of the logs. A space, smaller than the size of the stones, is left between the logs. This causes siltation and also takes fewer logs. Construction is usually carried out when the river falls to a low level or where it can be temporarily diverted. The bank may finally be planted with willow spars. Cribwork has been used in parts of Victoria, Australia. *See* HENCOOPING, STONEMESH CONSTRUCTION.

Crimp and Bruges formula This formula, which is based on MANNING'S FORMULA, was used to calculate the velocity of flow in a sewer:

$$v = 83 R^{0.67} S^{0.5}$$

where v is the velocity in m/s, R is the HYDRAULIC RADIUS in m, and S is the slope of the sewer in m/m.

critical concentration (irrigation) The limiting amount of an impurity in irrigation water beyond which it damages the soil and adversely affects crop growth and also its quality. *See* SALT BALANCE.

critical depth The depth of water in a channel corresponding to a recognized CRITICAL VELOCITY.

critical flow A flow condition in a channel whose mean velocity is equal to a recognized CRITICAL VELOCITY.

critical hydraulic gradient The HYDRAULIC GRADIENT at which a mass of sand becomes unstable or 'quick' and has a value of approximately unity. A sand becomes 'quick' if the upward flow of water is sufficient to cause flotation. *See* BOIL, QUICKSAND.

critical slope (1) The slope of a channel in which uniform flow occurs at the CRITICAL DEPTH, and in which the head lost due to friction, etc., per unit length is just equal to the fall of the channel in the same length. Under such conditions the energy gradient will be parallel to the channel invert. (2) A term sometimes applied to the ANGLE OF REPOSE.

critical tractive force (water) The TRACTIVE FORCE of water flow when bed material commences to move. *See* BED LOAD, LOAD.

critical velocity There are several critical velocities. Belanger's critical velocity is that velocity at which the SPECIFIC ENERGY of the flowing liquid reaches its lowest possible value while maintaining flow. The Reynolds critical velocity is that at which the flow of a liquid ceases to be streamline and becomes turbulent, and where friction becomes approximately proportional to the square of the velocity. Kennedy's critical velocity is that velocity at which a stream neither deposits nor picks up material when flowing over an erodible bed. *See* HIGHER CRITICAL VELOCITY, STREAMLINE FLOW.

critical velocity, lower *See* HIGHER CRITICAL VELOCITY, BERNOULLI'S THEOREM, SUPER-CRITICAL FLOW.

cross-section (stream) A section along a plane at right angles to the mean direction of flow, and bounded by the wetted perimeter and the free surface of the stream. *See* STREAM PROFILE.

cross-section, compound *See* COMPOUND CROSS-SECTION.

crown The topmost part of any arch construction, especially of the inside of a sewer or drain. *See* INVERT.

cruciform groynes Groynes formed by timber pieces about 2 m (6 ft) long, fixed together at about right angles to form a series of crosses. The crosses are spaced from 3 to 4.5 m (10 to 15 ft) apart, and placed vertically on the river bed and lined parallel to the flow. They are connected with long timber spars wired to the arms of the crosses. The lower ends of the crosses are driven into the river bed to give stability. The groynes are usually protected and supported by stonemesh sausages or drums filled with stone. These are placed on the upstream or downstream side and sometimes on both sides. If available, old rails may be used to construct the crosses. *See* CRIBWORK, GROYNES.

Crump weir An improved type of flow-measuring weir having a sharp horizontal crest, with upstream and downstream slopes of 1 to 2 and 1 to 5 respectively (Figure C.13). Flows are gauged by an arrangement for

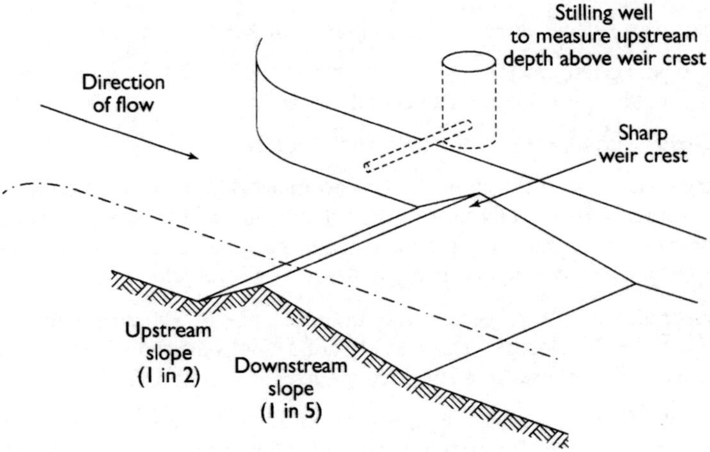

Figure C.13. Crump weir

upstream measurement usually in STILLING WELLS. This type of weir has the advantage of giving better results when the crest is partially drowned.

crustal-movement lakes Lakes that occupy depressions in the land surface as a result of earth movements such as the down-warping of an area. Faulting or subsidence hollows which impound bodies of water come into this class of lake. *See* ACCIDENTAL LAKE.

cryology The study of ice in all its forms and effects.

cryopedology The science dealing with intense frost action and ground which is perennially frozen. The term includes the study and application of engineering methods to minimize or avoid the difficulties involved. *See* PERMAFROST.

cryptoreic Subterranean streams. The term means 'hidden flow'.

culvert Any covered conduit, of box or circular section, for taking a watercourse, drain, or sewer under a railway, roadway or embankment and usually large enough for a person to pass through (Figure S.4). Also applied to a tunnel along which water is pumped to or from a dry dock. *See* BOX CULVERT.

cumec A flow of one cubic metre per second. *See* CUSEC.

cumulative runoff diagram A diagram in which the daily, weekly or monthly runoff figures are successively added, thus showing the total runoff from the commencement of the record. Cumulative diagrams may be used to obtain the minimum yield available from a reservoir during the period covered.

cumulative volume curve *See* DISCHARGE MASS CURVE, MASS DIAGRAM.

cup-type current meter A CURRENT METER incorporating a cup-type element which is rotated on a vertical axis by the water flow. *See* SPIN TEST.

current Running water; water flowing rapidly.

current curve A graph showing the flow of a TIDAL CURRENT. Rectangular co-ordinates are usually used, with velocity shown by the ordinates and time by the abscissae; the flood is taken as positive and the ebb velocity as negative. Broadly, the graph resembles a cosine curve.

current diagram A graph showing times of slack water and strength, and flood and ebb velocities over a stretch of a tidal waterway. The times are related to tide phases at a specified station.

current difference The difference between slack water and strength times in a locality and the time of corresponding phase of the current at a station for which predictions are given in a CURRENT TABLE.

current line A method of measuring velocity of current. A line is graduated and attached to a current pole, and the length of line carried out by the current pole in a specified interval of time indicates the velocity expressed in knots and tenths.

current meter An instrument for measuring water velocity, consisting of a wheel or propeller with vanes which are rotated by the water when submerged. The velocity is calculated from the speed of this rotation. Current meters are usually used in wide rivers, and velocity and depth measurements are made at regular intervals across it to obtain the MEAN VELOCITY. The volume of water flowing is equal to the mean velocity multiplied by the cross-sectional area of the flow. Current meter measurements are made mainly to calibrate STAGE DISCHARGE RELATIONSHIPS. *See* INTEGRATION METHOD, STANDARD CURRENT METER.

current meter rating Current meter tests under controlled conditions to determine the basic relationship between the actual velocity of water flowing past the meter, and the recorded velocity.

current table A list of daily predictions of velocities and times of tidal currents; additional predictions may be made for other places by the current differences and constants given.

curtain grouting Grouted foundation material formed beneath a dam. It consists of a deep, narrow, and continuous zone, and is generally carried out through the base of the CUTOFF WALL, as this helps to make the top of the grout curtain secure. Grouting is done at high pressure through a row of holes usually drilled vertically downwards. *See* GROUTING.

cusec One cubic foot per second, usually applied to water flow. 1 cusec = 0.0283 cumecs. *See* SECOND-FOOT DAY.

cuspate foreland A coastal feature which in effect consists of compound spits extending out to sea. Believed to be due to seas approaching the coast obliquely from two main directions and building up successive banks of shingle ridges, they sometimes enclose a small triangular lagoon.

cut *See* (1) CUTTING and (2) LOCK CUT.

cut and fill A type of canal construction or other works which is partly below ground in cut, and partly above ground in fill or embankment. *See* BALANCED EARTHWORKS.

cutoff (1) The new channel cut by a river, or excavated, through the neck of an oxbow curve. The natural cut-through often occurs during a flood. An artificial cutoff is often effective in providing a better overall alignment for a river; it diverts the current from bends where heavy bank erosion occurs; it increases velocities and thus promotes local bed scour; it reduces flood levels; and perhaps saves much cleaning and clearing work round a long bend. *See* PILOT CUTOFF. (2) A construction, such as a wall or collar, of impervious material placed below ground level to reduce water seepage.

cutoff collar (or **anti-seep diaphragm**) A collar, usually of steel or concrete, placed at right angles to a pipeline to prevent water percolating along the outside of the pipe (Figure O.1).

cutoff depth The depth to which CUTOFF TRENCHES are taken below the excavation level.

cutoff ratio The ratio of length of CUTOFF channel to original length of river around a loop or meander.

cutoff trench (or **diaphragm wall**) Any trench, wall or barrier of impervious material placed within and below an earth dam or embankment to prevent or reduce seepage; the construction may be of compacted clay, sheet piling, concrete or cementation (Figure O.1). *See* CORE (2).

cutoff wall A relatively watertight wall placed in a hydraulic structure, earth dam, or channel bank to arrest seepage.

cutter-dredger *See* SUCTION-CUTTER DREDGER.

cutting (or **cut**) Any excavation below ground level in the open for a pipeline, canal, trench or other works.

cutwater The wedge-shaped head of a bridge pier to streamline the oncoming flow of water.

cut yards *See* COMPACTED YARDS.

cycle of erosion The orderly succession of changes in the history of a land surface subject to erosion. A cycle of erosion may be described as young, mature, or old according to the stage reached in its development. Again, according to region and climate, various types of cycles may be recognized, such as the normal cycle, where flowing water is the dominant erosive agent, the arid cycle in desert regions, and the glacial cycle, in which ice is the dominant agent. *See* EROSION CLASS, GEOLOGICAL EROSION.

cycle of fluctuation (or **phreatic cycle**) The total time during which the water-table rises (due to replenishment) and the succeeding fall or decline (due to discharge). It may be daily, yearly or any other period.

cyclic depletion *See* CYCLIC RECOVERY.

cyclic recovery The gradual rise in level of the main water-table by recharge or replenishment of water in the zone of saturation. The recovery may extend over several years, following a period of cyclic DEPLETION.

cyclone A violent whirlwind or hurricane of limited extent; strong winds rotating round a DEPRESSION or centre of minimum barometric pressure. *See* TYPHOON.

cyclonic precipitation Precipitation associated with areas of low pressure or depressions. *See* FRONTAL PRECIPITATION, NON-FRONTAL PRECIPITATION.

cyclopean concrete Concrete containing AGGREGATE larger than 150 mm (6 in) and often used in mass concrete dams. *See* PLUM.

D

Dalmatian-type coast A coast with long narrow inlets and islands running roughly parallel with the coastline, as in Dalmatia on the eastern shores of the Adriatic. *See* RIA-TYPE COAST.

dam A massive wall or structure erected across a valley or river for impounding water; a bank of earth, mole or wooden frame to retard the flow or force of water; a natural obstruction or bar, such as ice or landslide debris, across a river. Some of the most commonly constructed types of dams are shown in Figure D.1. The deposits underlying a proposed dam for impounding water are tested for character and continuity to safeguard against the leakage of water underneath or around the dam. Leakage may be due to FAULTS, JOINT FISSURES, SOLUTION CAVITIES, BURIED CHANNELS or POROUS ROCKS. GLACIAL DRIFT is rarely homogeneous, and there may be lenses of permeable gravel or sand. If these occur near the structure, seepage of water is likely. The breaking or failure of a dam is sometimes caused by yield or slippage along shale or clay layers in the rock formation on which the structure rests. The entry of water into these clayey layers causes them to soften and slake, and they may become slippage planes for the foundation rock. The seepage of water along thin layers of sand, gravel or clay tends to increase considerably, and under the weight of the dam and the pressure of the water the solid rocks in the vicinity may become unstable and the entire structure be endangered. *See* DAM SITE INVESTIGATION, EARTH DAM, LARGE DAM.

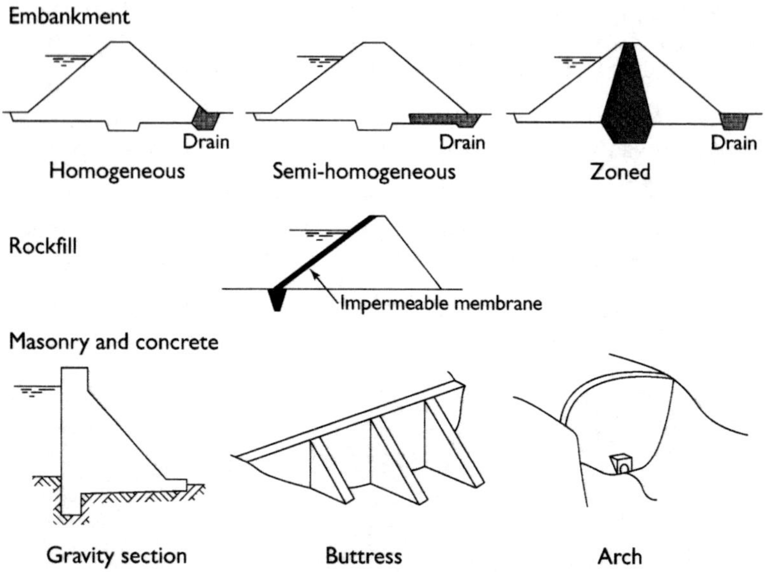

Figure D.1. Major types of dam

dam, axis of A reference line marked on plan or section along the length of a dam, used for measuring horizontal dimensions.

dam break wave *See* WAVE, DAM BREAK.

dam crest *See* CREST (1), Figure R.4.

dam foundation Broadly, the lower parts of a dam below ground level which transmit the weight of the structure and the forces from water pressure to competent layers in the abutments and the underlying soils and rocks in the valley floor. The soils and rocks forming the foundations must be either naturally of adequate strength and sufficiently watertight or treated to be made so. Both strength and rates of seepage loss can be respectively increased and reduced by many techniques, including compaction or grouting. Potential causes of the failure of dam foundations include the presence of naturally weak rocks or rocks which lose strength when they become saturated, slippage along shale or clay layers, and the removal of poorly consolidated deposits by seepage flows. Excessive leakage underneath or around a dam can be due to the presence of porous rocks, faults, joint fissures, solution cavities and undetected buried channels. As all of the above are initially concealed below ground, considerable skill and care are needed when investigating potential sites for foundations. *See* DAM SITE INVESTIGATION.

dam freeboard The vertical height from FULL SUPPLY LEVEL to the CREST of a dam required to contain FLOOD SURCHARGE and WAVE SURCHARGE, with also a margin for safety. *See* RESERVOIR RETENTION LEVELS, Figure R.4.

dam, groundwater *See* GROUNDWATER DAM.

damp air Air containing a large amount of moisture or possessing a relative humidity over 85%, as opposed to dry air with a relative humidity below 60%. *See* DEW POINT, DEW PONDS.

dam, raising of Increasing the height of a dam, either to increase the freeboard in the interests of safety or to increase reservoir capacity. Generally, gravity dams can be raised with greater safety than other types. In certain cases concrete can be added to the crest of a gravity dam at relatively low cost by securing it by vertical prestressing cables.

dam site investigation An investigation into the suitability of a potential site for the construction of a dam, its associated works and RESERVOIR area (Figure D.2). Before detailed investigations are undertaken it is usual to carry out preliminary studies of several alternative sites to compare their probable costs and benefits; normally these are desk studies based on existing hydrological data and geological and topographical surveys. Dam sites need to be carefully selected to ensure maximum benefits, minimum costs and adequate safety. Some of the important considerations include: (a) competent and stable soils and rock for the foundations; (b) valley sides that are steep and close together at the site of the dam, so reducing the volume of construction material and cost; (c) favourable topographic conditions, such as a shallow stream gradient and valley sides that widen out upstream of the dam site to provide a high STORAGE/EXCAVATION RATIO; (d) the site is such that the reservoir will not flood any areas of great historic, ecological or economic significance. Dam site investigations normally include detailed topographical surveys of the actual foundation area and the whole reservoir basin, hydrological studies and geological investigations. Methods commonly used include trial pits, exploratory drilling and geophysical techniques. See DAM, DAM FOUNDATION, EARTH DAM, LARGE DAM.

Figure D.2. Dam site investigation

Darcy's law The relationship between pressure gradient and the velocity of percolation of water through natural granular materials, which states:

$$V = k(h/l)$$

where V is the velocity of percolation, h is the difference of head under which flow occurs over a distance l, and k is the constant which varies with the nature of the material, and is calculated from field tests. Field tests may be made to determine the average velocity V. A GROUNDWATER TRACER, such as common salt, is introduced into a well and samples taken from neighbouring wells at timed intervals. Using the sample with the highest salt content, the average velocity of percolation through the ground between the two wells may be calculated. By using this value for V and the hydraulic gradient taken from the water levels in the wells, the value k in Darcy's law may be calculated.

datum Any permanent line, plane, surface or BENCH MARK used or specified as a reference datum for levelling or to which elevations are referred. *See* MEAN SEA LEVEL.

Deacon diagram A diagram developed by G. F. Deacon about 1902, giving reservoir storage and yield for different annual rainfalls; it was slightly modified in 1949 in C. F. Lapworth's Chart.

deadman (or **holdfast**) An anchor buried in the ground to hold wires or ropes for winches or for holding bank protective works such as stonemesh sausages, logs, etc. It may be constructed from heavy timber or concrete and well buried.

dead storage That volume of water which cannot be released from a reservoir under normal conditions.

dead water zone A stretch, area or region where water is temporarily restrained.

debacle The sudden breaking up of ice in a river; a rush of water carrying with it broken ice and debris.

debouch To issue from a stream into a plain or open ground, usually from a ravine or wood.

debris cone A fan-shaped heap of sand and debris deposited by a mountain stream when it reaches a plain or valley. *See* ALLUVIAL FAN.

debris dam A dam or barrier erected across a stream valley for retention of debris, driftwood, gravel and silt.

decantation Separation of a liquid from solids (or from a liquid of higher density) by allowing the heavier material to settle before drawing off the upper layer.

dechlorination The reduction of RESIDUAL CHLORINE by chemical or physical measures.

decompression sickness *See* CAISSON DISEASE.

decrement, groundwater *See* GROUNDWATER DECREMENT.

deep lead An ARTESIAN AQUIFER located beneath a PRIOR RIVER.

deep manhole A manhole or INSPECTION CHAMBER with a smaller access shaft above it. *See* SIDE-ENTRANCE MANHOLE.

deep percolation Rainfall that percolates downwards into storage in subsurface strata. *See* PERCOLATION, SHALLOW PERCOLATION.

deep well A well more than about 30 m (100 ft) deep, usually with a watertight casing down to 30 m (100 ft) or so, and thus largely unaffected by surface impurities. It commonly draws water from the zone of saturation and the water is relatively pure, but may be hard. *See* SHALLOW WELL.

deep well pump In general, a CENTRIFUGAL PUMP located at the bottom of a well and driven by a long shaft from a surface motor. *See* SHAFT-SINKING PUMP, SUBMERSIBLE PUMP.

defeated stream A stream that fails to maintain its original course because of major earth movements and is diverted, held back or even reversed. *See* ANTECEDENT STREAM.

deficiency The opposite of an excess, often applied to regions with inadequate rainfall or other hydrological quantities. *See* ARID.

deforestation The cutting down and clearing away of forests or wooded areas. Haphazard deforestation may be followed by excessive rain-wash and soil erosion, and in time the entire mantle of soil may be removed and the ground made useless for cultivation. *See* AFFORESTATION, SHEET EROSION.

degradation The lateral and downward erosion by streams; the process in which the land surface is worn down. *See* ALLUVIATION.

degree of saturation The volume of voids filled with water, expressed as a percentage of the total volume of voids in a soil or other mass. *See* POROSITY.

degrees of frost *See* FROST.

delayed junction A junction formed by a tributary that is forced to flow along the foot of a natural levee until it can effect a junction with the main stream.

delivery The volume of water delivered by a pump per unit time, usually expressed in gallons per minute or cubic metres per minute. *See* DISCHARGE.

delivery box (irrigation) A structure which controls and splits the available water in the agreed proportions to farm units.

delivery pressure head The residual pressure of the water at the delivery end of a pipe, or at the jet or nozzle in irrigation.

delta A type of river deposit, roughly semicircular in outline, and formed where a river loses its carrying capacity by flowing into a relatively quiet body of water such as the sea, lake, or reservoir. It consists of three main parts: (1) a broad, gently inclined upper deposit mostly above the stagnant water (topset beds); (2) the submerged and steeper front slope (foreset beds), and (3) the submerged silts and muds which extend outwards as a gently undulating layer from the base of the front slope (bottomset beds). The plan shape of a delta depends on many factors, such as the load carried by the river and the existence of any longshore currents in the body of water entered. Engineering efforts are often made to maintain one of the delta channels in a navigable condition out to sea. The channel selected may be improved by building parallel jetties and inducing the current to scour a deeper channel.

deltaic tract A PLAIN TRACT of a stream which has extended with the formation of a delta and deposition of fine sediments. *See* MOUNTAIN TRACT, VALLEY TRACT.

delta-plain swamps Swamps which owe their existence primarily to the flatness of the land at the mouth of rivers, where they are formed by sediments being deposited, and frequent innundation from the river at its maximum size.

demonstration area An area of land, usually a small drainage basin, on which water and soil demonstrations, and perhaps experimental work, are carried out by and between the local authority and the farmers.

dendritic drainage A treelike arrangement of rivers and streams, often seen over areas where structural controls are not exaggerated. *See* DRAINAGE PATTERN.

denitrification The anaerobic biological reduction of nitrites and nitrates to nitrogen gas.

density The mass of a substance per unit volume, expressed in kg/m^3 or lb/ft^3. *See* DRY DENSITY, SPECIFIC GRAVITY, UNIT WEIGHT OF WATER.

density current The gravity flow of a fluid above, below, or through a fluid of slightly different density, such as the flow of turbid water under clear water in a reservoir or lake.

density of snow *See* SNOW DENSITY.

dental A form of BAFFLE or tooth-shaped projection on a surface or APRON which breaks or deflects the force of flowing water.

dental treatment The filling of surface cracks, flaws, potholes, faults or weathered rock pockets in the rock foundations of a dam. Concrete and/or grout are used.

dentated sill A notched sill at the end of an APRON to break the force of a current of water, thereby reducing scour below the apron.

denudation The wearing away of land, rocks and soils by mechanical and chemical disintegration, particularly by rain, running water, wind and the action of frost, snow and ice.

deoxygenation Reduction of the amount of dissolved oxygen in a liquid either naturally through biochemical oxidation of the organic matter present or by the addition of chemical agents.

depletion The continued discharge or extraction of water from a well, basin, reservoir, pond, or stream at a rate in excess of replenishment. *See* ECONOMIC YIELD, GROUNDWATER DECREMENT, GROUNDWATER INCREMENT.

depletion (well) The exhaustion of a water well caused by: (1) interference or excessive pumping by neighbouring wells, (2) depletion of water in storage by pumping in excess of replenishment, and (3) defective casing allowing leakage of water into upper porous rocks. Well depletion may also occur during a prolonged period of drought.

depletion curve (groundwater) A curve indicating the loss of water from underground storage by springs or seepage into streams.

depletion, cyclic *See* CYCLIC RECOVERY.

depletion rate, annual *See* ANNUAL DEPLETION RATE.

depletion, stream flow *See* STREAM FLOW DEPLETION.

depression (1) An area of low atmospheric pressure usually surrounded by closed ISOBARS, associated with winds, rain and storm; in the northern hemisphere winds blow around the depression anti-clockwise and in southern hemisphere clockwise. (2) Low-lying land in undulating country within which water collects or flows. *See also* CONE OF DEPRESSION, DEPRESSION STORAGE.

depression, cone of *See* CONE OF DEPRESSION.

depression spring An outflow of water where the surface slopes down and intersects the water-table (Figure D.3). *See* VALLEY SPRING.

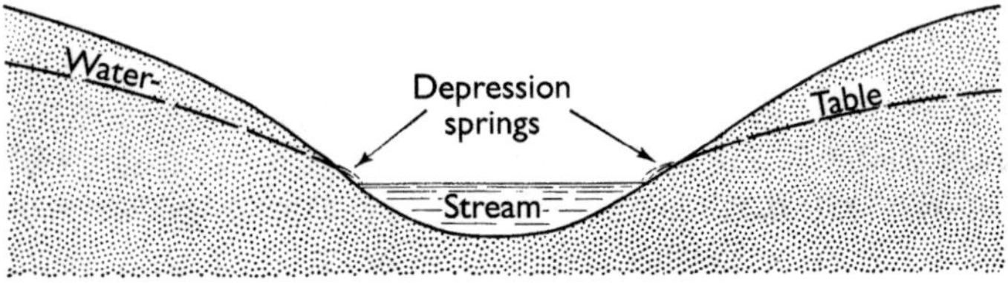

Figure D.3. Depression spring

depression storage The volume of water required to fill all natural depressions of an area to their overflow levels, expressed as depth for the entire water surface of the area.

depth area curve A graph showing the relation between progressively decreasing average depth of precipitation over a gradually increasing area, from the centre of a storm with maximum rainfall outwards to its limits.

depth–duration–frequency relationship The relationship between the average total fall of rain expected during a storm of given duration and RETURN PERIOD (Figure D.4). The same data are also often presented as INTENSITY–DURATION–FREQUENCY RELATIONSHIPS.

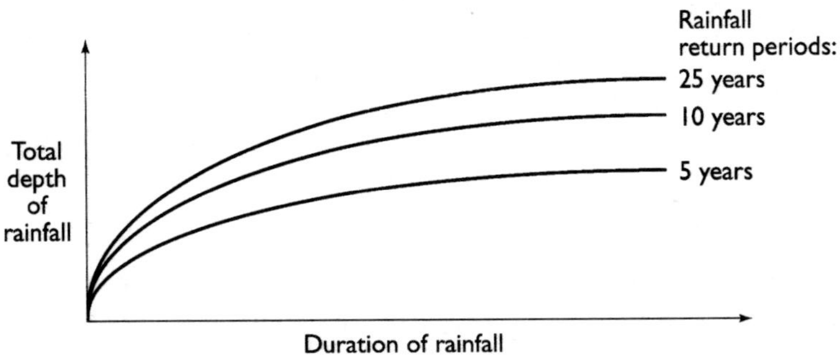

Figure D.4. Depth–duration–frequency relationships

depth duration graph A graph which shows the relationship between depth of precipitation and duration of storm.

depth of runoff Total runoff from a basin or drainage area, divided by the area; runoff may be given in depth or volume units per unit of the area.

depth velocity curve (or **vertical velocity curve**) A curve showing the relationship between velocity and depth of water at points along a vertical line at a specified cross-section of conduit or open channel.

desalination (1) The reduction in the salt content of sea or brackish water to a level acceptable for consumption. A relatively expensive process, it is most commonly used in countries where there are few alternative sources, or energy is cheaply available. (2) The removal of excess salt from irrigation lands. *See* LEACHING.

desert Usually, a widespread sandy plain, with an arid climate and largely waterless, treeless and uncultivated; vegetation is sparse and consists mainly of hardy, drought-resistant types that are able to flourish on the water from a single brief rainy period.

desert peneplain A PENEPLAIN resulting from long-continued desert erosion by wind and water.

desiccation A climatic change with a marked decrease in rainfall, drying up of streams and lakes, and loss of vegetation and surface soil.

design crest level The design level of the top of a dam, ignoring any extra for CAMBER (Figure C.1).

design flood The magnitude of flood, the statistics of which are used when designing a specific dam, barrage or other works for the control of a river. The design flood is not necessarily the largest flood possible, or even the largest recorded at the site; the choice is usually made after a study of the hydrology of the river, the damage liable to be caused and the economic factors involved. *See* FLOOD.

design flood hydrograph A graph of flow used for studies associated with FLOOD ROUTING and for design of dam or other construction.

design flood level The highest level to which the reservoir surface is estimated to rise in the event of a flood equal to the DESIGN FLOOD.

design recurrence interval The average number of years within which a given hydrological event, such as a flood, is equalled or exceeded, and which serves as a basis for the design of a dam or other hydrologic structure. For example, a RETURN PERIOD of 1000 years may be used as a basis for the design of a LARGE DAM. *See* DESIGN FLOOD, *N*-YEAR FLOOD.

design storm Measurements and estimations used in determining DESIGN FLOOD, mainly distribution and RATE OF RAINFALL over basin or drainage area.

desilting (farm storages) The removal of excess silt and mud from farm storages (ponds and reservoirs), usually carried out when water levels are low. Some form of mud scoop and two tractors may be used, with the material deposited on the downhill side. *See* DREDGING.

desilting basin (or **silt basin** or **settling basin**) A basin-shaped enlargement in a stream, where silt and debris carried in suspension may be deposited. *See* SAND TRAP.

desilting strip (or **filter strip**) A strip of land, located above a reservoir, stock tank, pond or cultivated fields, and used for the deposition of silt and debris carried by flowing water. The strip contains a vegetal cover of sufficient density to retard flow of runoff water with deposition of suspended material.

detachment (soil erosion) The disintegration and separation of particles or aggregates from a soil layer, usually by flowing water, raindrops or wind; later, this material forms part of the LOAD transported by streams.

detention dam *See* DETENTION RESERVOIR.

detention, initial *See* INITIAL DETENTION.

detention reservoir A reservoir for the storage of floodwater or stream flow for a relatively short period; it is retained until the channel or stream can carry released water from the reservoir plus its own flow.

detention, surface *See* SURFACE DETENTION.

detergent *See* SURFACTANT.

deterministic model One of the two main types of NUMERICAL SIMULATION MODEL, the other being the STOCHASTIC MODEL. These predict the value of a parameter to be modelled from a fixed relationship between it and observed historical data or other information. Deterministic models usually have the advantage of being based on relatively simple mathematical expressions, and are thus usually simple to use or set up on computers. However, as they use fixed relationships between the data, one set of input data will always give rise to identical predictions. Therefore, unless very long sequences of observed input data are available, they will yield little or no estimate of the probabilities to be assigned to the predictions they make.

detrital deposit A deposit of sand or gravel formed by river action. *See* ALLUVIAL DEPOSIT, ALLUVIAL FLAT.

detritus The coarse debris moved on the bed of streams and rivers, or carried by wastewater.

detritus chamber (or **detritus tank**) A tank in which solid matter from sewage settles to the bottom and is removed. *See* GRIT CHAMBER.

development (well) *See* WELL DEVELOPMENT.

developmental projects Schemes intended to improve land conditions, to increase productivity and gain additional revenue from such areas. These include plans to reduce or prevent flooding of lands bordering rivers, irrigation, and the drainage of swamps, etc.

dew Liquid water deposited by condensation of water vapour in the air, occurring especially during cloudless, calm nights.

dewatering The use of a SUBMERSIBLE PUMP or other means to remove water from a shaft or caisson; the removal of water from construction excavations or pipe trenches. *See* DEWATERING HOLE, DRAINAGE, GROUNDWATER LOWERING, PUMP, WELLPOINT.

dewatering (sludge) The removal of water from sludge to form a slurry or cake, using methods such as drying beds, pressure filtration and vacuum filtration. *See* CONDITIONING (SLUDGE), FILTER PRESS, SLUDGE CAKE, SLUDGE-DRYING BED, THICKENING, VACUUM FILTRATION.

dewatering hole A borehole put down to remove or drain away water from existing or proposed sites for cuttings, docks, harbours, tunnels or dams. The hole may drain water from a near surface deposit downwards through an impervious rock into storage at lower levels. *See* ARTESIAN WELL DRAINAGE, DRAINAGE WELL.

dew point The point at which DAMP AIR, when cooled, commences to deposit its vapour as water or dew.

dew ponds Shallow, often artificial ponds lined with cement or clay, fed by atmospheric CONDENSATION and sometimes rain or drainage of rain. They are constructed or occur usually on uplands, and form a drinking source for cattle and sheep.

diamond-head buttress dam *See* BUTTRESS DAM.

diaphragm An instrument or appliance for measuring water flow in pipes.

diaphragm pump A RECIPROCATING PUMP in which a rod moves a flexible diaphragm (of canvas, rubber or leather) to and fro. It can deal with very gritty water and is therefore useful in many civil engineering and water projects.

diaphragm wall *See* CUTOFF TRENCH.

diatomaceous earth *See* KIESELGUHR.

diatomite filter A FILTER which extracts fine particles of suspended matter and bacteria by passing the water through elements with a deposited coating of KIESELGUHR or diatomaceous earth. Often used in plants passing small water supplies.

differential erosion Applied to the action of flowing water in cutting into the softer rocks at a faster rate than in hard, resistant rocks, which remain as ridges or hills. *See* EROSION.

differential pressure The difference in pressure of a liquid at two different points, as, for example, on the two sides of a meter or device for measuring water flow. *See* GAUGE PRESSURE, MANOMETER.

differential settlement (dams) The settling of a dam in different places by different amounts. *See* SETTLEMENT.

diffuser (1) In pumping, the gradually increasing cross-section at the outlet of a centrifugal pump, which reduces the speed of the water and increases its pressure. (2) In sewage, the porous plate or material through which air is blown to stimulate sewage treatment. *See* CONTACT AERATOR.

diffuser chamber In a TURBINE PUMP, a chamber containing a number of fixed blades which guide the water outwards, on leaving the IMPELLER, with the minimum of turbulence.

digested sludge Sludge which has undergone either aerobic or anaerobic digestion. *See* AEROBIC DIGESTION, ANAEROBIC DIGESTION.

digestion The biological decomposition of organic matter in sewage with the formation of simple and relatively harmless compounds.

dike (or **dyke**) (1) An embankment, earth ridge or levee for confining water, especially along river banks to prevent flooding of lowlands. *See* POLDER. (2) A large drainage channel for land reclamation.

dike, closing *See* CLOSING DIKE.

dilution methods Measurements of water flow in open channels by the injection of solutions of known concentration at points upstream and ascertaining the degree of dilution at points downstream. *See* CHEMICAL GAUGING, CONSTANT RATE OF INJECTION METHOD, SALT VELOCITY METHOD.

dilutor (irrigation) An appliance fixed to an irrigation water supply pipe for diluting a concentrated stock solution of fertilizer. Usually the dilution rate is adjustable.

dimensionless unit graph (or **non-dimensional unit graph**) A graph plotted in non-dimensional units with respect to flow and time, useful for comparing UNIT HYDROGRAPHS resulting from different storm patterns or of different drainage areas. They are either constructed from a summation graph or obtained from an observed hydrograph of storm flow.

dip The maximum angle of inclination, or the inclination along a specified direction, of any surface which may be natural or artificial; the term is applied to batters, tunnels, conduits and other works. *See* GRADIENT, SLOPE.

Dippel's oil *See* BONE OIL.

direct-acting pump A RECIPROCATING PUMP in which the power cylinder (compressed-air or steam) and water cylinder are in line and at opposite ends of the same piston rod. *See* CENTRIFUGAL PUMP.

direct flood damage All damage caused by flood waters, mainly inundation of land, including stock and property. *See* FLOOD, FLOOD DAMAGE, INDIRECT FLOOD DAMAGE.

directional drilling The planned drilling of a borehole at an angle off the vertical.

direction float A wooden or metallic float with a small flag to indicate direction of flow of a river; the angle that the discharge section line makes with the direction of flow can be measured at an observation station. *See* FLOAT.

direction of groynes (relative to bank) A GROYNE may be built at right angles to the bank line, slightly upstream or downstream, of that direction. Opinions differ, but in general an IMPERMEABLE GROYNE is erected to point slightly upstream and a PERMEABLE GROYNE pointing a little downstream. A groyne pointing downstream induces siltation below itself, while one pointing upstream tends to cause siltation above and below itself. Groynes inclined slightly downstream are often used in series, so that the area of protection formed by each one extends to the next one downstream.

direct runoff Runoff that has not existed as groundwater or subsurface water since it was precipitated out of the atmosphere. Direct runoff may be expressed as depth of water (mm or in) spread over the entire drainage area, or in cubic metres or in acre-feet. *See* GROUNDWATER RUNOFF.

direct supply A supply-line drawing water at mains pressure with no intermediate storage cistern. The method is less costly in pipes and fittings.

disappearing streams Streams that emerge from the hills and flow for a short distance over the valley floor, then sink rapidly into layers of gravel and sand just below the surface, and continue as UNDERFLOW. They are fairly common in some arid regions and are found along the Coast Range of California. If the stream is long enough and the seepage and evaporation are abnormal, the entire flow may be used up, and the stream then comes to an end. *See* DRY VALLEY, SINKS, SWALLOW HOLES.

discharge (or **rate of flow**) The volume of water flowing past a DISCHARGE SECTION LINE in unit time; the delivery rate of a pump or pumping set. Discharge is expressed in litres per second, cumecs, gallons per minute or cusecs. *See* CURRENT METER, FLOAT MEASUREMENT, GAUGING STATION.

discharge area The cross-sectional area at a DISCHARGE SECTION LINE of stream, channel, or any other waterway.

discharge capacity The safe maximum volume of water which a channel, conduit or any other hydraulic flow structure is capable of passing regularly.

discharge coefficient *See* COEFFICIENT OF DISCHARGE.

discharge curve A curve showing the relation between the DISCHARGE and STAGE of a stream at a given section or station.

discharge head The height between the intake of a pump and the level at which the water is discharged freely to the atmosphere. *See* STATIC DELIVERY HEAD, TOTAL DELIVERY HEAD.

discharge hydrograph A graph in which discharge is plotted as ordinate and time as abscissa. *See* HYDROGRAPH.

discharge mass curve A curve in which accumulated values of recorded discharges are plotted against intervals of time; the average discharge is represented by the average slope of the curve. *See* MASS DIAGRAM.

discharge measurement The measurement of the volume of water flowing, usually in an open channel. *See* DISCHARGE.

discharge section line The line between two points fixed on either bank of a stream or channel along which velocities and depths of water are measured. *See* GAUGING STATION.

discharge site *See* GAUGING STATION.

discharge valve A valve in a pipeline for increasing or decreasing the discharge of water, frequently placed at the end of the pipeline.

disinfection The destruction of pathogens in potable water or wastewater, usually by chlorination or by boiling, ozonation or ultraviolet radiation. Chlorine has the advantage of carrying a free residual into supply. *See* RESIDUAL CHLORINE.

disintegration The break-up or separation of soils and rocks into small fragments by denudation. *See* DETACHMENT.

dispersion (1) The scattering or separation of soil aggregates into single grains, a factor linked with the erodibility of soils. (2) A SUSPENSION of very fine, often colloidal, particles in a liquid.

displacement (1) The volume of water displaced by a floating vessel or object. *See* ARCHIMEDES' PRINCIPLE. (2) The volume of water displaced by a pump ram or piston moving from one end to the other of its stroke.

displacement pump Broadly, any pump which incorporates a ram or piston, but the term is often restricted to an AIR-LIFT PUMP or a DIAPHRAGM PUMP.

dissected peneplain A PENEPLAIN which has been uplifted with respect to the sea level and a new cycle of erosion inaugurated. The streams cut their channels as young valleys below the peneplain, which in time is represented perhaps merely by the crests of the hills, which rise to about the same level.

dissection (erosion) The action of water and ice in cutting channels, ravines, and valleys in relatively flat ground.

dissolved solids The solids remaining after a filtered sample of a water has been evaporated. *See* TOTAL DISSOLVED SOLIDS.

distributaries Small channels or conduits for conveying irrigation water from the main channels to farms.

distribution coefficient, rainfall *See* RAINFALL DISTRIBUTION COEFFICIENT.

distribution curve A curve showing typical distribution of runoff as a percentage of TOTAL RUNOFF from a basin or drainage area.

distribution efficiency *See* IRRIGATION EFFICIENCY.

distribution reservoir *See* SERVICE RESERVOIR.

distribution system (1) An irrigation system comprising main channels, irrigation laterals, distributaries and associated structures for conveying water from the HEADWORKS to individual farms. (2) Any system which supplies water to a consumer and consists of any or all of the following: service reservoirs, mains, boosters, equipment and services.

ditch A long narrow channel to hold or conduct water as in drainage or irrigation; a boundary or waterway bordering fields; sometimes called a drain. A field side-ditch has the following minimum dimensions: depth 1 m (3 ft 6 in), top width 1.5 m (5 ft), bottom width 0.5 m (1 ft 6 in). Small dragline excavators, fitted with side-arms, and a specially shaped bucket are sometimes used for ditching work.

ditcher Mechanical equipment for cutting ditches or trenches for pipelines, etc. *See* TRENCH EXCAVATOR.

ditch, infiltration *See* INFILTRATION TUNNEL.

ditch outlet Another name for a BAY OUTLET or SLUICE BOX.

diurnal inequality (tidal currents) The difference in velocity between the two flood currents or the two ebb currents of each day; also the difference in height of the two high waters or of the two low waters of each day. *See* TIDAL RIVER.

diurnal range of tide, great *See* GREAT DIURNAL RANGE OF TIDE.

diurnal tides Tides with only one high and one low water in each lunar day; tides requiring a full day to complete their cycle. *See* DURATION OF TIDE.

diurnal tides, semi- *See* SEMI-DIURNAL TIDES.

diver A person skilled in underwater operations and normally supplied with air by a pipeline from the surface or a vessel. Equipped with copper helmet and conventional rubber suit, a diver can work for short periods at a depth of about 70 m. Aqualungs improve the mobility of divers. Underwater work includes inspection and repair of docks and harbours, dams and bridges. This often involves dangerous work; divers have been known to be sucked into dam outlet works. Another hazard is DECOMPRESSION SICKNESS. *See* CAISSON DISEASE, SIEBE DIVING HELMET.

diversion (1) A ditch or channel, with a confining ridge on the lower side, formed across sloping ground to intercept runoff and reduce soil erosion, or to prevent excess runoff from flowing over lands on the lower side. A series of diversions may be constructed in some areas. (2) A channel, cut or trench to divert water or a stream away or around swamps, dwellings or engineering works. It may be temporary or permanent. *See* CUTOFF, PARTIAL DIVERSION.

diversion cut *See* DIVERSION (2).

diversion dam (or **diversion weir**) A barrier constructed across a stream to divert all or some of the water into another channel or water supply conduit.

diversion requirement *See* GROSS DUTY OF WATER.

diversion terrace *See* DIVERSION (1).

diversion tunnel, river *See* RIVER DIVERSION TUNNEL.

diversion weir *See* DIVERSION DAM.

divide (or **drainage divide** or **watershed**) The path along the summit of a range of mountains at which surface water divides and flows down in opposite directions. A major divide would separate river systems rather than individual rivers. *See* WATERCOURSE.

diving bell A bell-shaped steel chamber, open at the bottom, which can be lowered to or raised from the river or sea bed by a powerful crane. People within the bell prepare foundations and other works which are necessary underwater. It is now obsolescent as DIVERS are more mobile, and also less costly. *See* HELIUM DIVING BELL.

divining (or **dowsing**) A method used to locate relatively shallow accumulations of water by holding a hazel fork, or other device, in the hands; the free end is claimed to bend downwards when positioned over the body of water in the ground. When submitted to impartial scientific tests, divining has not been successful. *See* DIVINING DEVICE.

divining device The device used by diviners when locating underground water. It may consist of a hazel, holly or willow twig, fencing wire, or watch spring or polished stone suspended by string (Figure D.5). Also called a dowsing rod, Aaron's rod, Jacob's rod, *baculus divinatorius* and *virgula divina*.

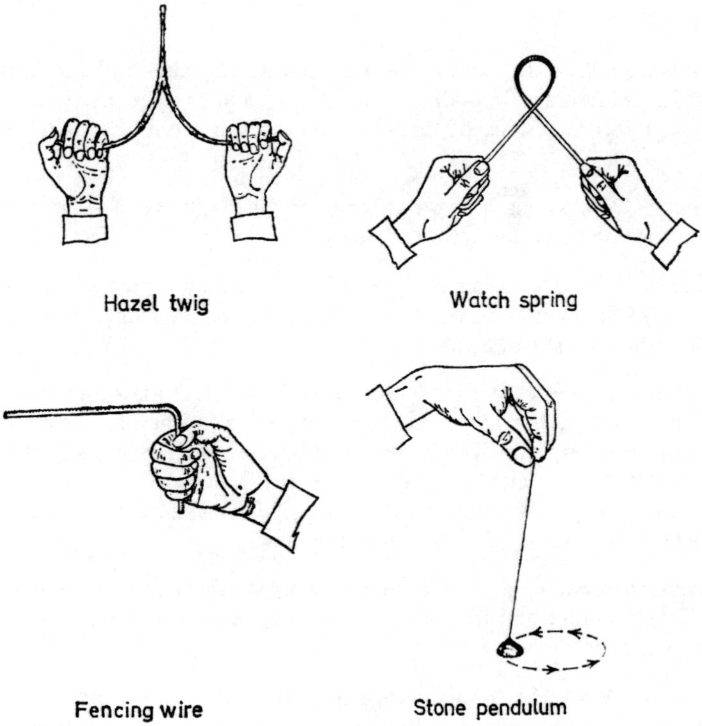

Figure D.5. Divining devices

division gate A frame or barrier which divides the water flow between two or more LATERALS.

divisor A device which continuously diverts a representative fraction of the runoff from a specified area. The fraction is weighed and sometimes analysed and used to estimate the total soil and water losses for the area (USA). *See* DEMONSTRATION AREA.

dock An artificially formed basin or area of water on the side of a HARBOUR or bank of a river, closed by gates for the reception of ships. *See* DRY DOCK, TIDAL DOCK.

docking blocks Timber blocks used in a DRY DOCK to support the underside of a ship's hull. There are rows of blocks along each side and also a central row of keel blocks. They are made of hardwood, such as oak, and often a softwood cap is fitted to prevent damage to the hull.

dockyards A naval establishment where ships are equipped, repaired and maintained.

doldrums Part of the ocean near the equator with calms and light variable winds, considerable cloudiness, warm moist air, frequent squalls and thunderstorms; a zone of relatively low pressure near the equator. *See* DEPRESSION.

dolines Depressions or basins formed by the collapse of large caves in limestone districts. *See* SWALLOW HOLES.

dolphin A fixed mooring in the open sea formed by a cluster of timber or steel piles driven into the sea bed; a guide for ships entering a harbour with a mouth. *See* BELL DOLPHIN.

dominant discharge A term with somewhat different meanings according to country or authority. It has been defined as (1) the discharge when a stream is at, or near, its PROFILE OF EQUILIBRIUM; (2) the discharge of a natural channel which determines its depth, width, radius and meander patterns, and (3) the discharge of a natural channel at its BANKFUL STAGE, where the channel is reasonably well defined.

dominant formative discharge (or **bed-building discharge**) A discharge of sufficient volume and frequency of occurence to affect formation of river bed and course.

Doppler effect The change in apparent frequency of a vibration due to movement of the source of the vibration relative to the position of the observer. This effect on sound is often noticed when a whistling train passes a stationary observer. It is named after Johann Christian Doppler (1803–1853).

Dortmund tank A type of upward-flow sedimentation tank employed in sewage treatment. A deep tank with a conical-shaped bottom. The waste-water enters through a long pipe going down to the bottom of the tank. The consequential upward flow helps the sludge to settle.

dosing chamber *See* DOSING TANK.

dosing siphon An automatic SIPHON attached to a DOSING TANK for discharging its contents.

dosing tank (or **dosing chamber**) A chamber which receives raw or partly treated sewage; after a certain quantity has collected it is discharged automatically for treatment.

double float *See* SUBSURFACE FLOAT.

double lock An arrangement of two parallel lock chambers in a canal. They are connected by a sluice which reduces the loss of water to half that with one lock.

double-wall cofferdam A COFFERDAM formed by two parallel lines of sheet piling, used at sites where the height of a single line would be too great for its stability. The space between the lines of piles may be packed with clay for watertightness or with other material if only stability is required.

downstream drains (earth dams) Drains made within earth dams to reduce (1) the pore water pressure in the downstream zone, and so increase the stability of its slopes, and (2) the seepage rate, and so prevent PIPING.

downstream total head (over a weir) A measurement at a point downstream of the elevation of the total head relative to the crest. *See* UPSTREAM TOTAL HEAD.

dowsing *See* DIVINING.

draft tube The metal casing through which water leaves a turbine. *See* DIFFUSER CHAMBER.

drain A DITCH (open drain) or a buried pipe (closed drain) for conveying surplus surface or subsoil water or sewage. *See* CATCHWATER DRAIN, HOUSE DRAIN, SUBSOIL DRAINAGE.

drainage The natural removal of water from land along its streams; the artificial removal of surplus surface water or subsoil water from land by drains; the removal of water by gravity flow from surface works or excavations or irrigation areas; the removal of water or sewage from works or towns. A 'well-drained' soil allows excess water to drain away rapidly, while a 'poorly drained' soil allows excess water to drain only at a slow rate, and interferes with plant growth and tillage. *See* DRAINAGE LEVEL, DRAINAGE TRENCH, DRAINAGE TUNNEL, SUBSOIL DRAINAGE, SURFACE DRAINAGE.

drainage (dams) The use of drains or pipes built into the base of a dam to conduct any water up and out through the downstream face. This relieves the upward pressure on the base of the structure due to water in cracks, joints and bedding planes in the rocks below the dam. The watertightness of the structure should not be jeopardized by extending the drains too far upstream under the dam.

drainage area *See* CATCHMENT AREA.

drainage basin *See* CATCHMENT AREA.

drainage basin yield *See* YIELD OF DRAINAGE BASIN.

drainage, closed *See* CLOSED DRAINAGE.

drainage coefficient Runoff from catchment area or district over a period of 24 hours, expressed in depth or other units.

drainage density The average length of a river per unit area of its CATCHMENT AREA.

drainage ditch A channel or ditch cut along the lower part of irrigated land for drainage of excess water. *See* DRAINAGE.

drainage divide *See* DIVIDE.

drainage equilibrium A hydrogeologic condition in which the volume of water lost to underground storage, from all sources, just equals the volume of water reaching the storage from all sources. A reasonable period of time is taken for the assessment. The position of the water-table is practically unchanged during a period of drainage equilibrium.

drainage, internal *See* INTERNAL DRAINAGE.

drainage level An underground roadway or tunnel with a very slight inclination to divert water and induce its gravity flow outwards and away from mine workings. The roadway is provided with an impervious seal or lining on the lower side to prevent the seepage of water to the dip. *See* DRAINAGE TRENCH.

drainage outlet The outlet or lower end of a drainage ditch or other drainage channel.

drainage pattern The arrangement or layout of the youthful streams draining a valley; it depends mainly on the nature and distribution of the superficial deposits and any surfaces of weakness, such as major joints and bedding planes. *See* DENDRITIC DRAINAGE, INSEQUENT DRAINAGE, RADIAL DRAINAGE, RECTANGULAR DRAINAGE, TRELLIS DRAINAGE.

drainage rate A charge or levy upon the owners or occupiers of all lands within a district where a government, state or authority is carrying out drainage works. The rate may be based on the rateable value of the land and may be of such an amount as will provide for interest on the cost of the works, as well as redemption, depreciation, maintenance and management.

drainage, subsoil *See* SUBSOIL DRAINAGE.

drainage, surface *See* SURFACE DRAINAGE.

drainage terrace A graded ridge or embankment of earth used primarily for hillside drainage. It is built with a relatively deep channel and a low ridge. *See* GRADED TERRACE.

drainage trench A channel or trench cut alongside an underground roadway or tunnel for drainage and to enable the proper ballasting of the rail track. The trench is graded to allow gravity drainage, and may be lined with pre-cast concrete sections on the underside.

drainage tunnel A horizontal or inclined stone drivage in a mine, used mainly for drainage purposes. It may be driven and graded to drain waterlogged rocks, old workings, or near-surface water, and thus prevent water seeping downwards to productive workings at lower levels. A central drainage tunnel may serve several mines in close proximity and by gravity flow reduce the need for pumps and pumping stations.

drainage valves Valves provided at all low points of pumping or similar hydraulic systems for allowing water to escape when repairs are carried out or as a frost precaution. The valves are left open when the installation is idle and empty.

drainage well A well put down to drain swampy ground or to remove wastewater or sewage at the surface. The water is admitted near the top of the well and discharged into a permeable deposit at lower levels. To be effective, a drainage well must enter a ZONE OF AERATION or a water-bearing rock with subnormal pressure head. *See* INVERTED CAPACITY, INVERTED WELL.

drainpipes Pipes for the removal of subsoil water, rainwater, wastewater, or sewage. They may be made of concrete, earthenware, cast-iron, asbestos cement or other material, and used with open or closed joints. *See* DRAINAGE, SUBSOIL DRAINAGE.

drain rods Rods pushed to and fro in a drain for cleaning and removing obstructions, made in sections which screw together. *See* LAMPHOLE, RODDING.

drain tile (USA) *See* AGRICULTURAL DRAIN.

drain well *See* ABSORBING WELL.

draw A ravine, tributary valley or coulée, where water discharge is confined to rainstorms; a small natural drainage way; a natural depression; the depth of water required by a ship to float.

drawbridge Any MOVABLE BRIDGE.

draw-door weir A WEIR in which the gates can be raised vertically. *See* FRAMED DAM, ROLLING-UP-CURTAIN WEIR, SLIDING-PANEL WEIR, SUSPENDED-FRAME WEIR.

drawdown The lowering or drop in water level of a reservoir; the lowering of the water-table in and around a well or borehole by pumping, usually expressed in metres or feet. *See* CONE OF DEPRESSION, SUDDEN DRAWDOWN. The term may also be applied when the surface slope of a stream exceeds the normal slope.

drawdown curve The theoretical curvature of the groundwater surface as affected by removal of groundwater by open ditch, tile or other methods (USA).

drawdown, sudden *See* SUDDEN DRAWDOWN.

draw-well A well from which water is extracted by a rope with a bucket attached.

dredger (or **dredge** [USA]) A vessel or floating PONTOON equipped with grab, bucket ladder, or suction dredging machine for underwater excavation. In general, harbour dredges are sea-going vessels. A STATIONARY DREDGE is a fixed vessel with a bucket ladder; the excavated material is discharged into a pipeline or a hopper barge. *See* BUCKET-LADDER DREDGER.

dredging The excavation, bringing up and removal of material from the bed of a river, lake or harbour by dredger, dragline or scoop. Dredging may be necessary to increase the depth of water in a river or to open up an intake channel. Draglines may be used from each bank of a river and the spoil placed directly on the banks. Lakes and other storages may be cleared by scoops pulled across by wire ropes attached to, and operated by, portable power winches.

dredging well The opening in a dredger through which the suction pipe or bucket ladder passes to the river or sea bed.

drift (1) The movement or speed of a current of water or any contained sediment or material. (2) The downstream movement of a discharge boat, anchored or not, during the period taken for velocity measurements. *See* MODIFIED VELOCITY. (3) *See* GLACIAL DRIFT.

drift barrier An open structure erected across a stream to catch driftwood; it may range from a simple wire fence or chains to massive piers supporting heavy cables between them. The driftwood is removed or burnt at low water. *See* DEBRIS DAM.

drift-dam lake *See* GLACIAL LAKE.

drift deposits *See* BURIED CHANNELS, ENGLACIAL MATERIAL, GLACIAL DRIFT, SHORE DRIFT.

drift, littoral *See* LITTORAL DRIFT.

drift map A geological map which shows the glacial drift or superficial deposits of an area in addition to solid rocks if they outcrop at the surface.

drift velocity The rate of movement in a specified direction due to drift. *See* DRIFT (2).

driller's log (or **operational log**) A form containing information entered by the drill supervisor for each shift. It covers drilling time, depth drilled, depth of groundwater standing in hole, nature of rocks, colour of sludge, caving of hole, size of casing and whether casing is following the drilling, and any abnormal features or mishaps.

drip irrigation *See* TRICKLE IRRIGATION.

dripping fault A fault along which small quantities of water seep downwards. Mine workings near such a fault often cause an increased flow because of subsidence and opening of the fault fissure.

driven well (USA) A well driven by a casing at the bottom of which a drive point is fixed. The well is usually between 25 and 75 mm (1 and 3 in) in diameter and there is no drilling, boring or jetting device.

drive pipe A length of piping, about 50 mm (2 in) in diameter, for sampling superficial deposits of crushed residues with no boulders or stones. The lower end of the pipe is sharpened to ease penetration and sometimes restricted in diameter, so that the sample material is retained within the pipe on withdrawal.

drizzle Fine, spray-like rainfall; droplets are so small that their individual impact on a water surface is not perceptible.

drop A structure in a channel for lowering the water level, usually vertically but an inclined chute may be used; a steep part of a pipe or channel. *See* CHECK AND DROP, CHUTE.

drop bars (Australia) *See* FLASHBOARD.

drop-down curve The curved longitudinal profile of a water surface upstream from a point where a sudden fall occurs in a conduit, river or channel.

drop-down section The part of the longitudinal profile of a river or channel along which a DROP-DOWN CURVE occurs.

drop-inlet dam A dam in which the overflow is carried off through a gently inclined pipe (or SPILLWAY) under the dam. The pipe is connected to a 'riser' or open-topped, vertical pipe at the pond side of the dam.

drop-inlet spillway A spillway consisting of a pipe built under or through the dam or embankment (Figure D.6). *See* DROP-INLET DAM, OVERSHOT SPILLWAY.

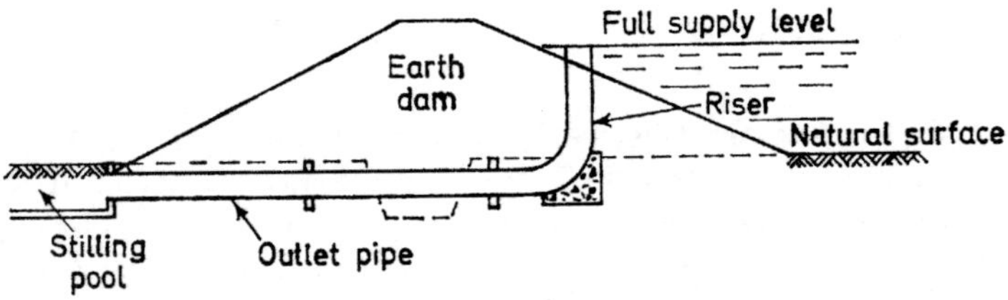

Figure D.6. Drop-inlet spillway

drop structure A structure built in a stream channel or gully to decrease the velocity of flow by lowering the slope or grade; the drop in velocity promotes the deposition of silt and prevents excessive channel scour. *See* CHECKDAM.

drought Continuous dry weather, that is, without significant rainfall. *See* PRECIPITATION (TRACE OF).

drowned flow The water flow through a flume, or over a weir, that is affected by fluctuations in the level downstream. *See* MODULAR FLOW.

drowned valleys Valley or river mouths which were submerged as a result of land subsidence (or rise of sea-level) and now form long narrow inlets or estuaries. If the valley mouths were broad, the sinking would give rise to wide shallow bays which, as a later phase, may be silted up to form land again. There are many examples of drowned valleys in Britain, as along the south coast of Devon and Cornwall, sometimes called rias. *See* COAST OF SUBMERGENCE.

drowning ratio *See* SUBMERGENCE RATIO.

drum gate A type of spillway CREST GATE in the shape of an acute circular sector in cross-section and controlled by valves (Figure D.7).

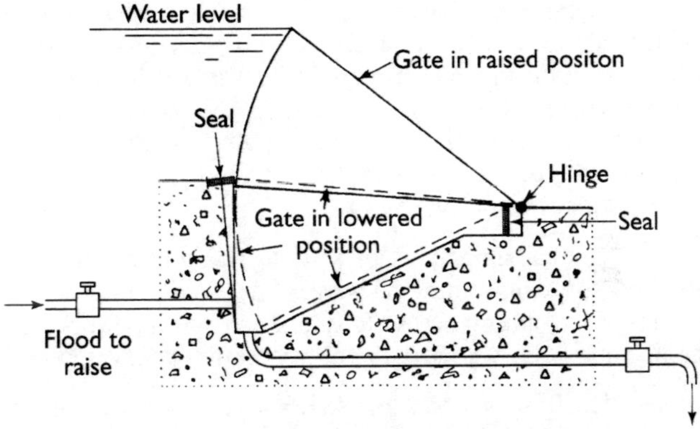

Figure D.7. Drum gate

drum screen A cylindrical or truncated cone-shaped screen, revolving on its centre line, and used in INTAKE WORKS or sewage treatment. *See* BAND SCREEN.

dry air *See* DAMP AIR.

dry density The mass of a dry sample of soil per unit volume. Before weighing, the sample is heated at a temperature of 105 °C (221 °F).

dry density/moisture ratio The relationship between DRY DENSITY and moisture content of a soil for a given amount of compaction. The values may be shown as a graph from which the OPTIMUM MOISTURE CONTENT may be determined.

dry dock (or **graving dock**) An enclosed basin from which the water can be pumped out. A ship is waterborne into the dock, the dock gates are closed, and the water is removed. The ship rests on DOCKING BLOCKS while undergoing repairs or cleaning.

dry farming Farming with no irrigation in arid or semi-arid regions.

dry risers One or more vertical pipes, for fire protection, in a building with outlets at different floor levels. They are normally empty, but standard fire hoses can be connected to their lower ends in an emergency. *See* FIRE HYDRANT.

dry salting The build-up of salt in the ROOT ZONE due to inadequate irrigation.

dry valley A valley without streams, caused in various ways. It is a feature of the British Chalk districts where the rainfall percolates downwards with no runoff. The original valleys were carved by streams when at higher levels. Other dry valleys were cut by melt waters from snow and icecaps which were situated on the uplands during the Ice Age. Again, a portion of a valley may become dry by RIVER CAPTURE. In the Mallee country of north-western Victoria, Australia, where the rainfall is low and the ground absorptive, there are few defined streams and all surface runoff is temporary. *See* UNDERGROUND DRAINAGE.

dry valve shaft A shaft forming part of a storage dam and containing a vertical standpipe for connecting the several draw-off pipes with the supply main. The shaft allows access to almost all the pipework and is generally adopted in relatively large dams. *See* INTAKE TOWER, VALVE SHAFT.

dry weather flow The flow of water in a stream during a dry period. Dry weather flow varies considerably in different areas, depending on the subsoil conditions and, for example, whether the stream is fed from springs. The term is also applied to the flow in a sewer measured during a 24-hour period of dry weather, which generally corresponds to the consumption of water in 24 hours.

dry well (1) (or **soakaway** or **rummel**, Scotland) A well into which surface water is drained and allowed to soak into the ground. It may be lined with bricks or stones without mortar and can be empty or filled with large stones. (2) That section of a pumping station where the machinery is housed, as opposed to the wet well or sump which contains water.

dual water scheme A scheme which utilizes both underground and surface water supplies. It gives greater flexibility and the differences in the seasonal fluctuations of the two sources may be exploited. For example, when the water level of an impounding reservoir is low, deep wells in the area often continue to yield substantial supplies.

dug well A well put down by manual labour; only picks and shovels are used. *See* WELL SINKER.

dumb barge *See* HOPPER BARGE.

duplex pump Usually a steam-driven pump comprising two water cylinders and two steam or power cylinders arranged side by side.

duplicate mains Often applied to the laying of SERVICE MAINS, one on each side of a wide concrete highway, to supply consumers on both sides. No communication pipes are required to cross the highway but capital costs tend to be higher.

duration curve A curve showing the frequency with which given values, such as flow and head of water in a channel, are equalled or exceeded during a certain period. For a river used for power generation the graph gives over a certain period (the abscissa) the percentage of total time, or number of times, that the flow was equal to or exceeded the ordinate. The area beneath the duration curve gives the total quantity (or runoff) which flowed in the specified period. The arithmetic mean flow is the average rating.

duration graph, depth *See* DEPTH DURATION GRAPH.

duration of rainfall In the UK, the duration of rain falling at a rate of not less than 0.1 mm (0.004 in) per hour, and obtained from self-recording rain-gauges.

duration of tide EBB TIDE, or falling tide: the interval of time from high water to the following low water; FLOOD TIDE, or rising tide: the interval of time from low water to the following high water. For a SEMI-DIURNAL TIDE these cover a period of 12.42 hours, but the presence of non-tidal flow may modify the duration. Duration of ebb tide in a river is usually longer than the duration of flood tide, owing to freshwater discharge.

duty of water *See* IRRIGATION REQUIREMENT.

dyke *See* DIKE.

dynamic delivery head *See* TOTAL DELIVERY HEAD.

dynamic similarity A principle of similarity which states that if a scale model of a hydraulic structure operates at a speed corresponding truly with that of the full size structure, then the resistances R, densities d, lengths l, and velocities v possess the following relationship:

$$\frac{R_1}{R_2} = \frac{d_1 l_1^2 v_1^2}{d_2 l_2^2 v_2^2}$$

dynamic suction head The friction head loss, usually measured in metres or feet, in the supply pipe leading to a pump from the intake sump to the pump centreline. *See* TOTAL SUCTION HEAD.

dysentery A general term for a range of infections of the intestine. The two basic forms are bacillary dysentery and amoebic dysentery. Both forms can be contracted from water that has been polluted by human faeces, but it is more commonly the result of eating or drinking contaminated foods or liquids. If suitable treatment is given both forms are now rarely fatal.

E

early youth A stage of the erosion cycle often characterized by (1) nearly all land surface well above the level of the major streams; (2) flat, wide inter-stream divide areas; (3) narrow, steep stream gorges, and (4) inefficient, poorly integrated drainage. *See* LATE YOUTH.

earth and rockfill dam A fill dam in which the impervious earthfill material exceeds 50%. The free-draining layers outside the earthfill core have an increasing perviousness towards the outside slopes. *See* EARTH DAM, GRAVITY DAM.

earth auger A hand-boring tool for putting down shallow test holes in soft deposits with no stones. It usually consists of a twist drill mounted on steel piping and rotated by turning a 'T'-piece attached to the upper end. The material adhering to the screw gives an indication of the ground penetrated. Depths down to about 6 m (20 ft) are possible. Earth augers may be used to fix bedrock or the water-table as a preliminary step to putting down shallow wells in areas which are more or less unexplored. The auger may be power-operated and mounted on a lorry or caterpillar tracks, when greater speeds and depths may be reached.

earth dam (or **earthen dam**) An artificial ridge or barrier for confining water, formed of earth, or rock and earth, with a core of clay, concrete, or other impervious material; it may be constructed with an impervious layer, such as steel plate or concrete, etc., on the water face. *See* HYDRAULIC FILL DAM, ROCKFILL DAM, SUPERFICIAL COMPACTION.

earthflow *See* LANDSLIDE, MUDFLOW.

earth-moving plant Mechanical equipment such as bulldozers, scrapers, trench excavators and loading shovels which are frequently used on earth dams, ditches, drainage and channel construction. *See* TRENCH EXCAVATOR.

earthquake A trembling of the earth's crust caused by the mass movement of rocks often along a pre-existing plane of weakness, such as a fault. Small local earthquakes or tremors may also be produced by new heavy loads on the earth's surface, such as the initial filling of a large reservoir. Earthquakes often cause severe damage to water mains, and the buckling of pipelines. Again, immense damage has been caused by coastal flooding resulting from earthquakes originating below or near the sea.

earthwork Any excavation, cutting, or artificial dumping or banking of SUPERFICIAL DEPOSITS, as during the construction of dams, canals, trenching and river improvement works. *See* BALANCED EARTHWORKS.

ebb channel The channel made in an ESTUARY by the EBB CURRENT, usually of 'S'-shape in a long estuary.

ebb current The return of tide water toward the sea; the movement of the tidal current down the tidal stream or from shore.

ebbing well A well in permeable rocks near a coast; the water level rises and falls with the fluctuations of the tides. Excessive pumping may cause salt water to enter the well and contaminate the supply.

ebb tide (or **falling tide**) The tide during the period between high water and the following low water. *See* DURATION OF TIDE.

echo sounder An instrument for measuring the depth of water by recording the times for sounds to be echoed back from the sea bed; times are converted into depths and automatically recorded by the instrument. It is also used to determine the depth of water in reservoirs. *See* SOUNDING.

E. coli *See* ***Escherichia coli***.

ecology The study of the relations between organisms and with their environment.

economic storage A storage which gives the cheapest water. The most economical size of impounding reservoir may be estimated by constructing curves of costs and gross yields for dams at different sites or catchments. Usually the cost of yield falls as the size of the reservoir is increased, reaches a minimum and then increases again. *See* STORAGE/EXCAVATION RATIO.

economic yield The maximum rate at which water may be extracted from a well or aquifer over a period without depleting the supply or causing a deterioration in quality. Deep wells in valleys have a relatively greater yield than those in upland areas, owing to greater depth in relation to the intake. *See* PUMPING TEST.

ecosystem A community of interdependent organisms within their environment, such as all life within a stream or forest.

eddy A small whirlpool; to move in a circular direction in flowing water, as a result of irregularities or obstructions in the bed or banks of a stream. *See* POTHOLES.

eddy, confined Eddy which is not free to move and spread owing to contact with a structure or obstruction in a channel.

eddy flow *See* TURBULENT FLOW.

eddy, free Eddy which is unrestrained and free to spread and move from a structure or obstruction in the channel.

eddy loss The energy lost by eddies, swirls and impact, as opposed to that lost by friction. The energy loss is converted into heat. *See* IMPACT LOSS.

effective area of an orifice The actual area of an orifice multiplied by its COEFFICIENT OF DISCHARGE.

effective porosity (or **effective drainage porosity**) The ratio of volume of water in a pervious mass (previously saturated with water), which can be drained by gravity forces, to the total volume of the mass. *See* SPECIFIC RETENTION.

effective precipitable water *See* PRECIPITABLE WATER (EFFECTIVE).

effective pressure (soils) The pressure between the points of contact of soil grains; it increases during consolidation of the soil to a maximum value when consolidation is complete. *See* FLOW NET.

effective rainfall The surface runoff portion of rainfall. *See* DIRECT RUNOFF.

effective size According to HAZEN'S LAW, the grain size larger than 10% by weight of the soil particles, as shown on the GRADING CURVE of the soil, described briefly as the D10 size. *See* UNIFORMITY COEFFICIENT.

effective snow melt That part of SNOW MELT that becomes runoff and reaches a stream. *See* SNOW DENSITY.

effective storage In certain large storages, the available water which is restricted to the quantity that can be drawn off by gravity through the outlet pipes. The water below the outlet pipe is known as the non-effective storage.

effective velocity (groundwater) The volume of groundwater passing through unit cross-sectional area divided by EFFECTIVE POROSITY of the ground or material. Also known as field, true or actual velocity. *See* AVERAGE VELOCITY.

efficiency of farm irrigation The percentage of the total irrigation water supplied to a farm which is available to plant roots. *See also* IRRIGATION EFFICIENCY.

effluent Flowing out; a stream that does not rejoin the main channel after separation; a stream flowing from a lake or underground storage; the flow of sewage from a process plant.

effluent seepage Seepage of water from the ground into a channel or stream (Figure E.1), as opposed to INFLUENT SEEPAGE.

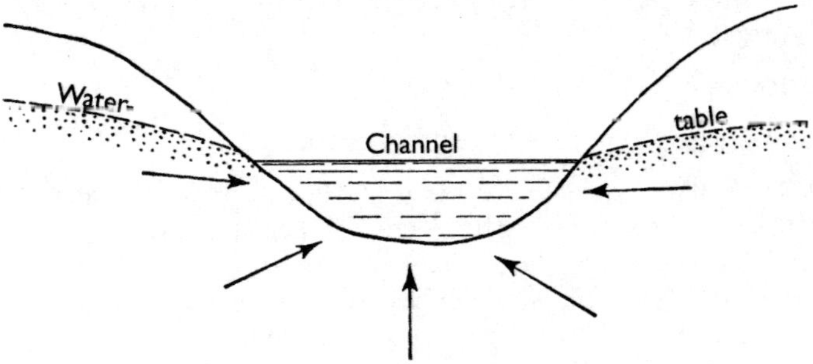

Figure E.1. Effluent seepage

effluent sewage The flow of sewage away from a treatment plant; the treatment may be complete or incomplete.

egg-shaped sewer An ovoid-sectioned sewer, often made of pre-cast concrete segments. They are placed with the small end downwards as this position ensures satisfactory flow when the sewer is nearly empty.

ejector An appliance for raising sewage to a higher level in which compressed air is injected into a chamber containing sewage and air pressure forces it upwards.

electrical resistance strain gauge A flat coil of fine wire used in investigations into the movements and strains in underground rocks, based on the change in electrical resistance of the thin wire when stretched under the influence of rock strain. In water engineering it has been used to investigate the strain developed in both longitudinal and circumferential directions in buried water supply pipes.

electrical tape gauge A gauge consisting of a graduated tape with a weight at the end which is lowered to touch the water surface. An electrical device indicates when contact is made with the liquid. *See* WIRE WEIGHT GAUGE.

electrochemical gauging Measurement of liquid flow in open channels. *See* CHEMICAL GAUGING, DILUTION METHODS, SALT VELOCITY METHOD.

electro-osmosis A method of lowering groundwater, particularly applicable to silts. It accelerates natural drainage away from surface works and excavations.

elevated ditch A large ditch or canal carried across a depression or canyon by earthfill crossing, similar to an earthfill dam, at normal grade; it is sometimes used for irrigation.

elevated flume A flume used to convey water across depressions or low-lying areas.

elevation head *See* POTENTIAL HEAD.

elevator dredger *See* BUCKET-LADDER DREDGER.

El Niño A term originally used to describe a warm South Pacific current that flows southward along the coast of Ecuador and Peru. It occurs around Christmas time, hence the name *El Niño* – the Child – after the Christ child. The term now applies to a more intense and prolonged equatorial Pacific warming, which occurs at intervals of several years and is associated with major changes in the climate system.

elutriation The washing of sludge with water or plant effluent to reduce alkalinity and thereby the amount of conditioning chemicals required. *See* CONDITIONING.

embankment *See* BANK.

emergency spillway A supplementary SPILLWAY which operates only when inflow floods exceed the capacity of the main spillway (Figure S.5), typically with a wide free overflow crest over which there are flows only during exceptionally large and rare floods. Emergency spillways can be constructed in tunnels, culverts or other structures, but most commonly they are side-open channels excavated to one side of a dam; the flow into these channels is controlled by a free-overflow weir crest at the upstream end. Emergency spillways are designed to be used rarely, if at all; should such usage be necessary, some damage is normally considered acceptable. *See* FUSE PLUG SPILLWAY.

encrustation (pipes) The formation of rust nodules or the deposition of calcium carbonate in imperfectly protected metal pipes, thus reducing their effective area and carrying capacity. *See* CORROSION (2).

end contraction A MEASURING WEIR is said to have end contractions when the sharp upstream edges of the notch sides cause the NAPPE to contract in width. *See* CONTRACTED WEIR.

energy dissipator Any structure or material used to disperse or destroy the energy of high-velocity waters. *See* CHECKDAM.

energy gradient (or **energy slope**) The gradient or slope of the ENERGY LINE.

energy head The head or elevation of HYDRAULIC GRADIENT line, plus the VELOCITY HEAD of average velocity of water, at any given section; any base may be used as datum.

energy line A line joining elevations of ENERGY HEADS of an open conduit.

engineer–hydrologist An engineer versed in the principles of hydrology and their engineering application. *See* FARM WATER SUPPLY ENGINEERING, HYDROGEOLOGY.

engineering geology Geological facts and principles that are usefully applicable to mining, municipal and civil engineering works. It includes undermined rock behaviour, tunnels, land drainage, irrigation, water supplies, river control, dams, docks, harbours, water power, sewage disposal and sewerage. When planning and designing these works, the geological aspects are investigated for safe and economic operation. *See* RIVER ENGINEERING.

engineering practices All water and soil conservation methods and engineering works which modify land slopes so that runoff may be reduced, controlled or diverted, and thus diminish soil erosion and improve irrigation. These include the use of DRAINAGE TERRACES, CONTOUR FURROWS, DAMS, DIKES, DIVERSIONS, etc. *See also* DRAINAGE, IRRIGATION, RIVER TRAINING, STOCKWATER DEVELOPMENT, VEGETATIVE PRACTICES.

englacial material The disintegrated rock material held and carried forward in the lower part of a glacier. Later, when the ice melts, the debris and boulders are left stranded on alien land. Englacial material of pre-existing glaciers often forms natural dams in valleys, allowing the formation of lakes which may serve as sources of water supply. *See* ESKERS, GLACIAL LAKE.

engrafted river system Streams that unite when flowing over the COASTAL PLAIN to reach the new coast.

enrockment *See* RIPRAP.

enteric viruses A mixed group of a large number of viruses infecting the alimentary tract which are excreted in vast numbers and are highly infective.

entrance head The head of water required to cause flow into a closed or open conduit or hydraulic structure. Entrance head is the sum of the ENTRANCE LOSS and VELOCITY HEAD.

entrance lock The lock providing entrance for ships to a dock in which the water level is different from that outside.

entrance loss The head lost at the inlet to a conduit or hydraulic structure due to friction and turbulence of water.

entry head *See* ENTRANCE HEAD.

entry loss *See* ENTRANCE LOSS.

envelope curve (1) A smooth curve covering all peak values of certain quantities, such as rainfall, runoff, etc., plotted against other values, such as time, area, etc. In this maximum envelope curve none of the peak values normally extend above the curve. (2) A smooth curve covering all trough values of the quantities as above. In this minimum envelope curve none of the trough points extend below the curve.

environment The air, water, land and other factors that support and influence the life of an organism.

ephemeral stream A stream with its channel positioned above the water-table; its water is derived from precipitation. *See* PERENNIAL STREAM.

epigene Pertaining to geological changes or processes active at or near the earth's surface. *See* HYPOGENE.

epilimnion The surface water above the thermocline in a thermally stratified lake; a turbulent region of relatively uniform temperature. *See* HYPOLINMION, THERMOCLINE.

equalizing bed A layer of concrete or ballast along the bottom of a trench in which pipes are laid.

equalizing reservoir (or **balancing tank**) An auxiliary reservoir or storage for excess river water that may be returned for use during a period of shortage.

equation of continuity *See* LAW OF CONTINUITY.

equilibrium moisture content The moisture content of a soil at a period when the moisture is stationary.

equilibrium pH The saturation pH of a water. *See* LANGELIER INDEX.

equilibrium, profile of *See* PROFILE OF EQUILIBRIUM.

equinoctial rains Rains occurring regularly about the time of the equinoxes in equatorial regions.

equinoctial tides The exceptionally high tides generated at the equinoxes (times at which the sun crosses the equator and day and night are equal). When these tides coincide with severe storms and heavy seas, the coarser deposits along the BACKSHORE are often piled up to form a storm-beach or ridge. *See* NEAP TIDES, SPRING TIDES.

equinox *See* EQUINOCTIAL TIDES.

equipotential lines Lines or contours of equal water pressure in the ground near or within an earth dam or other water-retaining structure. A model of the structure may be constructed with the water levels to scale. Glass STANDPIPES, connected to the soil mass, indicate by the water level in them the pressure at the points to which they lead. The equipotential lines can be plotted from the levels (Figure F.6). *See* FLOW LINES, FLOW NET.

equivalent pipe A pipe which can be used to replace a given system of pipes with equal head loss for a given flow. The calculation may be made using a pipe flow formula or a NOMOGRAM.

erodible Susceptible to erosion; the material eroded. *See* EROSIVE.

eroding bank A river, stream or gully bank which is undergoing active erosion by water flow; erosion may also occur by water and sediment flowing down the bank during heavy rains.

erosion The gradual wearing away of land forms, rocks and soils mainly by the action of flowing water and ice. *See* ACCELERATED EROSION, CREEP, CYCLE OF EROSION, GEOLOGICAL EROSION, SHEET EROSION, SHORE EROSION, SOIL EROSION, SPLASH EROSION, STREAM BANK EROSION, WAVE EROSION.

erosion class The erosion pattern of a region; a characteristic group or set of erosion conditions over a specified period and area. It often forms part of the land information given on a SOIL CONSERVATION SURVEY map.

erosion, stream bank *See* STREAM BANK EROSION.

erosive Pertaining to an agent causing erosion, such as flowing water; 'erodible' refers to the material eroded, such as soil or rock.

escapage Water which has been diverted into an irrigation system and escapes through spillways or is otherwise discarded.

escape A channel or wasteway along which the entire flow of a stream may be discharged. *See also* BY-WASH.

***Escherichia coli* (*E. coli*)** One of the species of bacteria living in the alimentary tract of mammals, including humans. It is not normally pathogenic but it is indicative of faecal contamination and hence the possible presence of pathogenic organisms.

eskers (or **eskars**) Long ridges of gravel and sand deposits which were left behind as the glaciers melted, parallel to the direction in which the ice is moving. *See* KAMES.

estuarine Pertaining to the tidal mouth of a river.

estuarine flat *See* MUD FLAT.

estuary The passage or mouth of a large river where the tide meets the river current and the water is brackish. *See* BAR, EBB CHANNEL, FLOOD CHANNEL, TIDAL CURRENT.

euphotic zone The upper layer of water where there is sufficient light to support photosynthesis effectively.

eustatic movements Elevations or depressions of sea level on a vast scale.

eutrophication A phenomenon in lakes with high nutrient levels leading to large quantities of planktonic algae. The low oxygen levels in the deeper layers kill animal life during summer months, often leading to formation of foul organic deposits and hydrogen sulphide. In time the process of eutrophication will cause most small lakes and reservoirs to become unable to support animal life and they are then completely filled and transformed first into marsh and then dry land; depending on local conditions this can take many centuries or only a few decades. *See* ALGAL BLOOM (Figure E.2).

Figure E.2. Eutrophication in a small artificial lake. The small dredger in the foreground is cutting the shoreline back to its position when the lake was built 250 years ago

evaporating basins Lakes in inland basins, as in south-western Victoria, Australia, with no visible outlet to the sea. These lakes may shrink in dry seasons and swell in wet seasons, but evaporation can usually cope with the inflow and they maintain a reasonably constant area. However, on rare occasions the rainfall may greatly exceed the evaporation rate, with flooding of marginal lands and properties. In some cases, outlet channels, cuts or tunnels have been made to minimize flooding hazards. *See* LAKE.

evaporation The conversion of surface water into vapour or ATMOSPHERIC WATER; the quantity of water converted into vapour under specified conditions. Evaporation varies greatly according to relative humidity, temperature, character of vegetation, wind velocity and nature of soil. It is more rapid from clayey than from sandy soils, and is greater in cleared areas than in areas covered with forests or vegetation. The amount of evaporation may be estimated from intake and outflow at reservoirs, from water levels in tanks, and from pans filled with soil and sunk flush with the ground. *See* ANTITRANSPIRANTS, EVAPORATION LOSS, MULCHING.

evaporation area (stream) The area along which evaporation of stream water occurs, consisting of the stream surface and the wetted ground along each side.

evaporation discharge The discharge of water, derived from the zone of saturation, by evaporation into the atmosphere; it may be SOIL DISCHARGE or VEGETAL DISCHARGE.

evaporation loss That part of rainfall that returns directly to the atmosphere by evaporation and by transpiration from vegetation.

evaporation opportunity The ratio of the actual rate of evaporation from water or land in contact with the atmosphere, to EVAPORATIVE CAPACITY under existing atmospheric conditions.

evaporation pan A large standard pan used to hold water for the accurate daily measurement by hook-gauge readings of the amount of evaporation at a given location. *See* EVAPORIMETER.

evaporation power An index of the degree to which a surface or area is favourable or unfavourable to evaporation; it is equivalent to EVAPORATIVE CAPACITY if referring to unit area exposed parallel to the wind and expressed in the same terms.

evaporative capacity (or **evaporativity**) The maximum rate of evaporation which can be developed by a given atmospheric environment from a unit area of wet surface exposed parallel to the wind, the temperature of the surface being at all times exactly equal to that of the surrounding atmosphere.

evaporativity *See* EVAPORATIVE CAPACITY.

evaporimeter (or **atmidometer** or **atmometer**) An instrument for measuring EVAPORATION.

evapo-transpiration The loss of moisture from a soil by EVAPORATION and plant TRANSPIRATION. Evapo-transpiration is viewed by some engineers as an unnecessary term.

excavated tank A TANK consisting of a storage basin excavated below ground level. *See* RING DAM.

excavated volume The volume of earth or rock excavated to build a water storage, such as a reservoir, canal, trench or foundation, or any works below ground level; usually expressed in m^3 or yd^3. *See* STORAGE/EXCAVATION RATIO.

excavator Any of a large variety of power-operated digging and loading machines, used widely in excavations for dams, canals, trenches, etc. *See* TRENCH EXCAVATOR.

exceedance interval The average number of years within which a given hydrologic event, such as a flood, is exceeded. *See* RECURRENCE INTERVAL, RETURN PERIOD.

excessive precipitation *See* PRECIPITATION (EXCESSIVE).

excessive storms *See* PRECIPITATION (EXCESSIVE).

excess water The water that runs off at or near the surface when precipitation exceeds infiltration plus absorption. *See* RUNOFF.

exit loss Head lost at the outlet of a conduit or hydraulic structure due to friction and turbulence of water.

expending beach A beach designed mainly for absorbing wave energy.

exposed intake A type of INTAKE WORKS often used in the UK to obtain water from a river or sometimes from a lake. The intake is usually sited on a straight stretch of river with no rapid currents, and its level set below the lowest known river level, while the floors of any buildings are formed above the maximum flood level. *See* SUBMERGED INTAKE.

exsiccation The drying up or draining dry of an area because of some change which causes loss of water or moisture without a reduction in rainfall. The drying up of marshy ground by drainage or seepage to subsurface excavations or cavities are examples. *See* DESICCATION.

extinct lakes Lakes which have been naturally obliterated in various ways. *See* OBLITERATION OF LAKES. The former existence of these lakes is indicated by a natural flat or basin filled with lake deposits, and perhaps by old beaches with indications of wave action. *See* LAKE-FLOOR PLAIN.

extratropical cyclone A cyclone exhibiting great variation in intensity, diameter and wind strength, found in latitudes higher than the tropics.

extreme value distribution A statistical distribution which describes the variations in the largest and smallest values in a process. Such distributions are important in assessing the risks from unlikely events, such as very large flood flows or rainfalls. Many types of extreme value distribution are available, of which the most commonly used is the GUMBEL DISTRIBUTION.

F

fabric, reinforcing *See* REINFORCING FABRIC.

fabridam A barrier of plastic sheet placed across a river and anchored to the river bed. When inflated with air or water it can rise to 6 or 9 m (20 or 30 ft); during heavy floods it is wholly collapsible, when it offers very little resistance to flow.

facing A layer or cover to protect cuttings, sea walls or dikes. *See* LINING.

facultative lagoon A lagoon or treatment pond in which aerobic processes occur in the upper section and anaerobic processes in the lower section.

faggot A fascine or long cylindrical bundle of brushwood. *See* FASCINES.

faggotting A covering consisting of brushwood FAGGOTS along the submerged areas of river banks where grass cannot be planted. They are weighed down with stones or other means. Thorn faggots are favoured, as they collect silt and also reduce scour. *See* FASCINES, MATTRESS, REVETMENT.

failure Broadly, there are two forms in water engineering: (1) structural, due to the inability of a hydraulic structure to physically fulfil its initial purpose, and (2) hydrologic, because of a defect in the hydrologic design, such as the failure of a reservoir to reach its designed capacity due to inadequate runoff from the catchment.

fairlead A metal fixture on a ship or quay to guide the ropes and enable ships to berth safely.

falaj A tunnel which taps groundwater and gravitates it to the surface for irrigation. Many falaj remain in use in the Sultanate of Oman.

fall (1) A sudden drop in bed level of a stream resulting in WATERFALLS. (2) The STREAM GRADIENT, expressed as a fall of so many metres per kilometre or feet per mile.

falling apron principle *See* UNDERWATER APRON.

falling tide *See* EBB TIDE.

falls *See* WATERFALL.

farm dam A small earth dam constructed on a farm to store water which is released and used as required for irrigation, or domestic or stock supplies (Figure F.1). *See* BY-WASH, DAM, EARTH DAM.

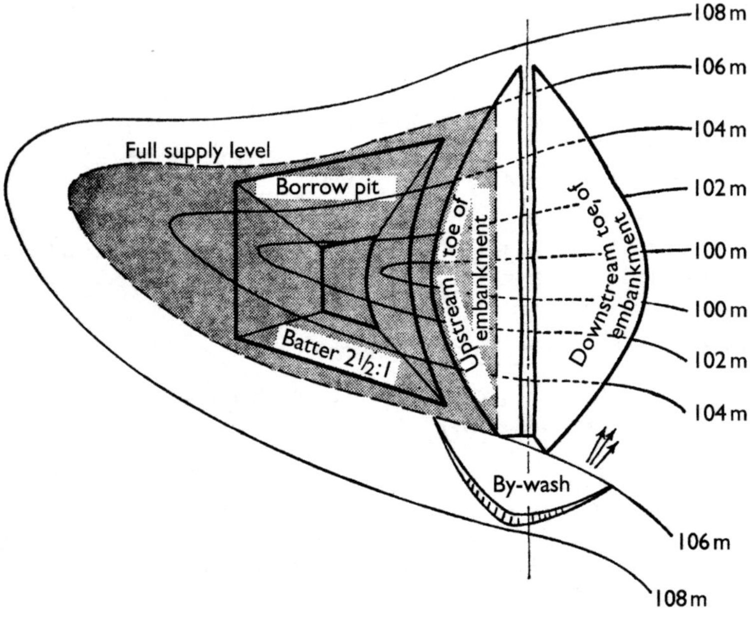

Figure F.1. Farm dam

farm drainage *See* DRAINAGE, FIELD DRAINAGE.

farm duty *See* NET DUTY.

farm pond (mainly USA) A relatively small body of water retained in a hollow excavated in the ground or held behind a dam for farm usage, water for livestock, etc. *See* GULLY DAM.

farm reservoir A natural or artificial lake near a farm to meet the water requirement of the property, such as irrigation and domestic and stock supplies. *See* STORAGE CAPACITY.

farm water supply engineering A specialized branch of civil engineering, mainly concerned with the development and use of the water resources available to farmers and landholders who live in isolated areas. Much of this branch of the engineering service involves supplies for domestic use, stock and small irrigation schemes. *See* WATER HARVESTING.

farm waterway A stream or constructed water channel which is wholly within the boundary of an individual farm.

fascine groyne A groyne, built with shingle and FASCINES, for diverting the current away from an eroding bank. A raft extending from the bank is secured by cables upstream. Shingle is loaded onto the raft until it sinks evenly. As it sinks, fascines are added along the sides and ends, and secured. The box-like structure is loaded with shingle until the raft comes to rest on the bed, and stone or weighed fascines are added to close any gaps. *See* WILLOWBOX GROYNES.

fascines Brushwood which has been rolled into cylindrical bundles and secured with wire, used to protect river banks and sea walls. The brushwood may be cuttings of ash, hazel or birch of a suitable age. The material is weighed down or otherwise secured to the bank. Willows or other vegetation are often used in conjunction with fascines. Another application of fascines is in filling ditches or to improve the bearing capacity of soft or waterlogged soils. *See* ANCHORED TREES, BRUSHWOOD.

fathom A British measure of length equal to 6 ft and used as a unit of marine depth; 1 fathom = 1.829 m.

fault When applied to geology, this refers to a break in the rocks along which displacement of one side relative to the other has taken place parallel to the fracture. Faults may form channels for the seepage of water or for the escape of water from a canal or reservoir. The fracture may be sealed off by grouting or by trenching across the fracture, and filling the trench with concrete or clay puddle. Faults often form outlets for heavy springs in excavations. Here again, grouting may be applied or a well sunk, with the spring water pumped away clear of the excavation. *See* EARTHQUAKE, SITE INVESTIGATIONS.

fault coast A coastal feature where the downthrow block of rocks of a fault is covered wholly or in part by the sea, and the fault scarp forms the coastline.

fault control (streams) *See* STREAMS (fault control).

fault-line valley *See* CONTACT EROSION VALLEY.

fault spring An outflow of groundwater at or along a fault plane. Usually a water-bearing bed is brought against impermeable rocks by the fault, and the water flows out at the point where the fault outcrops in a depression or valley (Figure F.2). *See* FISSURE SPRING.

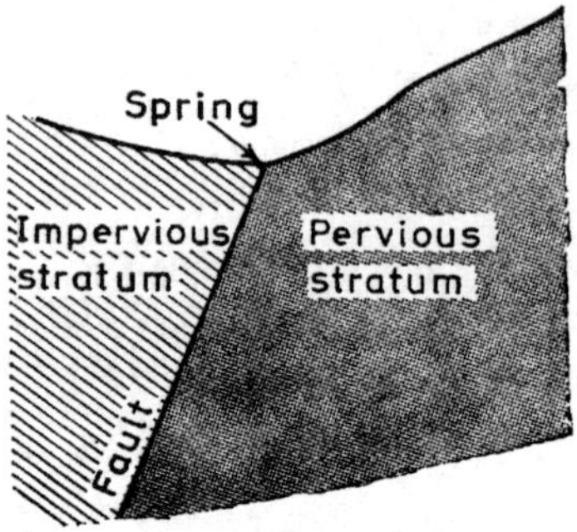

Figure F.2. Fault spring

fault trap (water) A geological structure in which water in a porous deposit has been trapped or sealed by an impervious deposit thrown opposite it by a fault. *See* STRUCTURAL TRAP.

feed canal *See* CANAL (MAIN IRRIGATION).

feeder A stream or channel which feeds or supplies water to a canal, reservoir or pond; a small flow of water from the surrounding rocks into a tunnel, excavation or works. *See* FAULT SPRING.

feed-pump A PUMP which supplies a steam boiler with feed-water; any pump which supplies a machine, appliance or plant with a regular supply of water of the required purity, quantity and pressure. *See* CIRCULATING PUMP.

feed-water The water pumped into a steam boiler, at boiler pressure, by a FEED-PUMP; the water is often purified, heated to almost boiler temperature, and de-aerated before entry into the boiler; any water which must be regularly supplied to a machine or plant.

Fellenius's circular arc method A type of landslide of homogeneous earth or clay where failure occurs on a circular arc. *See* SLIP SURFACE OF FAILURE. Research into this type of failure was undertaken in Sweden after a large portion of Gothenburg harbour slipped into the sea in 1916. The principle is used in the structural design of earth dams.

fender A shock-absorbing object (ball, rope mat or old rubber tyre, etc.) hung over the side of a vessel to prevent damage by striking against a pier, quay or dock, or another vessel.

fender pile Generally an upright, free-standing timber pile driven into the river or sea bed near a berth; it absorbs impact from a berthing vessel.

fender post or **guard post** *See* BOLLARD.

ferrous sulphate (copperas) A coagulant suitable for waters containing bicarbonate but not recommended for soft, coloured waters. *See* COAGULATION.

fetch The distance which the wind can travel, without obstruction, to any point when raising waves; that is, the distance from the nearest country near the shore in the direction of the wind. *See* HAWKSLEY'S FORMULA, STEVENSON'S FORMULA.

Fidler's gear Lifting tackle for handling and laying large stones or pre-cast concrete blocks at any angle in the construction of a quay or jetty wall below water level. *See* BLOCKWORK.

field capacity The maximum possible amount of water held by a soil against free drainage; the field moisture content of soil two or three days after heavy rain or irrigation. For economic irrigation, the objective is to restore the soil moisture content to field capacity, as any quantity beyond this is largely wasted. *See* MOISTURE REGIME.

field coefficient of permeability The COEFFICIENT OF PERMEABILITY at temperature of the water.

field drain *See* AGRICULTURAL DRAIN.

field drainage (or **farm drainage**) The removal of water from a field or farm in excess of that required for optimum plant growth and to make the ground firm enough for traffic and cultivation. Field drainage is usually a problem where the soil consists of clays or heavy silts with a low permeability. *See* DRAINAGE, MOLE DRAINAGE.

field geology The study and interpretation of rocks in their natural environment and in terms of the economic products sought. It seeks to correlate surface exposures with underground structures. It forms an important branch of HYDROGEOLOGY, particularly in areas which are more or less unmapped.

field moisture ADHESIVE WATER found above the water-table.

field moisture deficiency The amount or depth of water required to bring the moisture content of a soil up to FIELD CAPACITY.

field moisture equivalent The minimum soil moisture content at which a drop of water placed on a smooth surface is not absorbed immediately by the soil but spreads over the surface.

field test A measurement or experiment carried out under field conditions; control may not be equal to a formal test and the result may be less accurate.

field tile (USA) *See* AGRICULTURAL DRAIN.

field velocity *See* EFFECTIVE VELOCITY.

fill dam *See* EARTH AND ROCKFILL DAM.

filling Earth or waste rock deposited to bring construction to the desired formation level, as in embankments, earth dams, canals, rockfill dams and similar constructions. *See* BALANCED EARTHWORKS.

filling, run of the bank Soil and other material removed indiscriminately from a river bank for use as filling for structures or eroding river banks. The material may be good (e.g. stiff clay), but may also be wasteful and dangerous if it is not secured by protective works or vegetation before the onset of water attacks.

fill yards *See* COMPACTED YARDS.

filter A stage in a water treatment or purification plant which removes suspended matter. *See* COARSE STRAINERS, DIATOMITE FILTER, GRAVITY FILTER, PRESSURE FILTER, SAND FILTER, SLOW SAND FILTER.

filter bed *See* BACTERIA BED.

filter blocks Hollow blocks designed to support a BACTERIA BED, made of vitrified clay, and often salt glazed.

filter drain (dams) A drain, about 1.5 m (5 ft) wide, on the downstream side of the clay core of an earth dam to prevent seepage into the fill material.

filter material A layer of graded granular material which retains solids or suspended particles but allows water to pass through; the broken stone, clinker or strong metallurgical coke used in a BACTERIA BED. *See* GRADED FILTER.

filter medium Any material through which water or wastewater is passed for treatment. In the case of water treatment, usually fairly fine graded granular material on which a ZOOGLOEAL LAYER forms, and for wastewater biological filters, broken rock or clinker on which a ZOOGLOEAL FILM develops, comprising bacteria, fungi and protozoa.

filter press A mechanical press for dewatering sludge, comprising a series of ridged plates which enable the liquor to drain away from the compressed sludge cake. *See* SLUDGE CAKE.

filter run The interval between the cleaning and washing of a rapid sand filter.

filter strip *See* DESILTING STRIP.

filter well A well method of water-table lowering in the vicinity of cuttings or other works to remove water and strengthen the sides. Deep lowering of water may be achieved, since the pump can be submerged in the well. Graded material is provided around the pump suction to prevent fine material being withdrawn with the water. *See* GROUNDWATER LOWERING.

filter zone (or **protective filter**) A zone within an earth dam, consisting of a layer or layers of pervious material. Its purpose is to ensure drainage, but at the same time to prevent soil particles being transported by percolating water.

filtration The process of passing a water or wastewater through a filtering medium for the removal of suspended or colloidal matter, and also some bacteria. *See* FILTER MEDIUM.

fines Fine or small particles that settle slowly to the bottom in water. *See* SLIME, SUSPENDED LOAD.

fiord coast A coast which is deeply indented with long narrow inlets and steep valley slopes which emerge from deep water. In Norway this type of coast is formed by submerged glaciated valleys.

fire control (catchments) Fire protection of forested catchments, usually including fulltime firefighting teams, provision of a tower lookout system for early detection, good communications and also firefighting roads for access. Other firefighting authorities in the area often co-operate in the scheme adopted. *See* CATCHMENT MANAGEMENT.

fire hydrant A device or arrangement placed in water mains for firefighting purposes. It is controlled by either some form of valve or a wooden ball covered with rubber or composition. A good mains pressure is important. *See* DRY RISERS.

firn Loose surface snow on the heights of mountain ranges. As firn snow becomes older it hardens by a cement of ice. *See* NÉVÉ.

first bottom (USA) The normal FLOODPLAIN of a river. Areas which rarely flood are termed high bottom phase.

fish ladder *See* FISHWAY.

fish pass *See* FISHWAY.

fish screen A barrier placed across the inlet or outlet of a pond or channel to prevent the passage of fish.

fishway (or **fish ladder** or **fish pass**) A sloping structure or channel designed to enable fish to ascend or descend a dam, weir, or cataract. *See* SALMON LADDER.

fissure spring An outflow of groundwater at or along a fissure, joint or fault plane. Frequently, several springs occur along the same fissure. The source may be deep, in which case the water is often highly mineralized. *See* FAULT SPRING.

fissure water Water occurring along cracks, joints, and fissures; a relatively impervious rock may contain a large amount of fissure water. *See* CHALK.

fixed groundwater Water stored in rocks with fine voids. The water is permanently attached to the pore walls or moves so slowly that it is not available for extraction by wells and pumps. *See* MICROPORES.

flap trap *See* ANTIFLOOD VALVE, FLAP VALVE.

flap valve A valve which allows water to flow in one direction only. The term is usually used for open channels. *See* NON-RETURN VALVE.

flashboard (or **stop-log** or **stop-plank** or **drop-bars**) A board, plank, steel joist, or pre-cast concrete beam held horizontally between vertical grooves in end girders, piers or walls. The beams are placed horizontally on top of each other to close up or control the water level in a spillway or other water channel. *See* MOVABLE DAM, NEEDLE WEIR.

flashes Lakes occupying depressions in open country, often formed by irregular subsidence of the land by mining operations or removal of underlying salt by natural or artificial means.

flashy stream A YOUTHFUL STREAM which collects from the steep slopes of a catchment area; it has peaks of flood soon after rainfall and the flow subsides rapidly. *See* NULLAH.

flat roller *See* SMOOTH-DRUM ROLLER.

flats *See* RIVER FLATS.

flat-slab deck dam (or **flat-slab buttress dam** or **Ambursen dam**) A reinforced concrete slab, supported by parallel buttresses, with a flat upstream face inclined at about 45°.

flexible joint A joint which is sealed by rubber rings, having the advantage that the pipe can withstand earth movements and vibration. *See* RIGID JOINT.

flight of terraces A series of TERRACES or platforms formed by a degrading and swinging stream.

float *See* FLOAT GAUGE, FLOAT ROD, SUBSURFACE FLOAT, SURFACE FLOAT.

float, captive *See* CAPTIVE FLOAT.

float, direction *See* DIRECTION FLOAT.

floaters Isolated masses of rock or ore brought down by a stream from their parent bed and found on the land surface or stream bed. *See* COMPETENCE.

float gauge A small object made of cork, wood, or hollow zinc or copper, which rests and rides on the water surface; its movements are transmitted to a recorder or indicator.

float gauging The measurement of water velocity in an open channel by a float gauge and hence the discharge. *See* FLOAT MEASUREMENT.

floating crane A large crane, with a lifting capacity up to 200 t, carried on a heavy barge or pontoon. It forms an important appliance in most large ports and docks.

floating dock (or **floating dry dock**) A structure built up of steel plates which may be submerged or floated as required by means of air-chambers. It enables repairs to be carried out on a vessel in the dry. *See* DRY DOCK, SELF-DOCKING DOCK.

floating dry dock *See* FLOATING DOCK.

floating foundation *See* BUOYANT FOUNDATION.

floating harbour An arrangement of PONTOONS connected end to end to form a BREAKWATER.

floating on the system Pertaining to a reservoir which uses the same pipe to admit and release water.

floating pipeline A pipeline which conveys the semi-fluid spoil from a SUCTION DREDGER; the pipeline is carried on pontoons and the spoil material may be used in HYDRAULIC FILL DAMS or for land reclamation projects.

floating strainer A buoyant STRAINER or pump suction end which draws relatively clear water from the upper part of the sump or water lodge. A floating strainer is often used when pumping from ponds or rivers where the bed is muddy, but it must be kept below the water level to prevent drawing in air.

float measurement An approximate estimate of the volume of water flowing in a fast stream or open channel by timing a small floating object over a measured distance. A straight, uniform length of channel is selected for the test. The discharge is approximately 80% of the product of the average cross-sectional area of the channel multiplied by the velocity of the float.

float rod (or **velocity rod**) A rod with a weight at the bottom so that it floats in a vertical position; it gives the approximate mean velocity of the water between the bottom of the rod and the water surface, usually timed through a FLOAT RUN.

float rod correction The adjustment sometimes applied to the observed velocity to determine the mean velocity.

float run The known distance between two stations or cross-sections, along an open channel or stream, which is used for FLOAT GAUGING.

float, subsurface *See* SUBSURFACE FLOAT.

float, surface *See* SURFACE FLOAT.

float switch A switch operated by a float for starting or stopping a pump motor when the water level in the sump rises and falls. A similar device may also be used for automatically opening or closing CREST GATES. *See* AUTOMATIC PUMP.

float valves Valves installed on inlets to service reservoirs and tanks which close automatically at a predetermined height. *See* AUTOMATIC SELF-CLOSING VALVES.

floc Aggregates formed by the coalescence of fine colloidal particles.

flocculation A physical process of agglomeration of colloidal and fine suspended matter after COAGULATION by gentle mechanical or hydraulic mixing. After both coagulation and flocculation have been completed the floc can removed by filtration or settlement.

flood A flow of water in a stream or river channel which is beyond the capacity of the channel to carry, the excess overflowing the banks to form floodwater; water overflowing land not normally thus covered; any relatively high level or flow of water (Figure F.3). Floods may be caused by (1) excessive precipitation; (2) rapid melting of accumulated snow; (3) failure of reservoirs, banks, or levees, and (4) failure of ice jams or bars across rivers. The term flood when applied to rivers normally implies a rare event, and the magnitude of floods is often measured by a RETURN PERIOD. *SEE* ANNUAL FLOOD, DESIGN FLOOD, MAXIMUM PROBABLE FLOOD.

Figure F.3. Extensive flooding in a wide river valley

flood absorption The decrease in discharge by accommodation of water flow in a lake, reservoir, channel, or other natural or artificial storage.

flood absorption capacity The flood storage capacity of a reservoir between high flood level and normal reservoir level.

flood, annual *See* ANNUAL FLOOD, RETURN PERIOD.

flood axis The general direction of flow of flood current.

flood channel The channel formed by the flood tide in an estuary, generally wide at the downstream end and converging to a point at the upstream end. Where the estuary is narrow, the flood channels form a tree-like pattern in plan by branching upstream off the winding EBB CHANNEL.

flood coefficient An expression, given as a decimal, of the proportion of the PLUVIOGRAPH runoff that actually occurs as surface runoff.

flood control The regulation of flood waters to minimize or prevent the inundation of valuable property or land. Control measures include (1) the building of embankments or levees along the course of the river to confine the flood waters; (2) dredging to deepen the channel and increase its cross-section; (3) increasing the gradient of the river by cutting across loops, thus shortening its course, and (4) storing or diverting the flood water. *See* RIVER TRAINING.

flood control, complete The adoption of protective measures against large floods, such as MAXIMUM PROBABLE FLOOD. *See* PERMANENT BANK PROTECTION.

flood control storage A reservoir to reduce the floodwaters downstream of the dam. To impound maximum flood volumes, these special storages are kept as low as possible. The water is later released in controlled discharges which do not usually exceed the capacity of the river.

flood current, tidal *See* TIDAL FLOOD CURRENT.

flood damage The economic loss caused by floodwaters overflowing land and depositing silt and debris. *See* DIRECT FLOOD DAMAGE, INDIRECT FLOOD DAMAGE.

flood, design *See* DESIGN FLOOD.

flood forecasting The processes or methods of estimating the magnitude of future floods. The term is usually used only for individual events expected either in the near future or that have already begun. Flood forecasts are commonly made to give warning of high flows which are soon to be expected. The most reliable of such warnings are usually based on REAL-TIME FORECASTING which uses instrument observation of parameters, such as rates of rainfall and stream flow, which are observed within the catchment while the event is in progress.

flood formula A rule or statement in algebraic terms for estimating FLOOD INTENSITY. Some are purely empirical and derived from records of actual floods; others are based on more theoretical considerations; some again are based on a combination of both of these methods. There are a very large number of these formulae, but except for the RATIONAL RUNOFF FORMULA, or generally similar expressions, few give predictions of use to engineers except in the immediate regions where they were derived.

flood frequency *See* FLOOD, FLOOD PROBABILITY, RETURN PERIOD.

floodgate A barrier placed in a water channel to keep out tidal or flood water. *See* TILTING GATE.

flood hydrograph A graph showing the FLOOD INTENSITY plotted on a time basis (Figure F.4). If the stream is dry before the flood starts then the flood intensity will increase from zero to a maximum in the PERIOD OF CONCENTRATION. A large catchment will give a relatively low flood rate of discharge over a long period, and a small catchment a high flood intensity of short duration.

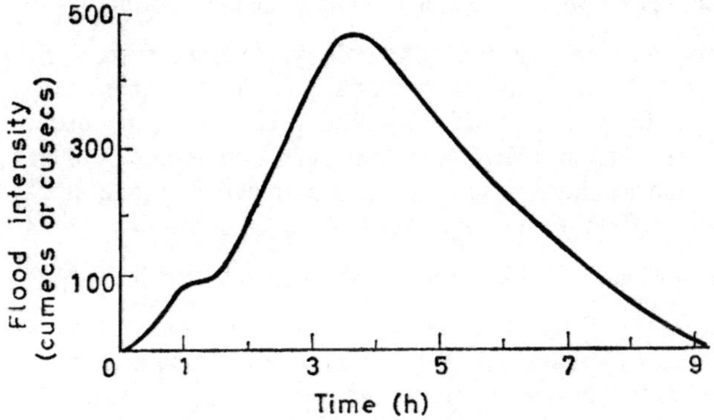

Figure F.4. Flood hydrograph

flood hydrograph, design *See* DESIGN FLOOD HYDROGRAPH.

flood intensity The rate of flood discharge at a given site, usually expressed in cusecs or cumecs.

flood irrigation (1) The application of wastewater on land so that it percolates through the soil for discharge through under-drains. (2) Any method of irrigation in which the water flows over the land surface. *See* SPRINKLER IRRIGATION, TRICKLE IRRIGATION.

flood level, design *See* DESIGN FLOOD LEVEL.

flood losses The water lost to surface runoff due to infiltration, interception and depression storage during a storm over a catchment area.

flood loss rate The average rate at which water is lost to surface runoff due to infiltration, interception and depression storage during a storm over a catchment area.

flood peak *See* PEAK FLOOD.

floodplain The flat lowlands bordering a river and which are partly or wholly covered with water during floods. In some cases floodplains have a width of many kilometres and extend upstream for great distances. *See* RIVER TERRACES.

floodplain clays Clays deposited along the lowlands bordering a river during periods of flood. The deposits usually taper out towards the valley slopes, and the sediment may be sandy or consist of alternate layers of sand and sandy clay. *See* ALLUVIAL FLAT.

floodplain discharge The water flowing over the floodplain.

floodplain terraces Natural TERRACES which border many rivers. They may consist of unconsolidated material, or of rock covered with a layer of floodplain deposit, or chiefly of solid rock; when unconsolidated, the material varies from sandy clay to gravel and sometimes boulders. In some areas the terrace material is sufficiently permeable but retentive to provide water for shallow wells. *See* INFILTRATION GALLERY.

flood plane The water level of a river during a specific flood.

flood prediction The forecasting of the magnitude of future floods. Usually these forecasts are of the peak rate of flow, and sometimes also their full HYDROGRAPH. Flood predictions can only be made on a basis of probability, often expressed as a RETURN PERIOD. The main techniques for flood prediction include: statistical analysis of field records and data, unit hydrograph techniques, rational methods and empirical formulas. Selecting the most suitable technique in any given case depends on the data available.

flood probability The likelihood that a flood of given magnitude will be equalled or exceeded in a given period, often measured by a RETURN PERIOD; an annual probability of 10% would be a 10-year flood return period, and similarly a probability of 1% would be a 100-year flood return period.

flood protection *See* FLOOD CONTROL, FLOOD CONTROL (COMPLETE), FLOOD RETENTION BASIN, FLOODWATER RETARDING STRUCTURE.

flood retention basin A dam and reservoir constructed on a river for flood storage. The reservoir is kept empty or partly empty under normal conditions and allowed to fill when a flood occurs. The impounded water is released during and after a flood in such quantities as not to exceed the capacity of the river channel below. Very few rivers contain natural sites favourable for such dams on an economic basis. *See* FLOODWATER RETARDING STRUCTURE.

flood routing The analysis of FLOOD HYDROGRAPHS as they are modified by flow down channels or rivers, and in and out of lakes and reservoirs (Figure F.5). The downstream passage flood hydrograph modifies the form, reducing the peak rate but extending the duration of flow, and, except for other losses, keeping the volume of the flood constant. Precise flood-routing calculations require extensive data and computers, but approximate estimates can be obtained quite simply.

flood series A table or list of flood occurrences over a specified period for a river or catchment area. The events are usually arranged in order of intensity or magnitude.

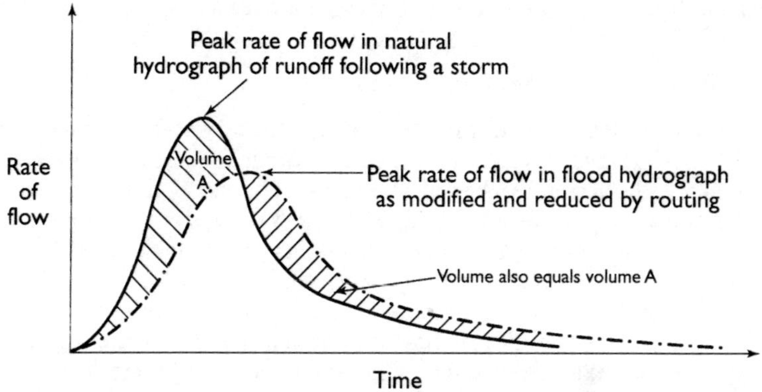

Note: (a) Total volume of flow in natural hydrograph and that modified by flood routing is the same.
(b) Volume A under the rising limb of the hydrograph is initially taken into storage in the channel or reservoir, and then later released.

Figure F.5. Flood routing

flood stage (1) The level marked at salient points on river banks indicating the elevation above which the river is considered to be in flood. (2) The water level at which a river overflows its natural banks. *See* BANKFUL STAGE.

flood, standard *See* STANDARD FLOOD.

flood storage *See* FLOOD RETENTION BASIN, FLOODWATER RETARDING STRUCTURE, NATURAL STORAGES.

flood strength, tidal *See* TIDAL FLOOD STRENGTH.

Flood Studies Report (1975) A report produced by the UK Natural Environment Research Council. It summarized and presented the results of an intensive study of the rainfall and river flow data then available for the British Isles, and put forward techniques of FLOOD PREDICTION which make use of this. Its publication was a milestone in engineering hydrology, and much important work arose from it. While many of the recommendations and most of the data in this very large report refer only to the British Isles, many of the methods and techniques described are of worldwide application.

flood surcharge The maximum rise of still water level above reservoir top water level during a design flood. (Surcharge water is not retained in the reservoir but is discharged as the flood passes, and the water surface returns to nominal retention level.) *See* Figure R.4.

flood tide (or **rising tide**) The tide period between low water and the following high water.

flood tide, duration of *See* DURATION OF TIDE.

flood warnings Previous information, notice or signal of the approach of a flood. On large and long rivers, a flood warning can often be given to those along the lower reaches many weeks beforehand, but with swift floods, notice may be very brief. Early flood warnings enable stock and goods to be moved to safety, and inhabitants in low-lying areas to be evacuated to higher levels. Warnings are usually based on river heights along the upper reaches, or on rainfall figures at stations in the upper part of the catchment. The information is collected and interpreted by one authority. Automatic river gauges may be connected by land lines to a central authority or to the telephone system.

floodwater retarding structure A dam built across a small stream or other channel designed to retard and decrease the discharge of floodwater. It has an uncontrolled outlet so that it temporarily impounds runoff from the catchment area, and releases it at a rate which prevents the channel overflowing below the dam structure.

floodway A channel, generally bordered by embankments or levees, designed to carry flood waters. *See* DIVERSION, LEVEE.

flow The quantity of water flowing in a channel, stream, conduit or pipe expressed in volume per unit time. *See* CUMEC, CUSEC, DISCHARGE.

flowage line (or **flow line**) A line generally used in connection with the acquisition of rights to flood lands for storage purposes (mainly USA). It may be a contour line around a pond, lake, reservoir or the water surface line along a stream, and represents a definite level, such as maximum, mean, low or spillway crest.

flow curve, integrated *See* INTEGRATED FLOW CURVE.

flowing well A well in which the hydrostatic pressure of the water is sufficient to cause it to rise and flow out at the surface. *See* NON-FLOWING ARTESIAN WELL.

flow lines (1) The lines shown in a FLOW NET indicating the paths taken by water flowing through an earth dam or the ground near a dam or cofferdam; the lines followed by solid particles in a current of water (Figure F.6). (2) FLOWAGE LINE.

flow meter An instrument to determine the quantity of water flow in conduits in unit time. *See* CURRENT METER, INTEGRATING METER, VENTURI METER, WATER METER.

flow net A net of EQUIPOTENTIAL LINES and FLOW LINES intersecting at right angles in a cross-section of a dam (Figure F.6). These lines indicate the relationship of head (or potential) and seepage of water through the ground near a dam. They give information regarding (1) ordinary seepage flow, (2) the increased pressure due to seepage after rain, (3) the NEUTRAL PRESSURE at any point from which the upward force or UPLIFT on the dam may be calculated, and (4) the EFFECTIVE PRESSURE on any plane required for dam stability investigations.

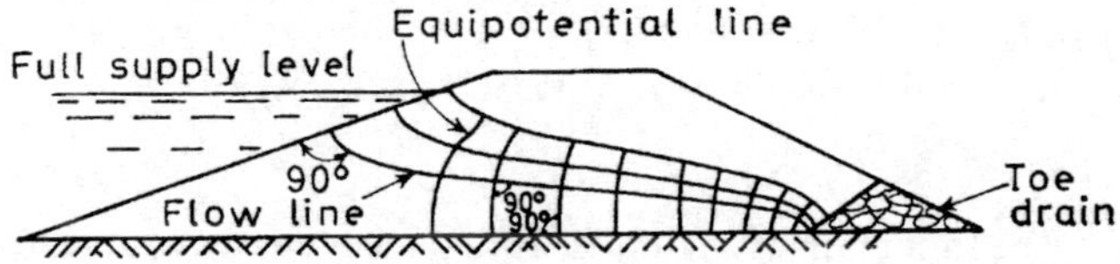

Figure F.6. Flow net

fluctuation of the water-table (or **phreatic fluctuation**) The alternate upward and downward movements of the water-table due to periods (usually seasonal) of intake and discharge of water in the ZONE OF SATURATION.

flume A wooden, sheet-metal or reinforced concrete trough or open channel for conveying water from a higher to a lower level or across depressions or other low-lying areas. *See* CONTROL FLUME.

fluoridation The addition of small quantities of fluoride – less than 1 part per million – into drinking water as a preventative measure against tooth decay.

flushing The rapid passage of water through a FIRE HYDRANT or wash-out to prevent the deterioration of water in 'dead-end' mains.

fluvial Pertaining to, or found in, rivers. *See* FLUVIAL HYDRAULICS.

fluvial hydraulics That branch of engineering science concerned with the motion of water in rivers. *See* RIVER ENGINEERING.

fluvioglacial deposits Deposits of detritus formed by GLACIAL STREAMS; material spread out along the margins of ice sheets or at the ice-front. *See* OUTWASH FAN.

fluviomarine Relating to the action of both rivers and sea, such as the deposits laid down in an estuary or estuarine clays. *See* MUD FLAT.

fluviomorphology Pertaining to the forms and structures of rivers. *See* LIMNOLOGY.

fluvioterrestrial The science concerned with land and its rivers.

fly-ash concrete Mass concrete in which pulverized fly-ash (obtained from coal-fired power stations) is used to replace about 20% of the cement content. The temperature rise of fly-ash concrete is about 15 to 20% less than the rise in concrete made with Portland cement. It is sometimes used in the construction of dams, particularly hydroelectric dams. *See* LOW-HEAT CEMENT, TRIEF PROCESS.

flying fox A short form of aerial ropeway, consisting of towers, pulleys and ropes, for conveying material from one point to another between the towers. Frequently used across a WET GAP or STREAM.

fly-off *See* INTERCEPTION.

foaming The formation of a foam or froth on wastewater treatment tanks caused by surfactants. *See* SURFACTANT.

fog Atmospheric obscurity caused by vapour suspended at or near the earth's surface. According to international meteorological practice, objects 1 km (0.6 miles) away are not visible in a fog; when visibility exceeds this distance, but is under 2 km (1.2 miles), the condition is called mist if caused by water droplets, or haze, if by fine dust or smoke. *See* CONDENSATION.

footing The widening at the base of dams, breakwaters and similar constructions to distribute the load and improve their stability.

foot irons *See* STEP IRONS.

foot valve A clack valve placed at the foot of a column of pipes, such as the suction line of a pump.

forced waves (or **seas**) The waves along the belt of water where they were first generated by wind action. *See* FREE WAVES.

force pump (or **ram pump**) A pump consisting essentially of a ram or plunger which forces the water to a level considerably higher than itself. On the up-stroke of the plunger, the SUCTION VALVE opens and the water rises in the SUCTION PIPE. On the down-stroke, the suction valve closes and the water is forced through the delivery valve into the discharge pipe.

ford A shallow stretch in a river or other body of water where a crossing may be effected by wading.

forebay A storage reservoir at the head of the penstocks to store water when the load on the hydroelectric plant is reduced, and also to supply water for the stages of an increasing load. *See* PUMPED STORAGE.

foreset beds *See* DELTA.

foreshore The shore area over which the water's edge migrates during normal tides. *See* BACKSHORE.

forest influences The effects of a large area of trees and plants on water, climate and soil.

formwork *See* SHUTTERING.

'fossil' water *See* CONNATE WATER.

foundation The lower part of a hydraulic structure below ground level which transmits the load to the rock or earth. In a dam foundation the term includes the valley floor and abutments.

foundation, dam *See* DAM FOUNDATION.

foundation investigation A branch of SITE INVESTIGATIONS, usually involving the testing by drilling into the deposits below a proposed foundation for a dam or other structure; the estimation of bearing capacities, shear strengths and settlements, and the design of a foundation to suit the prevailing ground conditions form part of the work. *See* DAM FOUNDATION, DAM SITE INVESTIGATION.

foundation soil The ground or SUPERFICIAL DEPOSITS directly carrying the load of a structure.

fountain An ornamental water jet; a spring or source of a river, or a public erection for a regular supply of drinking water, etc.

framed dam (1) Generally a timber dam framed to form a water face and supported by steel struts. (2) A type of movable weir built of steel or cast iron and timber; some types are lowered to the bed or hung above the water when in flood and propped up by struts against the river bed in low water. *See* SUSPENDED-FRAME WEIR.

frame weir *See* FRAMED DAM, SUSPENDED-FRAME WEIR.

Francis turbine A WATER TURBINE in which the water enters radially and leaves axially, with the turbine shaft usually vertical. It operates on a low-head to medium-head and has been installed in a number of HYDROELECTRIC POWER STATIONS.

freeboard The vertical distance between the maximum water level anticipated in flood design and the crest of a dam or other hydraulic structure. Freeboard provides that structures, particularly earth dams, are not overtopped, and this provision includes wave action (Figure O.1). *See* DAM (RAISING OF).

free chlorine *See* RESIDUAL CHLORINE.

free eddy *See* EDDY (FREE).

free flow The flow of water over or through a structure which is not affected by the level of the tail water. *See* FREE-FLOW WEIR.

free-flowing bore *See* ARTESIAN WELL DRAINAGE.

free-flow weir (or **free weir**) A weir or dam which is so high that flow over it is in no way affected by the tail-water level. *See* SUBMERGED WEIR.

free groundwater Groundwater which is not trapped or confined by an overlying impervious rock. *See* NON-ARTESIAN WELL.

free moisture The moisture in any soil or substance which can be expelled by normal drying.

free water Water which occupies cavities, fractures and voids in rocks, and is not chemically combined with minerals; water which is free to migrate laterally or vertically and is available for extraction in wells or removal by drainage. *See* INTERSTITIAL WATER, CONNATE WATER.

free water level The level or surface of a body of water at atmospheric pressure, that is, in contact or communication with the atmosphere.

free waves (or **swell**) Waves which have travelled some distance away from the area where they were first formed by wind action. *See* BREAKERS, FORCED WAVES.

free weir *See* FREE-FLOW WEIR.

freeze A condition when the surface temperature over a widespread area remains below 0 °C (32 °F) for a period long enough to form a characteristic feature of the weather. *See* SNOW LINE.

freezing rain Rain containing a portion which freezes to form a film of ice on exposed surfaces.

French drain An AGRICULTURAL DRAIN in which the pipes are bedded in some filtering material, such as gravel. *See* GRADED FILTER.

frequency curve In general, a curve which gives the number of occasions or times that a particular value, event or phenomenon has occurred during a given period. In hydrology, the curve may show the number of times that quantities such as runoff or intensity of rainfall have occurred or may occur. Usually frequency is plotted along the ordinate and quantity along the abscissa. The graph may indicate frequencies as percentages of the total values or observations available.

frequency diagram *See* HISTOGRAM.

frequency of rainfall, intensity *See* INTENSITY FREQUENCY OF RAINFALL.

frequent floods *See* RETURN PERIOD.

fresh (1) Referring to water containing less than 1 mg dissolved salt per litre. *See* BRACKISH. (2) A sudden rush of water in a stream.

freshet A flood or flow of fresh water into sea; river flood produced by heavy rains or melting snow; particularly a small rush or flow of water of short duration.

fretting The gradual wearing away of a river bank by water flow, also sometimes by runoff from the adjoining land mass. *See* EROSION.

friability The susceptibility of a soil or other mass to crumbling, having a bearing on its resistance to erosion. *See* ERODIBLE, EROSIVE.

friction The force that always opposes motion between two surfaces in contact. Where the surfaces are both of solid materials the frictional force is directly proportional to a coefficient times the reaction or force pressing the two surfaces together. Where one of the materials is a liquid the factors governing frictional forces are more complex. *See* HYDRAULIC FRICTION.

friction drag *See* SURFACE DRAG.

friction head (or **friction loss**) The energy or head lost due to HYDRAULIC FRICTION in a pipe or open channel; frequently eddy losses at bends and elsewhere are included.

friction loss *See* FRICTION HEAD, HYDRAULIC FRICTION.

friction slope The FRICTION HEAD or friction loss per unit length of channel or conduit.

fringe water Water in the ground along the CAPILLARY FRINGE.

frontal apron *See* VALLEY TRAIN.

frontal precipitation The precipitation that occurs at a FRONTAL SURFACE from the lifting of warm air over cool air. *See* NON-FRONTAL PRECIPITATION.

frontal surface A surface separating two air masses of different characteristics, commonly humidity and temperature, and usually associated with rain and cloud. There are three types: (1) cold front, along which colder air replaces warmer air, (2) warm front, along which warm air replaces colder air as the front moves, and (3) occlusion, which is formed when a warm front is overtaken by a cold front, appearing at the Earth's surface as either a cold or a warm front.

frost Freezing weather; frozen dew or vapour; temperature below the freezing point of water; 'degrees of frost' are the degrees that the temperature falls below freezing point. The two main methods of protection of low-growing crops against frost are (1) spraying during the time when frost is actually occurring and (2) applying irrigation not long before a possible night frost. Records of ground frost are obtained from the grass minimum thermometer, and earth temperatures by thermometers sunk in the ground at depths varying from 100 mm (4 in) to 3 m (10 ft).

frost action (1) The alternate freezing and thawing of water held in rock pores, cracks and fissures, resulting in the mechanical breakdown of outcrops. (2) Damage to new construction work, concrete and mortar when the contained water freezes and expands by about 9% of its volume. Fresh brick or concrete work should be protected from frost action by covering with tarpaulins or other material.

frost-active soil A soil liable to frost action; it undergoes changes in volume and bearing capacity. Some varieties of chalk, silt and fine sand are commonly frost-active. *See* PERMAFROST.

frost boil The loosening or softening of soil caused by thaw after FROST HEAVE.

frost heave The upward swelling of a soil or pavement by freezing and expansion of the contained water. A silty soil is most liable to frost heave, and layers of ice parallel to the ground surface are often formed in the mass. *See* UPLIFT.

frost line The line giving the depth of frost penetration into the ground.

frost valve A small automatic valve sometimes used if there is any danger of a hydrant becoming unusable by the formation of an ice block in the outlet. It ensures that the hydrant outlet or bend is drained and remains empty when not in use. It may be operated by weight or a spring.

Froude number (Fr) A dimensionless number expressing the ratio between influence of inertia and gravity in a fluid. In the case of a channel of rectangular cross section it is

$$\mathrm{Fr} = \frac{v}{(gy)^{0.5}}$$

where v = mean flow velocity, y = flow depth, and g = gravitational acceleration. *See* MODEL ANALYSIS.

FSL Full supply level.

full meander A stream MEANDER consisting of two loops, one flowing clockwise and the other anticlockwise.

full supply level (FSL or **top water level, TWL)** The water surface level when the reservoir is at maximum operating level, excluding periods of flood discharge. It is determined by the spillway crest level where an ungated spillway is provided.

full-tide cofferdam (or **whole-tide cofferdam**) A COFFERDAM in which the height is sufficient to exclude water at all periods of the tide. *See* HALF-TIDE COFFERDAM.

full-width weir A weir which extends across the full width of the CHANNEL OF APPROACH. *See* CONTRACTED WEIR.

furrow dams Small earth banks used to retain water in furrows. *See* DRAINAGE TERRACE.

furrow irrigation The use of furrows or small ditches for conveying irrigation water from a supply ditch or header. When the furrows are placed across the general slope, the method is called gradient irrigation.

fuse plug spillway A type of EMERGENCY SPILLWAY in which the upstream entrance is closed off by a temporary blockage, or fuse plug (Figure F.7). A fuse plug spillway is formed typically by an earth embankment (or by collapsible timber panels or other structures) with its crest lower than that of the main dam. If the water level in the reservoir ever rises to a pre-set critical level, the fuse plug collapses or is washed away, and excess water is rapidly discharged down the emergency spillway. When the reservoir surface has been lowered, it is then necessary to rebuild the fuse plug, but the very occasional cost of this work is often less than the cost of a permanent weir or other reliable outlet control.

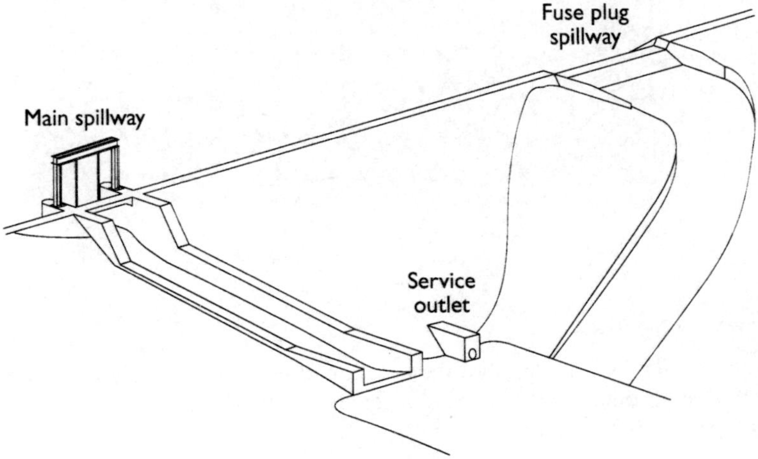

Figure F.7. Fuse plug spillway

G

gabion (1) A wire basket filled with soil or a small CELLULAR COFFERDAM. (2) A rectangular box made of wire mesh and filled with heavy stone, used for bank protection in river improvement (Figure G.1). If the water level and time allow, a series of steps are cut in the bank and the gabions built up in courses, each course being set back from the one below. Where water is deep, they are made on the bank and rolled down until they reach a footing.

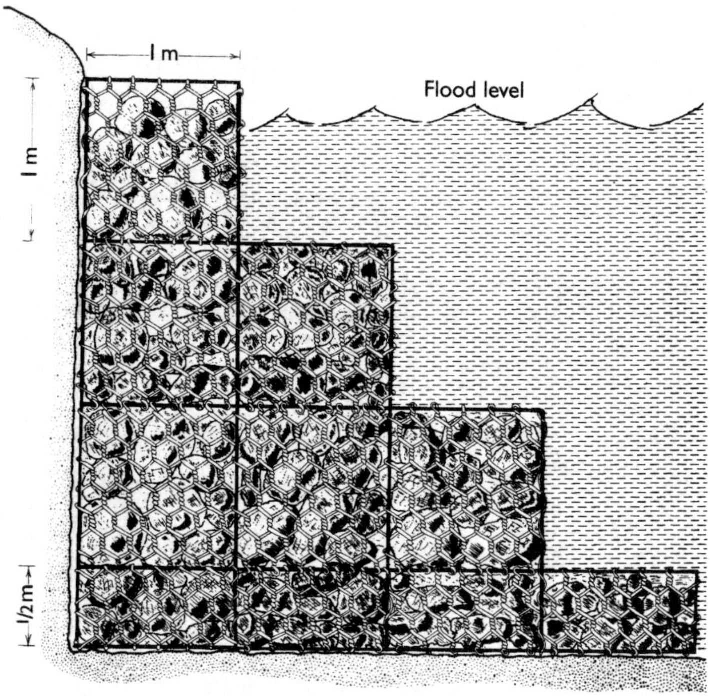

Figure G.1. Gabions

gage *See* GAUGE.

gallery (1) An underground reservoir or collector for percolating water. (2) A small tunnel constructed in a concrete dam for drainage, access and inspection. *See* HEADINGS, TUNNEL.

gallon A measure of capacity, usually for liquids. The British imperial gallon = 277¼ cubic inches = 4 quarts = 4.5459 litres = 10 lb of distilled water. The US gallon = 231.0 cubic inches = 0.833 imperial gallon. *See* UNIT WEIGHT OF WATER.

gang of wells *See* BATTERY OF WELLS.

gap A short valley across a ridge and connecting lowlands on opposite sides. *See* WATER GAP.

gas-lift flowing well A well in which natural gas is introduced to mix with the water and impart sufficient buoyancy to cause it to rise to the surface.

gas–water surface The surface between an accumulation of natural gas trapped under impervious strata and an underlying body of groundwater.

gate An adjustable frame or barrier to regulate flow in hydraulic constructions such as water channels and dam spillways, and over dam crests. *See* CREST GATE.

gate chamber (or **camber**) A space or recess provided in a lock wall to receive a SHIP CAISSON or other lock gate when it is open.

gate valve A STOP VALVE with a sliding plate which closes or regulates the water flow in a pipe. When open, the valve offers the minimum of resistance, since flow can occupy the full bore of the pipe.

gathering ground *See* CATCHMENT AREA.

Gauckler-Manning formula Robert Manning, an Irish civil engineer, presented a flow formula in 1889. It became known, incorrectly, as Manning's formula; a French civil engineer Philippe Gauckler had published the same formula 22 years earlier. It should be known as the Gauckler-Manning formula.

gauge (or **gage**, USA) An instrument for measuring the water level of a river (river gauge), or the amount of rainfall (RAIN-GAUGE). It may have an automatic recording device. *See* FLOAT GAUGE, HOOK GAUGE, STAGE.

gauge correlation (or **stage relation**) An empirical curve relating stage or discharge of stream at one or more upstream points to stage or discharge at a downstream point; it is often used at a downstream point to estimate stage flow at peak.

gauge datum The level of the zero of a gauge relative to a fixed datum or plane of reference. *See* BENCH MARK.

gauge height *See* STAGE.

gauge line A line in a fixed direction across a stream or channel and passing through a permanent gauge.

gauge, point *See* POINT GAUGE.

gauge pressure The difference between ABSOLUTE PRESSURE and atmospheric pressure; it may be positive (pressure gauge) or negative (vacuum gauge). *See* DIFFERENTIAL PRESSURE.

gauge reading, specific *See* SPECIFIC GAUGE READING.

gauge, tape *See* CHAIN GAUGE.

gauge well A chamber sunk in the bank, open at the top, and connected to the stream. It dampens the waves and surges, and allows the water level or stage of the stream to be measured in relatively steady conditions.

gauging *See* CLOUD VELOCITY GAUGING, COLOUR VELOCITY GAUGING, DILUTION METHODS, FLOAT GAUGING, SALT VELOCITY METHOD, STREAM GAUGING.

gauging station A site selected on a stream, river or open channel, fitted with a GAUGE and other means of making systematic measurements of water level and discharge. The station may be used to record flow conditions for some special purpose, or form one unit in a network to survey the runoff from an entire river basin.

Gaussian distribution *See* NORMAL DISTRIBUTION.

general base level of erosion The level of the sea into which a river flows. *See* BASE LEVEL.

Geographical Information System (GIS) A computer method for recording, manipulating and displaying spatial data. It is used for mapping objects and analysing events on the Earth's surface. This makes it suitable for all forms of data and information on the natural world. It is of particular use to planners, agronomists, earth scientists, engineers and others. *See* AERIAL PHOTOGRAPH, REMOTE SENSING.

geohydrology A term introduced in 1939 by Meinzer, who defined it as 'a study of groundwater'; it is an important requirement in the training of groundwater engineers and geologists. *See* HYDROGEOLOGY.

geological erosion (or **normal erosion**) The erosion of land forms, rocks, and soils under normal conditions and undisturbed by human activities. *See* SOIL EROSION.

geological map A map showing the main topographical features of an area, the exposed rocks, and their nature and general distribution. Maps with scales of 1:10 000 are useful for water supply surveys. *See* TOPOGRAPHICAL MAP, WATER PROSPECT MAP.

geological mapping The location and measurement of rocks, faults, folds and other geological details exposed at the surface, and marking their position on a large-scale topographical map of the area. *See* BASE MAP.

geology The science which deals with the rocks, and contained minerals and materials, which form the earth's solid crust, and also the agents which produce changes on the surface or underground. The investigation of these materials and processes provides the basic information for geological history.

geomorphology The science which deals with land forms, the distribution of land and water, and making geologic interpretations. *See* HYDROGEOLOGY.

geophysical prospecting Surface measurements by special equipment to reveal changes in the physical properties of rocks at depth. The information will help to elucidate hidden geological structures and locate, directly or indirectly, some mineral of economic value. The methods found useful are the RESISTIVITY METHOD, the SEISMIC METHOD, and sometimes the MAGNETIC METHOD. Geophysics may be used in the search for water, either by locating rock conditions and structures favourable for the accumulation of water, or by utilizing the fact that underground water has a direct influence on the electrical conducting properties of the strata. Again, geophysical surveys often make available data relevant to dam and other foundations.

geotechnical engineer An engineer with knowledge of geology and geohydrology, and their application to engineering projects and the use of construction raw materials. *See* GEOTECHNICAL PROCESSES.

geotechnical processes A term covering all processes that change the properties of soils, and also the associated water problems, such as GROUNDWATER LOWERING.

geyser An eruptive spring of hot water and steam; the discharge is usually intermittent.

Giardia A protozoan parasite capable of infecting humans and domestic animals, and causing diarrhoeal illness (giardiasis).

glacial cycle *See* CYCLE OF EROSION.

glacial drift (or **drift deposits**) Layers of sand, clay and debris transported and deposited directly or indirectly by glaciers. Since glacial drift may contain permeable beds, lenses or pockets of sand and gravel, the material is carefully tested at valley sites for dams or similar constructions in order to avoid or minimize leakage. *See* ENGLACIAL MATERIAL.

glacial drift well A shallow well sunk in GLACIAL DRIFT to obtain water mainly for domestic use and farms, although occasionally the supply is sufficient for municipal purposes. In some localities the structure is favourable for artesian wells. Since the drift deposits are somewhat erratic, the flow conditions tend to vary from point to point. *See* SHALLOW WELL.

glacial lake (or **drift-dam lake**) A lake formed by a glacier advancing across a river valley and impounding the water. It may also be formed by a TERMINAL MORAINE in a river valley cut prior to the Glacial period. The water is held back by a natural dam of glacial drift deposited across the valley. This type of lake is a hazard on account of floods following the sudden release of the waters. *See* ORIGINAL CONSEQUENT LAKES.

glacial stream A stream formed by the melting of ice near the SNOUT or termination of a glacier.

glacial valley A U-shaped valley with steep walls formed by vertical erosion by the moving ice mass; the floor shows smooth hollows and irregularities, and often contains boulders which are striated and faceted. It is probable that many glacial valleys were originally made by streams and later modified by ice. *See* CONTACT EROSION VALLEY.

glaciation (1) The wearing down of the earth's surface by moving ice, including the transportation of the material and its later deposition as GLACIAL DRIFT. (2) The process by which large areas of land are covered by ice or glaciers. *See* ICE AGE.

glacier A thick mass of ice, or snow and ice, formed in regions above the SNOW LINE, and moving slowly down valleys or mountain slopes. *See* CIRQUE GLACIERS, VALLEY GLACIERS.

glacier burst The sudden downstream rush of water by breakage and removal of glacial ice in a river valley. *See* ICE GORGE.

glacierets *See* CIRQUE GLACIERS.

glaciometer An instrument for measuring the movement of glaciers.

gland A washer or sleeve used in a pump to compress the packing in the STUFFING BOX.

Global Positioning System (GPS) A system of satellite navigation which makes use of a network of 24 satellites in fixed orbits such that at any one time and place it is possible for the signals sent out by 4 satellites to be received. Normally this is done with a small handheld receiver that can analyse the signals and compute the position.

global warming The steady rise in the average temperature of the surface of the earth, first observed during the last decades of the 20th century. All of the probable causes of this have not yet been established, but geological studies and other research work indicate that throughout its existence the temperature of the earth's surface has frequently changed, sometimes rapidly. Small changes in the temperature of the biosphere clearly do affect many natural processes, some of which are of very great importance to humankind, but these are far from clearly understood. *See also* GREENHOUSE EFFECT.

globe valve A water valve in the form of a circular metal disc, which is moved by a threaded spindle. It is open to flow when raised off its seating in the pipe, and flow stops when it returns. *See* GATE VALVE.

go-devil *See* PIG.

go-out A SLUICE formed in a tidal embankment to impound tidal water. At low tide the water can flow out through the sluice.

gorge A narrow channel or passage, with stream, between hills, cut by young streams which erode rapidly downwards and where weathering of the sides proceeds but slowly, as in areas of low rainfall. If the rock forming the sides is hard it will stand with steep or vertical walls. *See* NARROWS.

gorge, ice *See* ICE GORGE.

grab bucket A steel bucket consisting of two halves so hinged that they dig into the soil as they are lowered, and lift out this material when they are raised.

grade (1) The finished surface or bottom of an excavation or canal bed or top of an embankment. (2) The gradient or inclination of a sloping surface or of a line or any horizon, expressed as an angle measured in degrees from the horizontal or as a ratio or a percentage. *See* STREAM AT GRADE.

graded filter FILTER MATERIAL consisting of coarse gravel, fine gravel, coarse sand and fine sand, so layered that the fine material will not clog the coarse material. Graded filters are placed along FRENCH DRAINS, at the bottom of excavations which tend to BOIL, and at the downstream toe of an earth dam to improve stability while allowing drainage. *See* TOE DRAIN. It is often essential to place such filters within earthworks at points of very high hydraulic gradient to prevent PIPING. *See* BOIL, FRENCH DRAIN, TOE DRAIN, SAND BOIL.

graded reaches Stretches along a stream where the amount of material supplied is about equal to the amount the stream can transport. The stream is in early maturity and flowing over relatively soft rocks.

graded river floodplain A plain of river denudation with the bedrock concealed by a thin layer of river deposits. The plain is usually narrow and determined by the width of the meander belt. *See* AGGRADED FLOODPLAIN.

graded stream A stream which is neither building up nor actively cutting down its bed.The slope of the stream at this stage is said to be 'at grade'. When at or near this grade a stream is described as mature, and there are no irregularities such as pools or waterfalls. *See* PROFILE OF EQUILIBRIUM, REGIME.

graded terrace An embankment or ridge of earth having a constant or variable grade or slope along its length. *See* LEVEL TERRACE.

grader A towed or self-propelled machine with a row of cutting or digging teeth and a blade at the rear to spread and level the material. It is used for cutting topsoil and levelling the material.

grade-stabilizing structure A barrier or dam erected across a stream or gully to stabilize the grade and thus prevent lowering of the channel or further head cutting.

gradient *See* GRADE, HYDRAULIC GRADIENT, MOISTURE GRADIENT, SLOPE, STREAM GRADIENT, STREAM PROFILE, WATER-TABLE GRADIENT.

gradient irrigation *See* FURROW IRRIGATION.

grading curve A graph which shows the range and the relative amounts of each grain size in a sample of sand and gravel. The data are plotted with the particle size on a horizontal, logarithmic scale, and percentages are plotted on a vertical, arithmetic scale. The curve shows, at any particle diameter, what percentage by weight of particles in the sample is smaller in size than the given point.

grapevine drainage *See* TRELLIS DRAINAGE.

grassed waterway (or **vegetated channel**) A constructed or natural water channel for carrying surface water from cropland, generally shallow and broad, and covered with erosion-resistant grasses.

gravel An unconsolidated deposit consisting of roughly rounded rock fragments with a lower size limit of 2 mm. Conglomerate is the consolidated equivalent. Gravels may or may not be related to the present system of river valleys. The river gravels shown on a map usually occur as TERRACES, each of which is often at a constant height above the present river. FLOODPLAIN gravels, on the other hand, are level with the existing floodplain of the river.

gravel bank *See* SHINGLE BANK.

gravel filter well A large-diameter well in which graded pea-gravel is used as filter material between two slotted tubes in the well. The outer tube is often withdrawn and the gravel allowed to spread into the ground around the well.

gravel fraction That fraction of a granular mixture composed of particles between 2.0 and 60 mm (0.08 and 2.36 in) in size.

gravel packing (boreholes) The addition of gravel around the outside of a screen or slotted pipe to a thickness of about 75 mm (3 in). Often used in boreholes formed in unconsolidated ground, it has the effect of increasing the diameter of the hole and reducing the entry of sand and also the velocity of the water entering the pipe. *See* GRAVEL FILTER WELL.

gravel pump A type of CENTRIFUGAL PUMP suitable for raising a mixture of gravel and water. Rubber may be used as lining to the pump and pipes which lasts longer than iron or steel. The pump also has renewable impellers and a renewable lining. *See* BAILER.

graving dock *See* DRY DOCK.

gravitational water Water in soils and rocks above the water-table; water which seeps downwards under the influence of gravity. *See* VADOSE WATER.

gravity-arch dam A dam which maintains its stability by the combined effects of its own weight and arch action.

gravity dam A type of dam in which resistance to overturning, pivoting or sliding forces is provided by the weight of the dam alone. Such dams are usually of concrete or masonry, of triangular section, and very heavy and costly (Figure A.1). *See* PRESTRESSED GRAVITY DAM.

gravity filter A sand filter consisting of an open tank in which the difference in level between the filtered water channel and the water in the tank is sufficient to force the water through the sand bed. Since individual units can be built to deal with flows up to 13 500 m^3/d (3 million gallons per day), it is often used in major plants. *See* FILTER.

gravity groundwater The water which would drain out of a rock in the ZONE OF SATURATION, assuming the zone and the CAPILLARY FRINGE moved downwards for a period, no water entered the area and none was lost except through the force of gravity. The water discharged as springs, and that withdrawn from wells, is gravity groundwater.

gravity main Any pipeline from a reservoir which conducts water to lower levels by gravity, usually from the IMPOUNDING RESERVOIR downhill to the SERVICE RESERVOIR. The friction loss of maximum flow through the pipeline must not exceed the difference in level between the top water level of the service reservoir and the bottom water level of the impounding reservoir.

gravity scheme Usually applied to a water supply layout in which the bulk of the water flowing out of or into an IMPOUNDING RESERVOIR is by gravity and pumps are not required. *See* GRAVITY MAIN.

gravity spring Water discharged at the surface from permeable beds under the action of gravity; a spring at the outcrop of the water-table, usually in valleys. Also called DEPRESSION SPRING.

grazing management (catchments) On rangeland, the management of a continuing output of animal products while providing that the value of the area as a water catchment is not impaired. Where soil erosion has occurred or is likely to occur, grazing may be deferred for a period or the grazing pressure reduced. *See* CATCHMENT MANAGEMENT, RANGE.

great diurnal range of tide The difference in height between two low waters and mean higher high waters in each lunar day. *See* MEAN HIGHER HIGH WATER, SEMI-DIURNAL TIDE.

Great Ice Age *See* ICE AGE.

greenhouse effect The heating of the earth's atmosphere which has helped to maintain the earth's surface at a temperature suitable for sustaining life. The atmosphere acts as a filter controlling the exchange of energy between sun, earth and space. Heat-retaining (greenhouse) gases allow short-wave solar radiation to pass through this filter to the earth's surface while absorbing the long-wave infra-red radiation reflected back from the surface in the form of heat. *See* GLOBAL WARMING.

grey water Domestic drainage from washing and food preparation. Sometimes called sullage. *See* BLACK WATER.

gribble A crustacean found in warm waters that attacks timber by eating into it. *See* MARINE BORERS.

grid An open frame of timber beams resting on the foreshore which supports a vessel being repaired during low tide. *See* SLIP (2).

grip A shallow gutter formed along the margin of a road to carry rainwater from the road to a drain or ditch. *See* CATCHWATER DRAIN.

grit chamber A small DETRITUS CHAMBER for collection and removal of solids.

groins (USA) *See* GROYNES.

gross capacity (reservoir or tank) The volume of water stored between bed level and the sill level of the waste weir.

gross duty of water (or **diversion requirement**) The water diverted at the intake of a canal system and used for irrigation (USA). *See* NET DUTY (of water).

ground moraine The debris which accumulates in the sole of a glacier and is steadily pushed or carried forward with the ice; the debris deposited by a glacier when the ice melts. *See* ENGLACIAL MATERIAL.

groundwater Water in the ZONE OF SATURATION; SUBSURFACE WATER as opposed to SURFACE WATER. Groundwater has a number of advantages over surface water as supplies, e.g. better bacteriological and chemical quality, greater scope for well siting, and, commonly, lower cost of schemes.

groundwater artery A channel or passage saturated with ARTESIAN WATER and surrounded by a CONFINING BED. It is often a BURIED CHANNEL or depression filled with saturated gravel or other permeable material.

groundwater basin A basin-shaped group of rocks containing groundwater with geologic and hydraulic boundaries convenient for investigation and description. The basin commonly embraces both the recharge and discharge areas. *See* GROUNDWATER PROVINCES.

groundwater budget A statement or estimate of water resources, usually applied to a groundwater basin or province; it gives estimates of storage, discharge, and recharge of water and other assessments and forecasts. *See* WATER CONSUMPTION, WATER RESOURCES.

groundwater dam A subsurface mass or barrier (such as a fault or dike) which prevents or impedes the lateral movement of groundwater and results in a marked difference in level of the water-table on the two sides of the dam.

groundwater decrement A decrease in groundwater storage by withdrawal from wells, spring flows, infiltration tunnels and evaporation discharge.

groundwater depletion curve *See* DEPLETION CURVE, NORMAL DEPLETION CURVE.

groundwater discharge Discharge of water from the ZONE OF SATURATION into bodies of surface water or upon land. *See* EFFLUENT SEEPAGE, SPRING.

groundwater divide (or **phreatic divide**) The line of maximum elevation along a groundwater ridge where the water-table slopes downwards in opposite directions (Figure G.2). *See* DIVIDE.

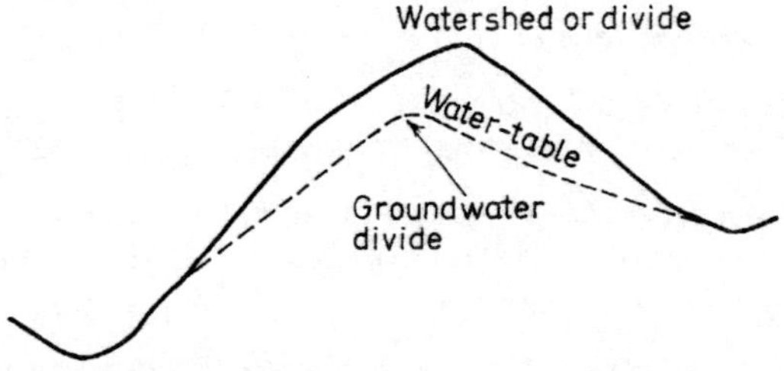

Figure G.2. Groundwater divide

groundwater equation The balance between water supplied to a basin and the amount leaving the basin. The information requires the collection of comprehensive data of water intake and discharge.

groundwater flow Part of the stream flow derived from the zone of saturation through seepage or springs. *See* EFFLUENT SEEPAGE.

groundwater increment *See* RECHARGE OF AQUIFER.

groundwater inventory A detailed statement or estimate of quantities of water forming RECHARGE OF AQUIFER, balanced against estimates of quantities forming GROUNDWATER DECREMENT for a particular area or basin. *See* GROUNDWATER BUDGET.

groundwater lowering Lowering the water-table locally so that excavations may be made in relatively dry conditions; it also helps to strengthen the ground excavation sides. The water may be lowered by wellpoints located along the fringe of the excavation, or by FILTER WELLS, which are preferred for deep lowering. The effective limit of a WELLPOINT is about 5.5 m (18 ft). *See* ARTESIAN WELL DRAINAGE.

groundwater modelling Techniques for predicting abstractions, yields, seepage flows, changes in water quality, movements in the water-table and other parameters related to water below ground. The first groundwater models were often analogue models making use of the similarity of water pressures in aquifers to electric potential. Numerical simulation models are now more commonly used because they can be readily run on computers.

groundwater mound A ridge or elevation formed in a body of groundwater by INFLUENT SEEPAGE.

groundwater provinces Districts, areas or basins throughout each of which the groundwater conditions are similar.

groundwater recession The general sinking or lowering of the water-table of a basin or area.

groundwater recharge *See* RECHARGE OF AQUIFER.

groundwater runoff Runoff which existed, wholly or partly, as groundwater since its last precipitation. *See* DIRECT RUNOFF, SHALLOW PERCOLATION.

groundwater storage The estimated quantity of water in the ZONE OF SATURATION; that stage of the HYDROLOGIC CYCLE when water is leaving and entering groundwater storage.

groundwater storage curve A curve showing the quantity of groundwater available for runoff at given rates of GROUNDWATER FLOW. *See* NORMAL DEPLETION CURVE.

groundwater tracers Dyes, chemicals or salts mixed with surface waters to trace the source of water seeping into tunnels, wells, shafts or deep excavations. Some radioactive isotopes form good tracers because of the high sensitivity with which they can be detected after seeping from the surface or other source to the point of issue underground. Tritium, a heavy isotope of hydrogen, has outstanding tracer qualities. *See* DARCY'S LAW.

groundwater trench A relatively narrow depression in the water-table caused by EFFLUENT SEEPAGE into a stream, channel or drainage ditch.

group action (USA) The planning, application and maintenance of land, soil and water conservation measures by a group of ranchers or farmers working with a leader or technician.

group enterprise The design and installation of a soil and water conservation scheme, which serves more than one ranch or farm, through the joint effort of those served. The scheme may involve the construction of a dam, canal or main drainage channel, or enlargement of the irrigation system.

grout curtain *See* CURTAIN GROUTING.

grouting (1) The injection, by pumps through boreholes, of cement slurry or chemicals into the ground around tunnels, excavations, shafts or dam foundations. When set, the grout reduces or prevents the inflow of water and also improves the mass strength and elastic properties of the rocks. Cement may be used to fill the larger fissures, and followed with chemical grout to seal the smaller cracks and voids. (2) The grouting of banks to form resistant masses to check the lateral erosion of important land areas. *See* BLANKET GROUTING, CLAY SEALING, CURTAIN GROUTING.

grouting (boreholes) The grouting of the annular space between the ground and the lining of the boreholes. It is often carried out for the first 30 m (100 ft) or so, mainly to retain the lining in position, to exclude unwanted water, and to support loose ground. The grouting is performed by pipes from the surface or from the bottom upwards.

groyne, filling for Filling material consisting generally of rounded stone, large gravel or shingle, depending mainly on the type of groyne and river conditions. Earth, sands, silts and small gravel are unsuitable as the material tends to be washed out through the groyne structure by the water.

groynes (or **groins**, USA) (1) River groynes (Figure G.3): elongated structures with one end, or root, in the bank of a river and the other end, the head or nose, projecting into the channel. They may be impermeable, semi-permeable or permeable. The construction material may be temporary (weighed FASCINES, etc.), semi-permanent (piles and sheeting or TIMBER SHEET PILING), or permanent (concrete, masonry, etc.). They are erected at right angles to the bank or current, or given an upstream or downstream slope. It is usual to use groynes in series so that the effect is continuous. A series of river groynes will check erosion and induce the deposition of silt in the sluggish areas between the groynes. (2) Coast groynes. Usually timber, but sometimes stone or concrete structures erected normal to the coastline or to the prevailing LONGSHORE DRIFT, to accumulate or retain beach deposits where there is LITTORAL DRIFT. In general, groynes are set at a distance apart about equal to their length, though this varies with local conditions. They are fixed sufficiently deep to avoid scour at low tides. *See* A-FRAME GROYNES, CRIB GROYNES, DIRECTION OF GROYNES, FASCINE GROYNE, SINKER GROYNE, STONEMESH GROYNES, WILLOW GROYNE.

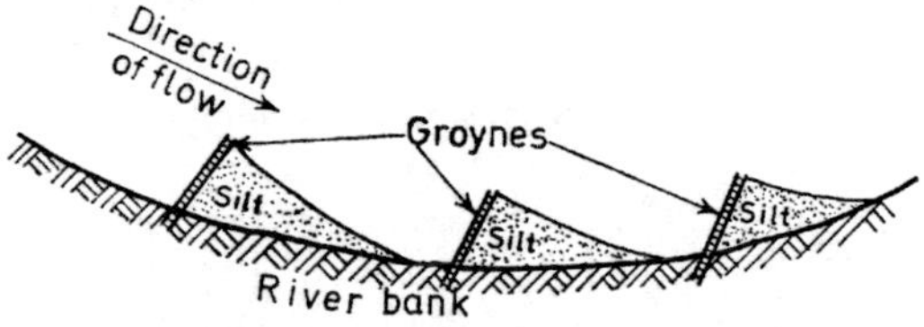

Figure G.3. Groynes

groynes, auxiliary protection *See* AUXILIARY PROTECTION.

groynework All works and practices in which groynes are used to control and protect rivers, coasts and banks. It includes the planning, design, location, construction and protection of groynes.

guard lock A lock which keeps a dock separated from tidal waters.

guard post *See* BOLLARD.

Guinea worm disease A waterborne infection from a source containing a small water flea (*Cyclops*) which carries the larvae of the Guinea worm (*Dracunculus medinensis*). Transfer of the thread-like larvae takes place where people use open wells or ponds for their drinking-water supplies.

gully A pit, chamber, or receptacle of concrete, stoneware, cast iron or other material, covered with a grating and placed on a roadside to collect water from a gutter or channel; a ditch, gutter or channel cut by water and along which water usually flows only after heavy rainfall or melting of snow or ice. *See* GULLY SUCKER, YARD GULLY.

gully control plantings The establishment or re-establishment of vegetative cover sufficient to control runoff and erosion, by planting seeds, cuttings or transplants in gullies. *See* COPPICING, TREE PLANTING.

gully dam A small FARM DAM constructed across a drainage depression or gully for the purpose of impounding water for irrigation or other farm purposes. *See* HILLSIDE DAM.

gully drainage (or **gully erosion**) Drainage which is concentrated in gutters or furrows. Gullying leads to rapid drainage, and erosion of valuable soil may also be rapid if vegetation is sparse or has been removed. *See* SHEET EROSION.

gully erosion *See* GULLY DRAINAGE.

gully stabilization structure (or **waterway stabilization structure**) A structure designed to prevent or minimize bed scour or other erosion in a channel or gully. The structure does not provide floodwater storage.

gully sucker A tanker lorry for removing silt and mud from road gullies; a pump sucks out the material and forces it into the tank. *See* YARD GULLY.

gully trap (or **yard trap**) A water seal fitted in a gully to prevent foul gases escaping from the drain.

Gumbel distribution An important statistical distribution which describes the occurrence probabilities of extreme values of many natural parameters. It is found to fit the extremes of observations sampled from data sets, for which the total population follows many distributions, including the NORMAL DISTRIBUTION and the LOG NORMAL DISTRIBUTION. Many water resource and hydrological parameters such as annual maximum flood flows or maximum rainfall intensities are often observed to follow a Gumbel distribution.

guttation Dropping or flowing in drop-like masses; the expulsion of water, in excess of transpiration, from uninjured leaves of vegetation.

gutter Any passage for water; a pipe or tube for conveying water from the roof or eaves of a building; a ditch or channel in the ground and lined with CLAY PUDDLE for conveying water alongside a canal, road, or street, to form into small longitudinal channels or gullies, usually by rain water.

H

hafirs Surface storage impounded by earth embankments in an area of natural run-off, sited in the Sudan. *See* RAINWATER HARVESTING.

hailstorm A STORM, often prolonged and severe, consisting largely of hail or frozen raindrops ranging in diameter from 5 mm to 10 mm or more; the ice particles are precipitated either separately or as aggregates of irregular size and shape. *See* SLEET, SNOW, THUNDERSTORM.

half-socket pipe An AGRICULTURAL DRAIN of which only the lower half is socketed. If made of concrete, the lower half may be impervious and the upper half porous.

half-tide cofferdam A COFFERDAM in an estuary or the sea, in which the height is not sufficient to exclude water at high tide, so that after every full tide de-watering is necessary. *See* FULL-TIDE COFFERDAM.

hand boring *See* AUGER.

hand-clean-up dam foundation A term used in some dam construction or similar earthwork contracts to indicate the requirement for the final cleaning of the rock top at the bottom of the excavation to be carefully done by handheld tools, air and water jets.

hand lead A sounding weight.

hand level *See* ABNEY LEVEL, BONING, CLINOMETER.

handpump A pump that is manually operated. It is a feature of the VLOM (village level operation and maintenance) concept introduced by the World Bank in field testing for the Water Decade Handpumps Project for Community Supply. Many pumps have been tested including the Afridev, India Mark II and the ROWER PUMP. *See* ROPE AND WASHER PUMP, TREADLE PUMP.

hanging glaciers *See* CIRQUE GLACIERS.

hanging valley A tributary valley which is discordant as to grade with its main valley. It may have been formed during the glacial period when the main valley was deepened by glacial erosion at a faster rate than that of its tributary, which formed a hanging valley when the ice disappeared. The considerable fall of water at the mouth of these valleys is sometimes used for power generation.

harbour A natural or artificially sheltered area of water where ships can anchor; it may have facilities for loading and discharging. *See* ARTIFICIAL HARBOUR, BREAKWATER.

harbour models Models of proposed harbours constructed to a scale ranging from 1 in 50 to 1 in 180, with outside waves about 20 mm (¾ in) high. Harbour models are useful during planning and design, and they also assist in solving problems of scour and silting. *See* MODEL ANALYSIS.

harbour of refuge A harbour which only gives ships shelter during storms and has no loading or unloading facilities.

hardness of water A feature of many waters due mainly to the presence of compounds of calcium and magnesium. Carbonate hardness is due to bicarbonates of calcium and magnesium that can be removed by boiling, whereas permanent (non-carbonate) hardness is due mainly to the presence of chlorides and sulphates of calcium and magnesium. 'Hard' water does not lather easily and forms an insoluble scum; it also leads to the deposition of scale in boilers, but removal for potable use is not to be recommended.

hardpan (or **pan**) A hard layer sometimes found below the topsoil which is relatively impermeable to water and does not soften when wet. The grains have been cemented together by iron or calcium salts leached from the topsoil. *See* BOUND GRAVEL.

Hawksley's formula An approximate formula used in the USA and UK to predict the height of waves in reservoirs:

$$H = 0{\cdot}025\sqrt{L}$$

where H is the height of wave in feet and L is the FETCH in feet.

haze *See* FOG.

Hazen's law A formula giving the approximate permeability of soils based on their effective grain size, and equals D10 size × 100 in cm/s. *See* EFFECTIVE SIZE, GRADED FILTER.

Hazen-Williams formula A formula for estimating the velocity of flow in pipes:

$$v = 0.849\,C\,R^{0.63}\,S^{0.54}$$

where v is the average flow velocity in m/s, C is the Hazen-Williams roughness coefficient for pipes, R is the HYDRAULIC RADIUS in m, and S is the FRICTION SLOPE in m/m. The Hazen-Williams coefficient, C, can vary between 150 for new smooth glass or plastic pipes to 80 for old riveted steel or concrete pipes.

head The potential energy of water due to its height above a given datum; it is usually given in metres or feet. *See* ARTESIAN HEAD, FRICTION HEAD, HYDROSTATIC LEVEL, HYDROSTATIC PRESSURE, PRESSURE HEAD, VELOCITY HEAD.

head bay The part of a canal lock immediately upstream of the lock gates.

head-control gate A type of gate often used in CHANNEL CHECKS to regulate the water level. The gate is set to act as an adjustable-level weir.

head, elevation *See* POTENTIAL HEAD.

head, entry *See* ENTRANCE HEAD.

header (1) A conduit or pipe which conveys water to or from another conduit or pipe system. (2) A stone or brick laid at right angles to the face of a wall.

head flume A flume placed at a junction with a terrace outlet to prevent cutting, or at the head end of a gully to prevent the extension of the gully.

headgate The upstream gate or control device through which water may enter a conduit. *See* TAILGATE.

headings (wells) Small tunnels, adits or headings excavated into the water-bearing rock formations to increase the yield of a well. The scheme may include (1) drilling large-diameter holes from the surface to intercept the headings; (2) enlarging fissures in the headings; (3) drilling small-diameter holes in the roof, sides and floor of the headings. *See* ADIT, INFILTRATION GALLERY.

heading up The upstream rise in water level from a construction because of regulation of flow as when gates at a barrage are closed. *See* HYDRAULIC BORE.

head, kinetic *See* VELOCITY HEAD.

head loss (or **lost head**) *See* FRICTION HEAD, HYDRAULIC FRICTION.

head of groyne *See* GROYNES.

head race A channel along which water flows to a turbine from a FOREBAY.

headwall A retaining wall, usually concrete, placed at the inlet and outlet of a culvert, drain or other hydraulic structure on water supply channels (Figure S.4).

headward erosion The erosion of the steepening hill slopes and the gradual extension of valleys and streams; the upstream movement of a scour or waterfall. Where necessary, the development of a deep scour or WATERFALL may be arrested by the use of heavy stone, stonemesh sills, or a concrete or timber sill. *See* CHECKDAM.

headwater control Flood control measures in a headwater drainage area. *See* FLOOD CONTROL.

headwaters The waters upstream of a structure; the upper reaches or source of a stream. *See* TAIL WATER.

headworks (or **intake heading**, USA) The diversion structure at the head of main irrigation canals and water supply conduits.

hearting The central or innermost zone of a masonry dam, located between the water face and the air face.

heave *See* FROST HEAVE, SWELL (2).

heel The part of the base on the earth side of a retaining wall, or the part of the base on the water or upstream side of a dam. *See* TOE.

heeling The traditional method of mixing wet clay and sand to make a watertight canal bed. The kneading of this mixture is done by labourers who 'heel' it with their boots. It was developed by James Brindley in the 18th century and is still used today on certain works. *See* PUDDLING.

heel post (or **quoin post**) A corner post of a lock gate. *See* HOLLOW QUOIN.

height of capillary rise *See* CAPILLARY RISE.

height of weir The height from bed to crest on the upstream side.

held water A term sometimes applied to CAPILLARY WATER; water retained in the ground above the STANDING WATER LEVEL.

helium diving bell A DIVING BELL in which a helium–oxygen mixture is breathed by the diver instead of ordinary air.

hencooping A form of CRIBWORK consisting of triangular cribs (or hencoops) filled with stone and planted with willow spars, which grow and hold the cribs in position along the river bank.

herringbone drain *See* CHEVRON DRAIN.

HHW *See* HIGHER HIGH WATER, HIGHEST HIGH WATER.

high-alumina cement A cement with the property of rapid hardening, prepared by fusing calcareous and aluminous materials to the completely molten state and crushing the resulting clinker to a fine powder. *See* PORTLAND CEMENT.

high bottom phase *See* FIRST BOTTOM.

higher critical velocity That velocity when eddies are first noted; lower critical velocity is that at which eddies in originally TURBULENT FLOW die out.

higher high water (HHW) The higher of the two high waters of any lunar day; when tide is diurnal, the high water occurring daily is accepted as higher high water.

higher low water (HLW) The higher of the two low waters of any lunar day.

highest high water (HHW) The highest known water level.

high-test hypochlorite (HTH) A combination of lime and chlorine in powder form, mainly consisting of calcium hypochlorite with a high, relatively stable, available chlorine content.

high water (HW) The highest level reached by a rising tide; the highest water level within any given period.

high waterline (tide) The line where the plane of mean high tidal water intersects the shore.

high water mark The highest level reached by water during a flood, as marked by debris or silt.

highway erosion control Including engineering and vegetative practices, the prevention and control of erosion at cross drains, in ditches, and on fills and road banks within a highway. *See* GUTTER.

hillside dam A small FARM DAM, usually curved or three-sided, built on the sloping side of a hill with no well-defined gully or depression. This type of dam may have storage ratios up to 3:1. *See* GULLY DAM, RING DAM, STORAGE/EXCAVATION RATIO.

hillside swamps Swamps on mountain sides often formed by a discharge of subsurface water into an area of poor local drainage, or by depressions or the downward percolation of moisture from melting snow. *See* COASTAL PLAIN SWAMPS.

histogram (or **frequency diagram**) A diagram or curve showing frequency distribution and so constructed that the area below the curve corresponds to the frequency. *See* FREQUENCY CURVE.

historic flood A flood whose date is known, and possibly its magnitude, but which occurred before the establishment of regular river gauging or recordings. *See* RETURN PERIOD.

HLW *See* HIGHER LOW WATER.

holdfast *See* DEADMAN.

hollow dam A dam built of reinforced concrete, plain concrete or masonry, in which the water pressure is taken on an inclined slab or vault carried by buttresses spaced at regular intervals. *See* MULTIPLE-ARCH-TYPE DAM.

hollow quoin Recessed masonry with a drilled hole in its upper surface carrying the HEEL POST of a LOCK GATE.

homogeneous dam An earth dam built more or less from one material (Figure O.1).

hooked spits SPITS terminating in a curve, formed by strong currents flowing past their ends.

hook gauge An appliance for measuring the elevation of the free surface of a liquid, consisting of a pointed hook attached to a vernier which slides along a graduated staff. The hook is lowered into the water and raised until the upward point just cuts the water surface. It measures water level with great accuracy. *See* POINT GAUGE.

hook groyne A type of groyne developed in New Zealand, consisting of a bank of shingle, curved in plan, with an armoured head pointing upstream. Above and below the groyne, heavy willow planting traps the silt, particularly on the upstream side.

hookworms Parasitic worms (nematodes) having mouth parts armed with hooks; in the adult stage they attach themselves to the wall of the intestine of mammals, including humans. Also called roundworm.

hopper barge (or **dumb barge** or **scow**) A barge for transporting dredged material from a DREDGER to a site for dumping. It may be equipped with power and discharging facilities. *See* FLOATING PIPELINE.

hopper dredger A DREDGER built to contain and carry its dredged material to the dumping ground.

horizontal alignment *See* REALIGNMENT.

horizontal drainage blanket A DOWNSTREAM DRAIN frequently placed in moderately high earth dams. It is located in the base of the dam and covers about a third of the downstream area. It may consist of 0.3 m (12 in) compacted layers of cobbles, gravel, and sand built up to a thickness varying between 1.5 and 3.0 m (5 and 10 ft).

hot springs Springs with a constant discharge of hot water which is sometimes used for domestic purposes, as in Iceland. *See* GEYSER.

house drain (or **collection line**, USA) A drain or drains taking all sewage or wastes from a house. In the UK it extends from the dwelling to the point of connection with the local authority's public sewer. *See* COLLECTING SYSTEM, COMBINED SYSTEM, SEPARATE SYSTEM.

HTH *See* HIGH-TEST HYPOCHLORITE.

humid Damp, moist; pertaining to climate or land where agriculture can be carried out without irrigation because the moisture, when distributed normally throughout the year, is adequate.

humidity A measure of the water vapour content of the atmosphere; it may be expressed as RELATIVE HUMIDITY or as ABSOLUTE HUMIDITY. Relative humidity is the ratio, expressed as a percentage, of the water vapour in the air to that in saturated air. Absolute humidity is the mass of water vapour present in a unit volume of air. *See* HYGROMETER, SATURATION DEFICIT.

humid, semi- *See* SEMI-HUMID.

humus tank The final settling tank from which sewage effluent passes out to the land or into a stream.

hurdle groynes Groynes used for checking erosion; they consist of light, driven piles with saplings secured horizontally to them, and also brushwood occasionally. They are placed about 20 m (65 ft) apart along the bank (Western Australia).

hurdle work (or **wattle work**) A low fence on a river bank, consisting of osiers (species of willow) interlaced with vertical sticks, to retard scour and encourage silting. Often cheap and effective along smaller streams with a sandy bed, as well as along the margins of lakes to check erosion by wave action, it has been used successfully in Victoria, Australia.

HW High water.

hydrant (irrigation) A small box or chamber containing a valve and coupling, with a removable cover flush with the surface; used for connecting irrigation spraylines or sprinkler laterals to MAIN PIPES. Any pipe with a nozzle at which water may be drawn by means of a hose. *See* STANDPIPE.

hydraulic Relating to the flow or conveyance of liquids, especially water, through pipes or channels; relating to the use of water as motive power. *See* HYDRAULICS, HYDRODYNAMICS.

hydraulic bore A wave in an open channel which advances upstream from a point where flow has stopped or diminished because of obstruction. *See* SURGE, TIDAL CURRENT.

hydraulic cement A special type of cement which possesses the property of hardening or setting under water.

hydraulic check nut A fitting used for the protection of pressure gauges against sudden pressure releases.

hydraulic conductivity A term sometimes used for the COEFFICIENT OF PERMEABILITY.

hydraulic discharge The loss of groundwater by discharge through springs, pumping from wells, etc. *See* GROUNDWATER DECREMENT.

hydraulic dredger *See* SUCTION DREDGER.

hydraulic ejector (or **silt ejector**) A pipe arrangement for removing mud, sand or silt from the chamber at the bottom of a pneumatic caisson. Water is injected into the bottom end of the pipe by means of a pump at the surface. The removal of mud and silt depends on the same principle as the ejector.

hydraulic elements The values that affect the flow of water in a conduit, such as velocity, area, depth, wetted perimeter and so on.

hydraulic elevator An arrangement widely used in various goldfields, towards the end of the 19th century, for lifting gravel and sand up to the drainage or discharge level. A jet of water creates a powerful suction in a hopper, and the gravel and water are carried up a pipeline and then discharged along sluice boxes. *See* HYDRAULIC EJECTOR.

hydraulic excavation The use of powerful water jets to break down deposits of coal or gravel containing gold, tin, etc. *See* HYDRAULIC MINING.

hydraulic extraction (or **hydro-extraction**) A general term for the various processes of excavating and transporting coal, metals, minerals, and gravels by water power.

hydraulic fill The filling of waste spaces, embankments or construction cavities by sediment or debris carried by water through a pipe or a flume. The filling of waste spaces in mines by waterborne material is known as hydraulic stowing.

hydraulic fill dam An earth dam built up from gravel and silt carried into place by pipelines or sluiceways located along the outer slopes of the dam, with the waterborne material flowing inwards towards midstream (Figure H.1). The coarse material settles on the outer shell (towards the banks) of the dam, while the fines are carried into the central part to form a more or less impervious core or core wall.

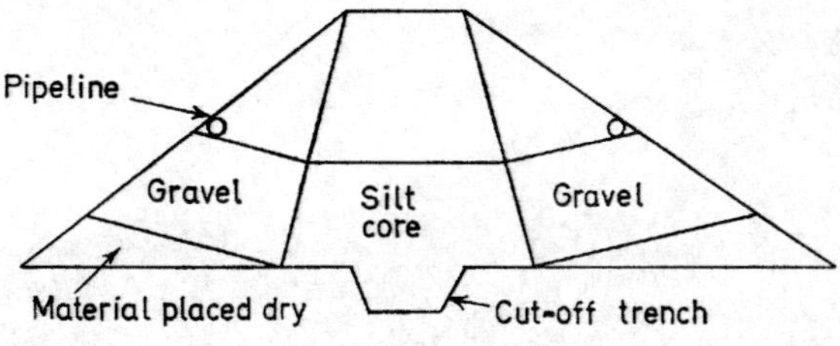

Figure H.1. Hydraulic fill dam

hydraulic flume transport The use of flowing water to transport pulp, broken coal, mineral or gravel in rectangular or semicircular channels. The channel slope should not be less than 3°. Movement of broken coal in flumes commences at a water velocity of about 1 m/s (3 ft/s), but in practice velocities of at least 2 m/s (6 ft/s) are used.

hydraulic friction (or **loss of head**) The friction or resistance to flow of the wetted surface of a channel, conduit or pipe; eddies and cross currents normally associated with TURBULENT FLOW are included. Whenever possible, the effect of obstructions, excessive channel variations, curvatures, impacts, etc., are not included in hydraulic friction. *See* ROUGHNESS COEFFICIENT.

hydraulic grade line (USA) *See* HYDRAULIC GRADIENT.

hydraulic gradient In a closed conduit, an imaginary line connecting the points to which water will rise in vertical open pipes extending upwards from the conduit; in an open channel, it is the free surface of flowing water. The gradient along a closed conduit may be expressed as the drop in metres per kilometre or feet per mile or as a ratio.

hydraulic jump A rapid and often turbulent flow of water from low stage below CRITICAL DEPTH to high stage above critical depth; the CRITICAL VELOCITY changes from HYPER-CRITICAL FLOW to SUB-CRITICAL FLOW (Figures H.2 and H.3).

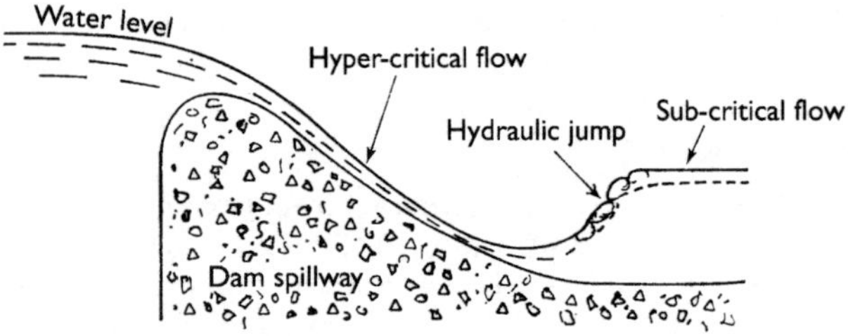

Figure H.2. Hydraulic jump

Figure H.3. Shallow hyper-critical flow immediately downstream of a weir which then reverts to sub-critical flow downstream of a well-pronounced hydraulic jump

hydraulicking The use of a strong jet of water from a nozzle to break down and remove a face of alluvial wash or other deposit, and the subsequent recovery of the gold or other valuable mineral so set free in sluice boxes or other appliances. The method is adopted only in undeveloped regions and where water is plentiful.

hydraulic loading The sluicing or flushing of broken material, obtained by hydraulic excavation, along the floor and into FLUMES or other appliances convenient for loading the material. In the case of broken coal, the material flows towards the flume if sufficient water is available and the floor slope is not less than 6–7° in favour of the flow. The flushing operation may be accelerated by flexible hoses with 1.05 to 1.4 N/mm^2 (150–200 lbf/in^2) water pressure.

hydraulic main A main pipeline which supplies water under high pressure to subscribers for driving hydraulic machines, such as lifts, cranes and so on. *See* MAIN PIPES.

hydraulic mean depth *See* HYDRAULIC RADIUS.

hydraulic mining (1) Surface: the use of high-pressure jets of water for breaking down relatively soft deposits of gravel or sand containing gold or tin or other valuable material. The washed debris flows along flumes with ribs or riffles in the bottom which trap the gold or other heavy mineral. (2) Underground: mainly the excavation and extraction of coal by high-pressure water jets. The jets impel the coal fragments along the floor to the point of collection. Tests indicate that water at a pressure of about 4.2 N/mm^2 (600 lbf/in^2) may be adequate to break down the softer coals, while pressures up to 8.4 N/mm^2 (1200 lbf/in^2) may be required for the harder coals. In the latter it may be necessary to pre-crack or loosen the seam by shot-firing or water infusion before applying hydraulic mining.

hydraulic model A scale representation of a hydraulic structure which is geometrically similar at all solid–liquid boundaries; a close general resemblance to its prototype is also valuable. The type of flow in the model must be the same as in the full-size plant. Model experiments ensure considerable accuracy in the solution of many hydraulic problems. *See* FROUDE NUMBER, HARBOUR MODELS, MODEL ANALYSIS.

hydraulic permeability The capacity of a rock or soil for transmitting water under pressure; it may be different in different directions.

hydraulic press *See* HYDROSTATIC PRESS.

hydraulic pressure snubber A SNUBBER used in self-contained hydraulic units to protect pressure gauges against sudden fluctuation of pressure and pressure ripple.

hydraulic profile (aquifer) A vertical section of the PIEZOMETRIC SURFACE, or the surface to which water will rise under its full head, from a given aquifer.

hydraulic radius (or **hydraulic mean depth**) The cross-sectional area of a stream divided by its WETTED PERIMETER; used in formulas to predict the velocity of water flow in an open channel. For a given channel and slope, the greatest hydraulic radius gives the largest flow.

hydraulic ram pump A self-acting pump which uses the momentum of a fall of water in a stream to force a small quantity of water to an elevation considerably higher than the initial stream fall. If correctly installed it will operate for many years without maintenance.

hydraulics The engineering application of the principles of HYDRODYNAMICS. In mining, the applications include wet coal or ore treatment plants, drilling fluids, water infusion, hydraulic stowing and hydraulic power. In civil engineering, it is applied to the flow of water in rivers, open channels, irrigation, drainage, seepage, dam construction and water supply.

hydraulic stripping The removal of overburden or waste rock overlying an economic deposit at the surface, by hydraulic methods.

hydraulic test (1) A water test to locate any leakage of water in newly laid drains. They are closed at the lower end and then filled with water and kept to a maximum head of 2.1 m (7 ft) for an hour. The test is satisfactory if no fall of level occurs during the hour. (2) A general test for pipes, pressure vessels, boilers, etc., which are filled with water and subjected to the designed pressure or slightly in excess.

hydraulic transport The conveyance of broken or crushed material by means of water flowing in flumes or pipes.

hydrodynamics The scientific study of the flow of fluids, particularly water flowing through pipes, in open channels and orifices and over weirs and notches, including the problems relating to energy and pressure. *See* HYDROSTATICS.

hydroelectric power Electrical power obtained from a water-driven dynamo. *See* HYDROELECTRIC POWER STATION.

hydroelectric power station A building in which turbines are rotated to drive generators, by the energy of natural or artificial waterfalls. *See* PUMPED STORAGE, PUMPED-STORAGE HYDRO STATION.

hydroelectric scheme A major and complete project for using water power to generate electrical energy; such schemes usually include dams, spillways, tunnels, buildings, roads, bridges, housing accommodation and many other works, services and facilities. *See* DAM, RESERVOIR.

hydro-extraction *See* HYDRAULIC EXTRACTION.

hydrogenesis The natural condensation of moisture in the air voids in soils and surface deposits.

hydrogeological map A map which depicts, from observed facts, the surface distribution of water and the principal aquifers, together with their thickness, lithology and outcrop belts in the area covered. It may also show structure contours on the aquifers and perhaps the ISOPACHYTES of the formations. The sites of the major abstraction wells and all surface springs and water supplies are shown. It may include sections across important aquifers and catchment areas. Most of the hydrogeological maps now being published are scaled at between 1:50 000 and 1:200 000.

hydrogeology The science concerned with the occurrence and movement of both surface water and groundwater, including a considerable amount of geologic orientation. *See* GEOHYDROLOGY, HYDROLOGY.

hydrograph A graph showing level, velocity or discharge of water in a channel or conduit plotted against time. The most common graph is discharge against time, with time expressed in hours, days, weeks or months.

hydrograph, flood *See* FLOOD HYDROGRAPH.

hydrographer One who surveys and draws maps of the seas, lakes, rivers and other waters including the adjacent shores; one who measures and records rainfall, runoff, water level, water flow and associated quantities.

hydrographic datum A datum used as a reference for measurements of heights of tides, levels of rivers or canals, or depths of water in wells and underground storage.

hydrograph, recession (or **normal recession curve**) The curve obtained from specified lengths of hydrograph that represent discharge from channel storage or a natural valley after subtracting BASE FLOW; the curve showing the decreasing rate of flow in a stream channel.

hydrograph time base The duration of flow of the runoff hydrograph following a distinct rainstorm (Figure H.4). Where the flow is conveyed in a watercourse which is otherwise dry, this is simply the time interval between flow commencing and ceasing. Where there is a continuous BASE FLOW, separating the duration of the flow resulting from an individual storm from that from other sources may require more complicated analysis.

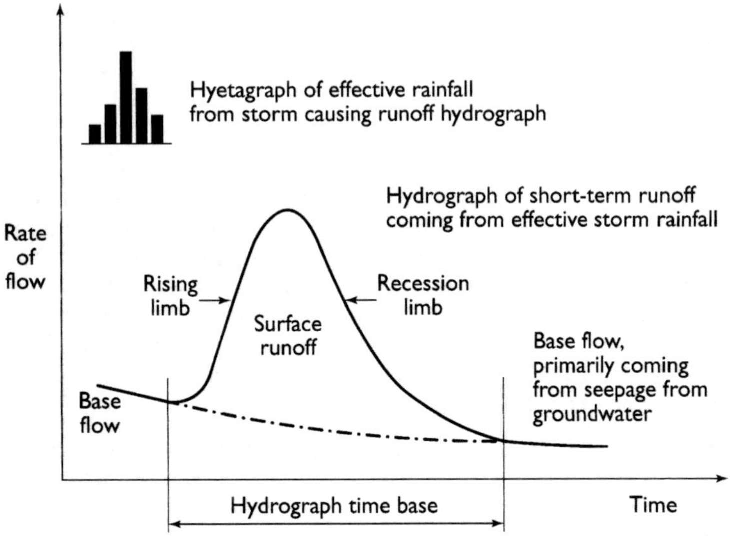

Figure H.4. Hydrograph time base

hydrograph, well *See* WELL HYDROGRAPH.

hydrography The study and measurement of seas, lakes, rivers and other waters, including their marginal land areas, and the preparation of maps, charts and publications for the use of navigators in such waters. *See* OCEANOGRAPHY.

hydroisobaths Lines connecting points of identical depths of the water-table below ground surface. *See* WATER-TABLE CONTOUR PLAN.

hydroisopleth map A map showing fluctuations in position of water-table with respect to time.

hydrologic cycle (or **water cycle**) The series of transformations in the circulation of surface waters to atmosphere, to ground as precipitation, and back to surface and subsurface waters (Figure H.5).

hydrologic processes Processes which may be divided into two main phases: (1) the wetting phase, when precipitation is the dominant process, and (2) the drying phase, when evaporation is the dominant process with no precipitation. During the wetting phase, evaporation diminishes and the catchment gains water from rainfall through infiltration and depression storage. *See* MOISTURE STATUS.

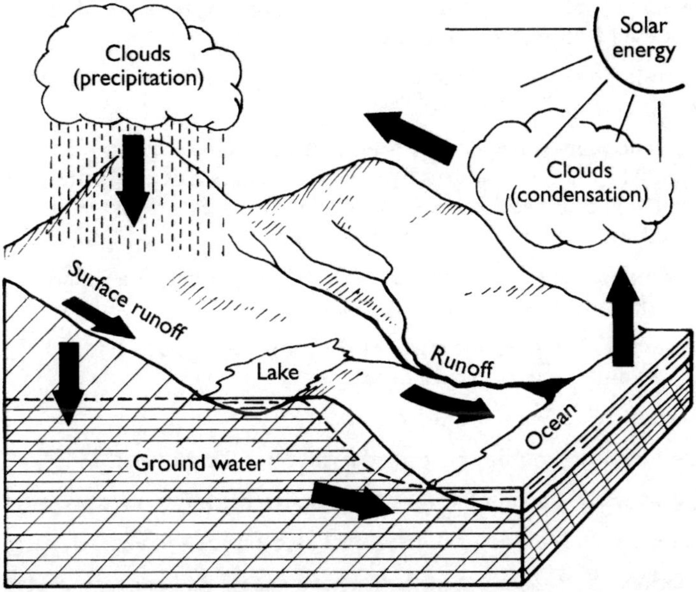

Figure H.5. Hydrologic cycle

hydrology The science concerned with the occurrence of water on and in the earth, its physical and chemical properties, and the changes involved from its period of precipitation until its discharge into a sea or other waters or return to atmosphere, that is, during a complete HYDROLOGIC CYCLE.

hydro-mechanization The use of modern hydraulic methods of excavating and moving coal or other minerals in surface or underground operations.

hydrometeorology The study of METEOROLOGY in relation to hydrology.

hydrometer An instrument for measuring the density or specific gravity of liquids. The ordinary type consists of a slender graduated stem fixed to a weighed bulb; when floating vertically in the liquid to be tested, the specific gravity is indicated on the stem. In liquids of low density a shorter length of stem is exposed than in liquids of high density.

hydrometry Concerned with, or determination of, the SPECIFIC GRAVITY or density of liquids. The term is sometimes used in relation to the measurement of velocity or discharge of waters as from reservoirs, rivers and other waters.

hydrosphere The zone containing all the waters of the earth's crust, including the ATMOSPHERE and LITHOSPHERE.

hydrostatic joint A spigot-and-socket joint in a water main. A hydraulic ram is used to force lead into the socket.

hydrostatic level (or **static level**) The level to which water will rise in a tube, pipe or well under its full PRESSURE HEAD.

hydrostatic press (or **hydraulic press**) An appliance, based on PASCAL'S LAW, in which a force applied by a piston over a small area is transmitted through water to another piston having a large area; by manipulating the small piston high pressures may be obtained.

hydrostatic pressure The pressure at any given point in a liquid at rest, equalling its density multiplied by the depth. *See* PRESSURE HEAD.

hydrostatics The scientific study of the properties of, and particularly the pressures and forces in, water at rest.

hydrostatic test *See* HYDRAULIC TEST.

hyetal Relating to rain, or to its distribution. *See* ISOHYETALS.

hyetal interval The rainfall difference represented by two ISOHYETALS.

hyetal regions Major world divisions based on rainfall characteristics.

hyetograph A chart or graph showing average rainfall or average intensity of rainfall; a recording rain-gauge.

hygrograph A self-recording HYGROMETER.

hygrometer (or **psychrometer**) An instrument for measuring the RELATIVE HUMIDITY of air. In the whirling type two thermometers (a dry-bulb thermometer and wet-bulb thermometer) are fixed side by side on a brass frame, and fitted with a loose handle so that it can be whirled in the air to be tested. After whirling for about one minute at approximately 200 revolutions per minute, the readings on both thermometers are recorded and by referring to Glaisher's or Marvin's hygrometrical tables a consistent and precise figure can be obtained for the relative humidity.

hygroscopic coefficient The amount of moisture a dry soil will absorb in nearly saturated air (98% RELATIVE HUMIDITY) at 25 °C (77 °F). The result is given as a percentage of the dry weight of the soil.

hygroscopic moisture The moisture contained in air-dried soil but which is driven off if the soil is dried at 105 °C (221 °F).

hyper-critical flow (or **super-critical flow**) An OPEN CHANNEL in which the velocity is greater than the CRITICAL VELOCITY and the depth of flow is less than the CRITICAL DEPTH.

hypogene Relating to changes, processes or forces within the earth's crust, such as earth movements.

hypolimnion The bottom layer below the THERMOCLINE in a thermally stratified lake; a region of constant temperature. *See* EPILIMNION.

I

ice Water in the solid state. Water freezes at 0 °C (32 °F) and expands and becomes less dense than water. *See* FROST ACTION, GLACIER, SNOW.

ice age The Pleistocene geological period, covering some hundreds of thousands of years, and ending only a few thousand years ago. The glaciers and ice sheets covered large continental areas, including at times nearly all Britain north of the Bristol Channel and the Thames Valley. Extensive deposits of ice-borne material were left, and around some coasts new land was added; many of the countries affected by the glacial period have published papers on these deposits, which are of value to water engineers as they give information as to thickness of drift and presence of BURIED CHANNELS, etc.

ice apron (or **ice breaker**) A structure for breaking ice and thus protecting a bridge pier; it consists of a ramp on the upstream side of a bridge, sloping upwards from below water level.

iceberg A huge floating mass of ice. When a glacier reaches the sea, it begins to float and break up into icebergs of varying shape and size.

ice caps Thick and widespread sheets of glacier ice in regions where the SNOW LINE is low.

ice cover A mantle of ice over an area of land or water, and particularly its thickness.

ice gorge (or **ice jam**) The piling up of ice against a restriction or obstruction in the channel of a stream, forming a temporary bar or dam. The sudden release of water when the obstruction collapses often causes flooding of any low-lying areas in the valley.

ice-pushed terrace A terrace formed along some lake shores by the thrust of ice up the beach. In northern latitudes, similar terraces have been noted on the shores of fiords and inlets of the sea.

ice ramparts (lakes) Terraces or ridges of gravelly material along the shores of lakes, caused by the contraction and expansion of the ice cover during cold weather. Sometimes the ice thrust causes damage to structures along the shore of such lakes.

igneous intrusion An igneous rock which, while still in its molten state, has invaded the surrounding sedimentary rocks and has solidified below the surface.

igneous rock A rock created by the solidification of molten matter from the Earth's interior.

image well An imaginary well which, if assumed to exist, can greatly simplify the analysis of hydraulic flow within an aquifer (Figure I.1). Image wells are usually introduced into the analysis of flow in aquifers near streams and rock boundaries.

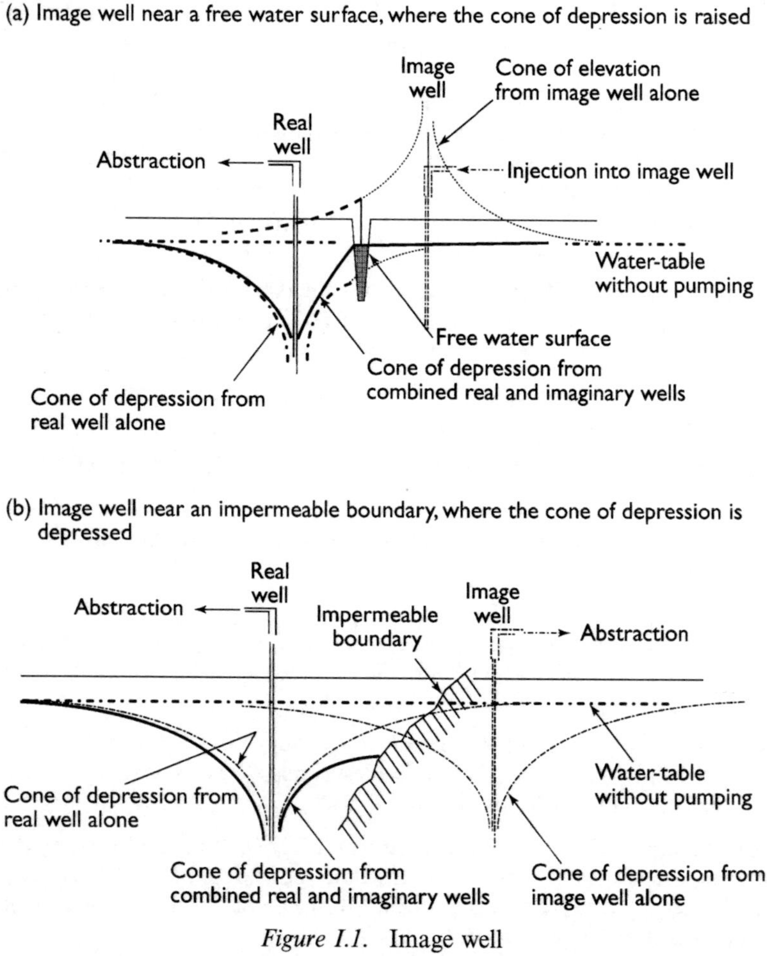

Figure I.1. Image well

imbibition Absorption of water by plants from the soil; absorption of water into the pores of a rock or soil.

Imhoff cone A graduated glass cone for the laboratory measurement of the proportion of settleable solids in a wastewater.

Imhoff tank A two-storeyed tank for the fermentation of sewage; methane gas is formed and the sludge which settles out is collected and dried.

immediate runoff *See* SURFACE RUNOFF.

impact loss The energy or head lost by impact or collision of particles of water; it is included in EDDY LOSS.

impact pressure The pressure exerted during impact or collision between a solid and a fluid.

impeller The rotating curved blades of a CENTRIFUGAL PUMP or TURBINE PUMP. *See* DIFFUSER CHAMBER.

impeller pumps *See* CENTRIFUGAL PUMP, SUBMERSIBLE PUMP, TURBINE PUMP.

impermeability factor (or **runoff coefficient**) A factor which enables runoff to be calculated, and is the ratio of the DIRECT RUNOFF to the average rainfall over the whole drainage area for any storm.

impermeable (or **impervious**) Relating to a soil, rock or other substance which permits the passage of water only at a very slow rate. *See* HAZEN'S LAW, DARCY'S LAW.

impermeable groyne A type of GROYNE often used for confining, straightening or narrowing a river; it induces scouring along specific lines to eliminate shallow, wide and meandering lengths. These groynes are usually long and, to keep down costs, often made of earth or shingle with a head reinforced with stonemesh crates or concrete blocks. Their crests are built well above flood level to avoid overtopping and erosion. Many of the river control groynes used in New Zealand are of this general type and known as Ashley type groynes. *See* PERMEABLE GROYNES.

impervious rock A rock that will not permit the passage of water, oil or natural gas, or only at a very slow rate. Impermeable rocks include shales, clays, compact limestones, marls and unfissured igneous rocks. *See* AQUIFUGE.

impounding reservoir River flows vary considerably throughout the year and also from year to year. To fully utilize these flows it is necessary to construct reservoirs so that water collected during wet seasons is available for use in dry periods. These are known as impounding reservoirs.

impounding schemes (planning) Field investigations and research so that the impounding reservoirs make full use of the catchment area. In planning, it may be decided (1) to construct a single dam to the full height required to develop the catchment. This is often adopted where only one site is favourable, or (2) to construct the first dam in such a position to enable the building of further dams at later stages of the entire scheme. In general, it is considered advantageous to exploit to the maximum the possibilities of a single dam site. *See* DAM SITE INVESTIGATION, WATERWORKS PLANNING.

impregnation Any method which prevents or reduces water seepage through ground by saturating it with a waterproofing liquid. *See* GROUTING.

improved Venturi flume An earlier term for a PARSHALL MEASURING FLUME.

impulse turbine A WATER TURBINE (e.g. the PELTON WATER WHEEL) which acquires its rotative force from the impulse of a jet or body of water, rather than by a fall in its pressure. *See* REACTION TURBINE.

incineration Combustion of volatile matter from sludge or solid waste to reduce their volume, while producing hot gas emissions and dry inorganic ash.

incised meanders A feature formed when a river, after it had reached a meandering stage, is subjected to REJUVENATION and cuts downwards once more, and so carves a gorge-like valley along the old meander channel. *See* VALLEY-IN-VALLEY STRUCTURE.

incised river A river whose channel has been formed by DEGRADATION. *See* ALLUVIAL RIVER, SUBSEQUENT STREAM.

inclined gauge A sloping gauge or plank graduated to record vertical heights above a permanent line or plane of reference. *See* STAFF GAUGE.

incoherent A descriptive term applied to a rock or deposit that is loose or unconsolidated and unable to hold together firmly.

increment (groundwater) *See* GROUNDWATER INCREMENT.

index contour An important CONTOUR LINE which is marked more heavily than others and labelled at intervals along its length by a number which denotes its elevation above mean sea level. *See* STRUCTURE CONTOURS.

index map (or **location map**) A map or plan showing the main surface features (towns, roads, rivers, etc.) of the area embraced in a geologic or hydrogeologic survey. The outline of any proposed works may also be shown. An index map may accompany a geological or engineering report. *See* SITE PLAN.

index, moisture *See* MOISTURE INDEX.

index, zero moisture *See* MOISTURE INDEX (ZERO).

indicator plants Plants that indicate, broadly, the underlying soil conditions, such as its qualities, deficiencies, salinity and alkalinity. The plants may also indicate the approximate depth to the zone of saturation and the climatic conditions of the area. *See* MARITIME PLANTS.

indirect flood damage All flood losses resulting from the interruption to services and production of goods, etc., as opposed to DIRECT FLOOD DAMAGE.

induced recharge of aquifer Recharge of an aquifer by inflow of stream water.

industrial water Water used by industry for all purposes except that used by employees and public health. Industrial uses include steam raising, cooling plants and air-conditioning. *See* POTABLE WATER.

infiltration The slow movement of water through, or into, the pores or interstices of a soil or other mass; the absorption of water by the soil. It may be determined by an INFILTROMETER.

infiltration area (well) The area or extent of water-bearing rocks penetrated in a well, and which discharge water into it. *See* GRAVEL FILTER WELL.

infiltration capacity The maximum INFILTRATION RATE of a soil or other porous material under specified conditions.

infiltration coefficient The ratio of INFILTRATION to PRECIPITATION for a soil or substance under specified conditions.

infiltration ditch *See* INFILTRATION TUNNEL.

infiltration diversion A form of stream water diversion by the use of pervious conduits, or perforated pipes, which are set under the bed of the stream.

infiltration gallery A construction to obtain a supply of water from a river where the banks consist of porous gravels or sands. A gallery or tunnel is driven close to and parallel with the river. Variations of the method consist of pipes placed in a trench or small tunnel dug parallel to the river, or in sinking wells in the porous banks. *See* INTAKE WORKS.

infiltration index The INFILTRATION RATE, as determined from rainfall and runoff records. There are several different indices, depending upon the method of calculation.

infiltration pond A pond for the purpose of retaining surface runoff, or water from other sources, so that it is able to seep downwards towards the WATER-TABLE. Infiltration ponds are used either to dispose of excess water or to increase groundwater recharge.

infiltration rate The rate at which water is absorbed, or seeps into or through the interstices of a soil or other porous material; it may be stated in millimetres or inches per hour. It varies with the INFILTRATION CAPACITY of the soil and the rate at which water is applied. As an approximation for irrigation purposes, gravelly soils may take in water at the rate of 25 mm (1 in) in an hour, whereas a heavy clayey soil cannot absorb more than 8 mm (1/3 in) per hour.

infiltration tunnel (or **infiltration ditch**) A tunnel, ditch, or gallery driven into the zone of saturation and along which water flows outwards by gravity into a well or water lodgement, or to the surface. *See* INFILTRATION GALLERY.

infiltration water Groundwater that has percolated into a sewer. *See* AIR TEST.

infiltrometer An instrument for measuring infiltration. The water during a test is applied artificially by either sprinkling or flooding. *See* PERCOLATION GAUGE.

inflow hydrograph (reservoir) A graph showing inflow into a reservoir.

inflow–outflow method A SEEPAGE LOSS test in a water supply channel. A test section of the channel is taken and measurements made of flow into it and the corresponding flow out of the section. The seepage loss is the difference between the two flows, with adjustments for rainfall and evaporation over the test period. During the test, channel levels are kept at the operating supply levels to ensure the normal seepage opportunity. *See* PONDAGE METHOD.

influence basin (of well) The basin-shaped depression in the water-table around a well due to withdrawal of water by pumping. *See* AREA OF INFLUENCE, CONE OF DEPRESSION.

influence of well, area of *See* AREA OF INFLUENCE.

influent seepage The seepage of water from a stream or channel outwards into the ground (Figure I.2); movement of GRAVITATIONAL WATER towards the ZONE OF SATURATION. *See* EFFLUENT SEEPAGE.

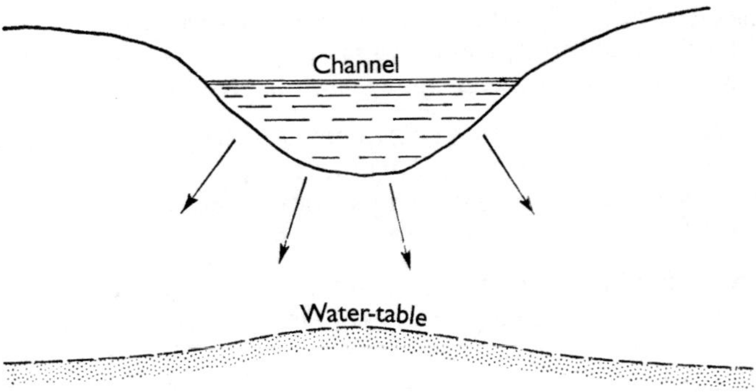

Figure I.2. Influent seepage

influent stream A stream which continuously loses water by seepage to ground storage.

infra-red photography Photography on special films which are more sensitive to infra-red rays than visible light rays, sometimes used in air surveys during misty weather. Can be used to locate waterlogged land and also the presence of freshwater intrusions in coastal waters.

ingrown meanders The series of meander-like loops formed by a YOUTHFUL STREAM when it enlarges pre-existing bends.

initial abstraction The part of precipitation forming DEPRESSION STORAGE and INTERCEPTION.

initial detention The part of rainfall which does not appear as surface runoff or as infiltration during a period of precipitation; includes evaporation, interception by vegetation, and depression storage.

initial rain Rainfall during the initial period of a storm before DEPRESSION STORAGE is completed. *See* RESIDUAL RAIN.

injection station The point or cross-section at which an indicating solution is injected into a stream or channel to measure velocity and discharge. *See* DILUTION METHODS.

inlet (1) The upstream end of a conduit, or any structure through which water flows. (2) A structure at the diversion end of a channel. (3) A surface connection to a closed drain. (4) A recess or bay in a lake, large river or the shore of the sea.

innings An area of land reclaimed from a waterlogged or marshy lowland or from the sea. *See* LAND ACCRETION.

insequent drainage A drainage system in which young streams follow the irregularities of the ground, often on a nearly flat plain or upon superficial deposits of indefinite structure. *See* DRAINAGE PATTERN.

insequent stream A stream whose course is determined by factors of a minor nature which may be obscure; the term is often applied to streams formed on the valley slopes of CONSEQUENT STREAMS and SUBSEQUENT STREAMS.

inspection chamber (or **manhole**) A pit from the surface down to a sewer, large water pipe or siphon, which a person can enter for inspections and repairs. Deep manholes are often made of 0.9 m (3 ft) diameter pre-cast concrete pipes fitted with step irons. Manholes may not be necessary for sewers larger than 0.9 m (3 ft) diameter through which a person can pass.

insulated stream A stream which neither receives water from the ZONE OF SATURATION nor contributes water to it; a stream which is separated from the zone of saturation by an impermeable deposit. *See* PERCHED STREAM.

intake (of a well) The voids, cracks or fissures in a water-bearing rock through which water passes into a well. *See* OPEN-END WELL.

intake area of aquifer The extent or area of outcropping permeable rocks from which an aquifer is fed with surface waters, such as streams and DEEP PERCOLATION. *See* RECHARGE OF AQUIFER.

intake heading *See* HEADWORKS.

intake of groundwater *See* GROUNDWATER INCREMENT.

intake tower (or **valve tower**) A concrete, stone or cast-iron tower forming part of the outlet works of a storage dam. All water released from storage passes through the intake tower, which also houses the outlet control gates and valves of pipes drawing off water at different levels. *See* DRY VALVE SHAFT.

intake works The structures, channels, pipelines, etc., used to obtain a supply of water from a river, lake or other source. Intake works are located where the banks, ground and flow conditions are reasonably stable, with the minimum risk of erosion or silting. *See* EXPOSED INTAKE, SUBMERGED INTAKE.

integrated flow curve A flow graph showing accumulated volume of runoff at various discharges.

integrating meter An instrument for recording the total quantity of water which flows past it. *See* FLOW METER.

integration method (velocity) A mean velocity measurement of a stream at a vertical depth by means of a current meter. The meter is slowly lowered from surface to bed of stream and back to surface, and the total number of revolutions of the wheel and time are recorded. *See* CURRENT METER.

intensity curve (rainfall) A curve showing the relation of RATE OF RAINFALL to its duration.

intensity–duration–frequency relationship The relationship between the average intensity of rainfall expected during a storm of given duration and the RETURN PERIOD. The same data are also often presented as DEPTH–DURATION–FREQUENCY RELATIONSHIPS.

intensity frequency of rainfall The RATE OF RAINFALL which, on average, is exceeded or equalled once in a given period of years. Also called rainfall recurrence interval or recurrence interval (rainfall).

intensity of rainfall *See* RATE OF RAINFALL.

interception The process by which precipitation is caught or retained by foliage and vegetation before reaching the ground.

interception channel (or **interception ditch**) A ditch or channel excavated near roads, at the foot of slopes or at the top of earth cuts, to intercept and collect surface water; a CATCHWATER DRAIN. *See* GRIP.

interceptometer A device to determine rainfall lost by interception; it is placed under trees or foliage and its catch of water compared with that of a rain-gauge placed on open ground without vegetal cover.

interference of wells *See* WELL INTERFERENCE.

interflow *See* SUBSURFACE RUNOFF.

interfluve A ridge of land between drainage valleys or between streams. *See* DIVIDE.

intermediate belt That part of the ZONE OF AERATION which lies between the CAPILLARY FRINGE and the BELT OF SOIL WATER; the contained water is sometimes termed intermediate vadose water.

intermediate vadose water The water contained in the INTERMEDIATE BELT.

intermittent filtration (or **land treatment**) A method of sewage disposal in which the sewage effluent is spread over land, the water being removed by field drains. *See* BROAD IRRIGATION, SPRINKLER IRRIGATION.

intermittent spring A spring from which the flow of water is variable and the discharge ceases during certain periods of the year.

intermittent stream A stream which ceases to flow during certain periods of the year; it is usually perched above the water-table and flow can usually be correlated to the seasons or periods of rainfall.

internal drainage A basin or area where runoff has no direct outlet to the sea. *See also* CLOSED DRAINAGE.

internal water A term sometimes applied to any water contained in the lower regions of the earth and below the ZONE OF SATURATION.

interrupted protection (river banks) In river engineering, any form of intermittent bank protection. Alternate lengths of river bank are protected, leaving the intervening portions only lightly protected or even unprotected. The objective is to secure the maximum possible bank protection for a specific outlay. *See* SADDLEBAG GROYNES.

interrupted stream A stream which does not flow regularly throughout its course; a stream with perennial stretches with intervening intermittent or ephemeral stretches. *See* EPHEMERAL STREAM, INTERMITTENT STREAM.

interrupted water-table A water-table with a marked difference in level in the vicinity of a fault or dike or other obstruction to lateral flow of water. *See* GROUNDWATER DAM.

interstices *See* VOIDS.

interstitial ice *See* SUBSURFACE ICE.

interstitial water The water contained within the voids or interstices of a rock or soil. *See* CONNATE WATER.

intertropic front (or **intertropic convergence zone**) A zone of heavy rains, local thunderstorms and strong squalls, which develops when an outbreak of cold air penetrates the DOLDRUMS, and this calm belt is replaced by a narrow zone of violent interaction with equatorial air; location of severe tropical weather.

intrapermafrost Layers of water within a mass of PERMAFROST.

inundation canal A canal which depends for its supply upon the water level in a stream or river. It may or may not have some form of head regulator.

invert An inverted arch, as at the bottom of a sewer, culvert, drain, tunnel or channel (Figure S.4). *See* SOFFIT.

inverted capacity The maximum rate at which an INVERTED WELL can remove surface or near-surface water by discharge through openings into deposits at its lower end. *See* DRAINAGE WELL.

inverted siphon (or **sag pipe**) An irrigation structure which carries water discharges under railways, rivers, roads or ground depressions. The inlet and outlet structures are usually of reinforced concrete, with a connecting tube or tubes of similar material. *See* SIPHON.

inverted well A well in which the water flow is downwards; the water enters at or near the top and is discharged into a permeable deposit through openings at lower levels. *See also* RECHARGE WELL.

invert level The level of the lowest part of an INVERT; the level which defines the elevation and slope of a channel, sewer or drain.

ion-exchange process A water softening process in which calcium ions in solution (which cause hardness) can be exchanged for sodium ions by being passed through special resins.

iron bacteria Bacteria that assimilate iron from ferruginous waters and excrete its compounds, such as ferric oxide.

iron-clad catchment *See* ARTIFICIAL CATCHMENT.

irrigable land Arable land sufficiently low for irrigation. It may include areas and constructions not previously irrigated. *See* NON-IRRIGABLE LAND.

irrigating head The flow of water required or available for irrigation; the flow rotated among a group of irrigators; the flow of water in a single farm lateral, or distributed at a single irrigation.

irrigation Spraying or causing water to flow over arable land for farming and to benefit crops. The work may involve excavating canals, laterals and ditches, and constructing small dams and associated works. The rate of irrigation should not be excessive, or lime and nutrients in the soil may be lost by leaching. *See* WATER HARVESTING.

irrigation district An irrigation area, with definite geographic boundaries, under the control of a local authority, state department or other body. *See* GROUP ENTERPRISE.

irrigation efficiency (or **distribution efficiency**) Usually expressed as the percentage of the total inflow into the channel system that is recorded as deliveries to farms. The quantity of water that is actually delivered should be carefully measured to obtain reliable results. *See* EFFICIENCY OF FARM IRRIGATION, WASTE.

irrigation intensity That proportion of any given area which is irrigated.

irrigation lateral A ditch or channel branching off the main canal and carrying water to the farm ditches.

irrigation layout plan A scale drawing showing the entire arrangement for the irrigation of a farm; it may show the quantity of water used or required, ground slopes and elevations at key points, etc.

irrigation requirement (or **duty of water**) The amount of water required, including wastes (but excluding rainfall), for farming and production of crops; usually expressed as rate of flow per unit area, or as depth of water in a given period of time.

irrigation structure Any structure or appliance required for the regulation, conveyance and measurement of water within an IRRIGATION DISTRICT.

isobars Lines on a map or weather chart connecting points or places of equal barometric pressure at a standard level.

isobath (water-table) A line on a plan connecting points at the same height or elevation above an aquifer or water-table. *See* WATER-TABLE CONTOUR PLAN.

isochion (or **isonival**) A line on a map or ground connecting points of equal water content of snow, or of equal snow depths.

isochrone A chart or map of a sewer system or river drainage showing a series of time lines; from the origin to the outlet of the system, the transit time of water is indicated.

isohyetal map A map showing rainfall depth contours.

isohyetals (or **isohyets**) Lines of equal rainfall. The lines are marked at intervals by interpolation on maps showing a large number of stations at which rainfall observations are made. In water engineering schemes, local records of rainfall are important; commonly one gauge per 200 ha (500 acres) of gathering ground is used. *See* BRADFORD GAUGE.

isomeric values Rainfall values that are ISOPERCENTALS in which the monthly average is given as a percentage of the annual average.

isonival *See* ISOCHION.

isopachyte A line on a geological map connecting points of equal thickness in a specific rock stratum, aquifer or other deposit; a contour of equal thickness. The depths or thicknesses for constructing isopachytes are obtained from boreholes, wells, outcrops and perhaps geophysical surveys.

isopercental A line on a map connecting points of equal percentage of rainfall; each rain-gauge station shows annual or monthly rainfall as a percentage of annual average values for that station over a long period.

isopiestic line An imaginary line connecting points which possess the same static level; a contour of the PIEZOMETRIC SURFACE of an aquifer. *See* PIESTIC INTERVAL.

isopluvial line A line connecting all localities with the same PLUVIAL INDEX over a given period.

isotherm A line on a map or weather chart connecting points or places of the same mean annual temperature.

J

jet *See* ORIFICE.

jetting A hydraulic method of inserting wellpoints or piles into sandy material, and used where a pile hammer might damage structures in the vicinity. *See* WELLPOINT.

jetty (1) A ridge, dike or construction of rock, piles or other materials that jets or projects into a stream or into the sea at the mouth of a river, for bank building or protection, or to induce scouring. *See* GROYNES. (2) A landing stage or deck carried on piles from the water's edge.

joint-sealing material Mastic or bituminous material used for filling expansion joints in concrete flumes, siphons and other hydraulic columns and structures. A pressure gun may be used when applying the material.

Joosten process A GROUTING method in which chemical fluids are injected into the ground through pipes in boreholes. The chemical grout strengthens and seals the soil, sands or gravel, and prevents the seepage of water, air or gases into or from tunnels, shafts or other works. Sodium silicate is used to react, in the voids of the rock, with calcium chloride to form calcium silicate, which acts as the sealing agent.

juvenile water (or **magmatic water** or **plutonic water**) Water derived from magmas or molten masses of igneous rock, during their crystallization, or from lava flows in the form of steam.

K

kames Ridges of gravel deposited from the ice along the margins of a glacier. They lie parallel to the ice limits and indicate pauses in the retreat of the ice. Kames frequently enclose small lakes and some subsequently become peatbogs. *See* ESKERS.

Kaplan turbine A propeller-type water turbine with blades of a pitch that can be automatically adjusted with the load to improve its performance. *See* FRANCIS TURBINE.

karst topography A region where limestone underlies the soil and the land is characterized by many SINKS, with irregular divides between them, disappearing streams, and solution valleys. *See* SWALLOW HOLES, TUBULAR SPRING.

karst water A term sometimes applied to water with a high capacity for dissolving carbonates, as when seeping or flowing through limestone rocks. *See* CHEMICAL WEATHERING, SINKS.

keel blocks The central row of DOCKING BLOCKS in a dry dock.

Kelvin tube An appliance for determining the depth of water, consisting of a lead tube containing a chemical, placed inside a glass tube. The tube is dropped into the water and taken out immediately it touches the bed. The depth of water, given on a scale, determines the colour of the chemical changes.

Kennedy's critical velocity *See* CRITICAL VELOCITY.

kerb inlet A kerb with openings to convey storm water to a drain or gully.

kid A FAGGOT or FASCINE.

kidding *See* FAGGOTTING.

kieselguhr (or **diatomaceous earth**) Sometimes, an unconsolidated deposit found in ponds and lakes, and consisting of the skeletons of siliceous organisms, such as diatoms. A very porous and absorbent substance, it is often used in waterworks as a filtering material. *See* DIATOMITE FILTER.

kinetic head *See* VELOCITY HEAD.

Kirpich formula An empirical formula used to estimate of the TIME OF CONCENTRATION in small agricultural catchments:

$$t_c = 3.97\, L^{0.77} S^{-0.385}$$

where t_c = time of concentration (minutes), L = length of the channel from divide to outlet (km), and s = average slope in m/m. The formula may be used to calculate the duration of STORMS and so RAINFALL INTENSITY when predicting flood flows using the RATIONAL METHOD.

knick point *See* NICK POINT.

Kutter's formula An empirical formula expressing the value of the coefficient C in CHÉZY'S FORMULA, in terms of hydraulic mean depth R, FRICTION SLOPE S, and ROUGHNESS COEFFICIENT N.

Metric units:

$$C = \frac{23 + 1/N + 0{\cdot}001\ 55/S}{1 + \left(N(23 + 0{\cdot}001\ 55/S)/\sqrt{R}\right)}$$

Imperial units:

$$C = \frac{41{\cdot}65 + 0{\cdot}002\ 81/S + 1{\cdot}811/N}{1 + \left(N(41{\cdot}65 + 0{\cdot}002\ 81/S)/\sqrt{R}\right)}$$

See MANNING'S FORMULA.

L

lacustrine Pertaining to lakes.

ladder The bucket ladder of a machine for underwater excavation. *See* DREDGER, BUCKET-LADDER DREDGER.

lag (time) *See* TIME LAG.

lagoon (1) A shallow stretch of salt water parted from the sea by a ridge or barrier of sand. *See* ATOLL, BARRIER BEACH, COASTAL LAGOON. (2) A shallow natural storage connected to a river or sea. (3) A large pond used to detain wastewater during sedimentation and biological oxidation.

lagtime (storm) The time interval between the beginning of rainfall excess and the maximum rate of runoff during a particular storm. *See* DESIGN FLOOD, PRECIPITATION (EXCESSIVE), STORMWATER, TIME LAG.

lake A large inland body of water. It may or may not have a single direction of flow, and the water may be either fresh or brackish. Natural and artificial lakes often provide water for domestic and industrial use. *See* CRUSTAL-MOVEMENT LAKES, EVAPORATING BASINS, LANDSLIDE LAKES.

lake-floor plain The exposed floor of a lake from which the water has drained or evaporated away. The deposits forming the floor are usually evenly stratified and fine-grained. *See* PLAYAS.

lamella settlers or **separators** A series of thin plates or tubes placed in a sedimentation tank to improve the efficiency of solids removal.

laminar flow *See* STREAMLINE FLOW.

laminar velocity The velocity of water in a particular channel below which STREAMLINE FLOW always exists, and above which the flow may be either streamline or turbulent.

lamphole A small pit put down over the centre of a sewer to enable a lamp to be lowered on a string. Someone looking along the sewer towards the lamp from a manhole can detect any sewer damage or flow obstruction. *See* INSPECTION CHAMBER.

land Any portion of the solid surface of the earth's crust. It includes soils, rocks, watercourses, climatic conditions, and any earthworks and man-made improvements. *See* HYDROSPHERE.

land accretion The reclamation of land from swamps, low-lying boggy areas, or the sea. The methods employed include drainage and pumping, elevation of ground by groynes, dumping of soil or mud, and the planting of reeds or maritime plants to induce the deposition of silt. Large-scale reclamation schemes are usually carried out by local authorities or state departments. *See* INNINGS.

land drain *See* AGRICULTURAL DRAIN.

land drainage *See* DRAINAGE, INTERNAL DRAINAGE.

land drainage (pumping) *See* PUMPING.

land management The management and planning of all measures for the protection and prudent use of land and soil for the maximum long-term benefit to agriculture, industry and the national economy. *See* AFFORESTATION, LAND PREPARATION, SOIL CONSERVATION SURVEY, WATER REQUIREMENT.

land preparation Making the ground surface capable of more intensive and uniform application of irrigation water by levelling and reshaping irregularities and slopes. *See* SOIL CONSERVATION.

land reclamation *See* LAND ACCRETION.

landslide (or **slip** or **slide**) The downward movement of rock and soil on embankments or hillside slopes; it may be slow or sudden. Movement may start because of an increase in water content, deep cuttings, and heavy blasting along the base. The usual remedies are drainage of slopes, lowering of the water-table by pumping, and planting of trees or shrubs. Wet patches indicate accumulations of water, and trenches are often cut across such areas and the water removed. A site with steep slopes may become a major problem when a reservoir has been filled, owing to changes in groundwater conditions and levels. *See* DAM, FAULT.

landslide lake A lake formed along the drainage line in a valley which has become blocked by debris from a landslide. *See* ACCIDENTAL LAKE, GLACIAL LAKE.

land spring A spring, the water from which is derived from a surface stratum. The water is affected by local variations of rainfall and is liable to be polluted. *See* SPRING.

land treatment *See* INTERMITTENT FILTRATION.

Langelier index The difference between the actual pH value of a water and its saturation pH (equilibrium) value. Waters with a positive index are supersaturated with calcium carbonate, and will tend to deposit it as scale. Conversely, waters with a negative index will tend to be aggressive and dissolve calcium carbonate.

Lapworth's chart *See* DEACON DIAGRAM.

large-area floods The flooding of a relatively large area by low-intensity storms, with a duration ranging from a few days to several weeks. *See* SMALL-AREA FLOODS.

large bore *See* OPEN PIT.

large dam According to the International Commission on Large Dams (ICOLD) World Register of Dams, any dam above 15 m in height (measured from the lowest point of foundation to the top of the dam), or any dam between 10 and 15 m in height that meets at least one of the following conditions: (a) the crest length is not less than 500 m; (b) the capacity of the reservoir formed by the dam is not less than 1 million cubic metres; (c) the maximum flood discharge dealt with by the dam is not less than 2000 cubic metres per second; (d) the dam had specially difficult foundation problems; (e) the dam is of unusual design.

late maturity (river) A stage of the erosion cycle characterized by: (1) major rivers with a MEANDER BELT roughly equal in width to the VALLEY FLOODPLAIN; (2) nearly all land at about the level of river grades; (3) rounded hilltops of low relief and poorly drained, gently rolling areas. *See* EARLY YOUTH.

lateral A small irrigation channel branching off a main supply channel. *See* DIVISION GATE.

lateral abrasion (stream) The cutting into and wearing away of the banks of a stream, especially active along the outer banks at curves. Along the inner banks water flow is sluggish, and deposition of silt often occurs.

lateral canal A canal excavated more or less parallel to a river that is unsuitable for navigation owing to high velocity or another reason. *See* CANALIZATION.

lateral-flow spillway (or **side-channel spillway**) A spillway used on dam sites located in narrow valleys or canyons. The water passes over the spillway crest into a channel and then flows out parallel to the crest; that is, the initial and final flow are approximately at right angles to each other.

lateral moraines *See* MORAINES.

lateral pipes (irrigation) Portable pipes which convey the water from a HYDRANT or STANDPIPE on the main to the positions where the water distributors are located. The remote end of each pipe is closed by a 'stopend' or plug. A shut-off valve is often used to enable part of the system to be moved while irrigation continues along other sections. *See* MAIN PIPES, RING MAIN.

lateral storage *See* BANK STORAGE.

lateral stream A stream which flows in the depression along the edge of a lava flow in a valley. *See* TWIN LATERALS.

late youth A stage of the erosion cycle often characterized by (1) efficient and well-integrated drainage of the area covered by the stream system, (2) alternating spurs in most stream valleys and (3) pronounced relief and sharp inter-stream divides. *See* LATE MATURITY.

lavants Streams formed by VALLEY SPRINGS. *See* BOURNES.

law of continuity (flow of water) The principle that the inflow of water to a given reach of a river or storage during a given period is equal to the outflow during the same period, plus or minus any changes in storage volume.

lay-flat tube irrigation Irrigation by means of jets of water issuing from small holes punched in thin-walled lay-flat polythene tubes. The tubes are laid down along the centre of beds, one end connected to the mains and the other end closed. The throw of the jets can be varied by a standpipe tap which adjusts the head of water. Water is applied at a high rate. *See* LOW-LEVEL SPRINKLER IRRIGATION.

leachate Water that has been contaminated with dissolved or suspended matter when percolating through solid waste or other permeable material, such as in landfill sites.

leaching The flushing out of excess salts from irrigation land, usually by applying large amounts of water over the affected area for extended periods. *See* SALTING.

leaching requirement The proportion of the applied water that is used to leach the salt out of the soil.

lead line *See* SOUNDING LINE.

leat (or **mill stream**) A stream, or any open watercourse, used for power purposes in a mill, mine, farm, etc.

ledges Platforms formed along a coast, composed of soft and hard beds, by the attack of storm waves; they are formed on hard rocks by the accelerated erosion of the overlying soft beds.

leech *See* LIMPET.

left bank The left-hand bank of a river facing downstream (Figure A.4).

legionella Any of over 30 species of bacteria of the genus *Legionella*, most of which have been implicated in human disease, notably *L. pneumophila*, the cause of an acute pneumonic infection of the lungs (legionnaires' disease).

length (stream) The linear dimension measured along the course of a stream; usually between two specified points or cross-sections. *See* REACH, STREAM PROFILE.

length-of-run (irrigation) The length along which irrigation water is conveyed in furrows or by flooding from the head channel.

levee (or **levee banks**) (1) An embankment to protect land from inundation during river floods or to control or confine river flow. An artificial levee may be made wide enough at the crest to carry vehicles or to provide access in flood time; access ramps and turning places may also be provided. (2) A NATURAL LEVEE (Figure L.1) or river bank formed by the deposition of silt during flood periods. It forms elevated ground bordering the floodplain of a river. Natural levees are often increased in height by artificial means to protect the plain from overflow during river floods. *See* BANK, RING LEVEE.

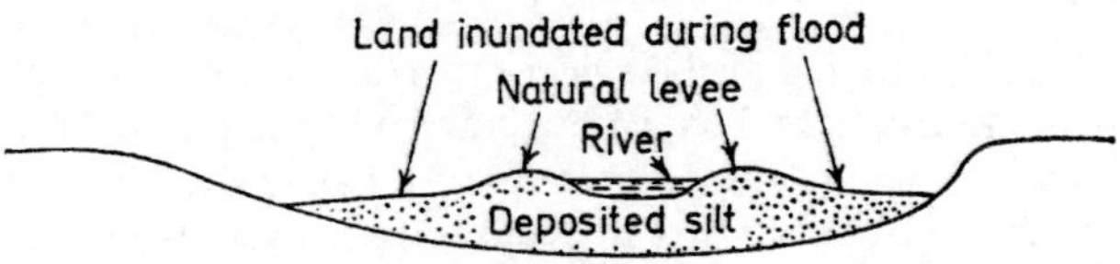

Figure L.1. Levee

level course *See* STRIKE.

level recorder An automatic device which records the water level in a stream or canal, operated by pressure or by a float. *See* GAUGE.

level terrace An embankment or ridge of earth formed along the absolute contour of the ground. It is generally used where outlet channels are not practicable, or on permeable soils where water conservation is very important. *See* GRADED TERRACE.

LHW Lower high water.

life buoy A device to keep a person afloat when he or she is in danger of drowning.

life of reservoir The period in years that a reservoir yields useful supplies of water. Mainly, it varies with the amount of silt deposited by streams, etc., and the remedies applied. Some storages in India, China, and the USA will last only about 50 years. *See* SILTING.

lift (1) The vertical distance through which the water must be raised from a river or channel. (2) A power-operated hoist which elevates or lowers vessels from one reach to the next when a lock is not available. (3) The vertical distance a vessel is elevated or lowered when passing through a lock.

lift gate A LOCK GATE which opens by moving vertically upwards. *See* PENNING GATE.

lift pump An earlier type of pump consisting of a column of rods operating a piston within a cylinder or working barrel. A reciprocating motion causes the water to be raised to the surface, or a higher level, by a series of lifts. The piston on its down-stroke (with the flap valves open) admits water to the upper section of pipe. On the up-stroke the valves close and the water is lifted in the pipe column. *See* CENTRIFUGAL PUMP, FORCE PUMP.

lighthouse A tower or structure supporting and protecting a recognisable light signal at the top, which is transmitted to serve as a guide and warning to ships at night; it is erected at some important point on a coast, at the entrance of a port or offshore.

lightship A ship moored at a site near the coast where it is not possible to build a LIGHTHOUSE. It is fitted with a powerful light and foghorn to warn or guide ships at sea. Modern lightships also have radio direction-finding apparatus and radar. The ships are manned or unmanned, the latter being fitted with automatic lanterns. The lanterns are normally visible for a distance of at least 16 km (10 miles).

limestone A general term for a widely occurring type of sedimentary rock, which mainly consists of calcium carbonate. Limestones were laid down on the beds of ancient seas by the accumulation of the shells of marine organisms. The properties of different types of limestone vary enormously, CHALK being quite soft, while Portland Stone is hard and durable. Many of the forms have important practical applications, including use as building materials, a source of calcium carbonate and other minerals and providing AQUIFERS.

limestone sinks *See* SINKS.

limits of oscillation The valley area or width within which a river has ranged within historic times. *See* MEANDER BELT.

limnology The branch of hydrology pertaining to ponds, lakes and inland waters; the study of life in such waters.

limpet (or **leech** or **limpet dam**) A small OPEN CAISSON designed to fit against a dock wall. It provides convenient access to underwater repair work.

limpet dam *See* LIMPET.

lined ditches (or **cement ditches**) Ditches with linings usually 50 mm (2 in) thick and made of a sand–cement mixture in the ratio of 3:1 to 5:1. They are usually trapezoidal in cross-section, with side slope in the ratio of 1:1 or 1½:1, and the depth of water is about equal to the width of the bed (Figure L.2).

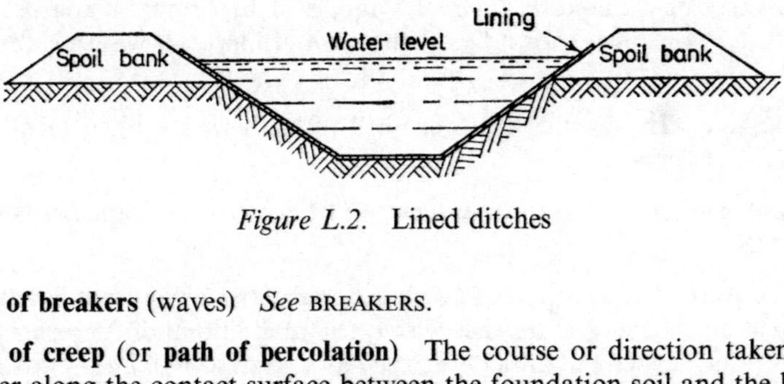

Figure L.2. Lined ditches

line of breakers (waves) *See* BREAKERS.

line of creep (or **path of percolation**) The course or direction taken by water along the contact surface between the foundation soil and the base of a dam or other structure. *See* PIPING, TUNNELLING.

lining A layer or cover over the whole or part of the perimeter of a conduit or a reservoir to minimize seepage losses, resist erosion, withstand pressure and in general improve flow conditions. On canals, a lining of clay, concrete, polythene or other material may be used. *See* REVETMENT, RIPRAP.

lip (1) A small structure, such as a wall, to break the flow on the downstream end of the APRON. (2) Part of a dam.

liquefaction The transformation of a cohesionless soil, such as saturated sand, into a fluid mass. It may be caused by a sudden shock or vibration. Liquefaction is a potential cause of failure of foundation and earthworks in areas of seismic activity, and should always be considered in planning and construction. *See* QUICK CONDITION, QUICKSAND.

liquid level recorder *See* WATER STAGE RECORDER.

liquid surface profile *See* SURFACE PROFILE.

liquor Wastewater or effluent, usually after separation from sludge or main solids. *See* MIXED LIQUOR.

listening methods Methods used to detect underground leakage of pipes. The devices used include listening rods and stethoscopes fitted with earphones. *See* WASTE.

lister An earth-moving machine in which the blades throw the soil in opposite directions, thus forming a series of alternate ridges and furrows.

lithological map A GEOLOGICAL MAP showing the main rock types of an area, their general distribution, and the principal lines of faulting and folding. Where local water supplies are involved, the arrangement of the pervious and impervious rocks and any fault lines are important. The rock types may be denoted by signs or symbols or by colour. Water-bearing rocks are sometimes shown by a light wash of blue. *See* WATER PROSPECT MAP.

lithosphere The Earth's crust or shell of hard rock with a mantle of superficial deposits.

littoral current A flow of water parallel to the seashore. *See* SHORE CURRENT.

littoral drift The transport of beach material along a shore belt by marine current and waves. Interference with the normal littoral drift by engineering works may lead to difficulties, such as heavy erosion. *See* LONGSHORE DRIFT.

littoral zone The shallow zone of a body of water where light penetrates to the bed, so that rooted vegetation can occur.

livestock reservoir *See* TANK (EARTH), FARM RESERVOIR.

LLW *See* LOWER LOW WATER, LOWEST LOW WATER.

load (stream) The weight or volume of material carried or transported by a stream per unit of time. The load may be moved (1) in SUSPENSION when consisting of mud, silt, or sand; (2) in chemical or colloidal solution, when consisting mainly of calcium and magnesium carbonates and small quantities of other salts, such as sodium chloride; (3) by traction when pebbles and boulders move along the bed of the stream, and (4) by saltation or the intermittent movement of heavy particles by turbulence and eddy currents. *See* TRANSPORT COMPETENCY.

loaded filter (or **reversed filter** or **weighed filter**) A GRADED FILTER placed at the foot of an earth dam or similar hydraulic structure. Its weight stabilizes the toe of the dam and its permeability prevents any water being trapped under pressure.

local base level The level of a lake or body of water into which a stream flows. *See* TEMPORARY BASE LEVEL.

local inflow A flow of water into a stream between two gauging stations.

location map *See* INDEX MAP.

loch A lake; a bay or arm of the sea (Scotland).

lock A chamber between two stretches of a canal, with an upper and lower gate, through which small ships or barges can pass in either direction. *See* DOUBLE LOCK, HEAD BAY.

lockage The water needed to fill a lock, and so that which flows from the upper to the lower reach of a canal during the passage of a vessel through the lock.

lock bay *See* CHAMBER.

lock cut (or **cut**) A short length of canal excavated near a river to enable boats to bypass a weir and pass through a lock.

lock gate A gate or movable barrier which separates the water in an upstream or downstream stretch from that in the lock chamber.

lock paddle A SLUICE to enable a lock chamber to be filled or emptied. *See* PADDLE.

lock sill (or **clap sill** or **mitre sill**) The step in the floor of a lock CHAMBER forming a stop against which the gates bear when closed.

lodge A pumping station or the storage sump near the pit-bottom of a mine or other pumping centre. *See* SUMP.

log A float and line for measuring the speed of a ship; a record of the rate of a ship's speed. *See* INTEGRATION METHOD.

log chute (or **log-way**) A bypass maintained around or through a dam for the passage of driftwood or logs.

log line (or **sounding wire**) A length of flexible wire with a weight at one end for measuring the depth of water in streams, canals, etc. *See* HAND LEAD, SOUNDING LINE.

log normal distribution A statistical distribution in which the natural logarithms of a data set, and not the actual data, are found to follow a NORMAL DISTRIBUTION (Figure L.3). Many water resource and hydrological parameters, such as monthly rainfall totals or stream flow totals, are often observed to be log normally distributed.

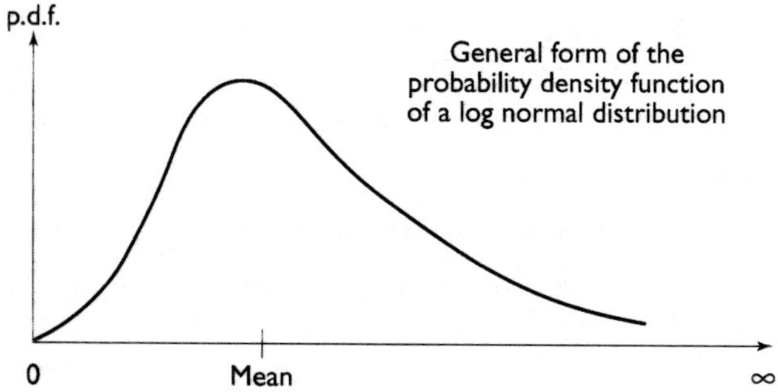

Figure L.3. Log normal distribution

logs, bush *See* BUSH LOGS.

log training wall A TRAINING WALL to prevent erosion of river banks, consisting of pyramidal trestles built of logs about 150 mm (6 in) thick and about 3.7 to 6 m (12 to 20 ft) apart, with other logs wired from trestle to trestle on the river side. The wall induces the deposition of shingle, especially during floods, and an embankment is formed. *See* WOOLF SYSTEM.

log-way *See* LOG CHUTE.

longitudinal crack A form of crack in an earth dam or embankment which develops more or less parallel to the axis of dam. It is not as dangerous as a TRANSVERSE CRACK but can be troublesome. *See* SETTLEMENT.

longitudinal section (or **longitudinal profile**) A section drawn vertically through the centreline of a canal, conduit, dam or pipeline. It shows levels along the original and final ground. *See* PROFILE.

longshore drift The transport of material by waves running obliquely up the beach. The dominant direction of the drift is determined by the prevailing winds, of moderate intensity, blowing obliquely on the beach. *See* LITTORAL DRIFT.

loose rock dam A rubble dam; a dam constructed with rock material without the use of mortar. *See* ROCKFILL DAM.

loose stone Large or small stones which are not restrained by wire mesh, timber work or vegetation. When used for groynes and river bank protection, they are usually secured to prevent dislodgement by water flow and waves. *See* STONEMESH CONSTRUCTION.

losses, flood *See* FLOOD LOSSES.

loss of head *See* FRICTION HEAD, HYDRAULIC FRICTION.

loss rate, flood *See* FLOOD LOSS RATE.

lost head *See* FRICTION HEAD, HYDRAULIC FRICTION.

lower critical velocity *See* HIGHER CRITICAL VELOCITY.

lower high water (LHW) The lower of two high waters of any lunar day.

lower low water (LLW) The single low water, occurring daily during periods when the tide is diurnal; the lower of the two low waters of any lunar day.

lowest low water (LLW) The lowest known water level.

low-heat cement A cement with a lower heat generation during setting than ordinary Portland cement. It is often used in concrete dams because the lower temperature rise reduces the risk of consequent shrinkage cracks and leakage. Low-heat cement is a coarse-ground low-lime cement. *See* TRIEF PROCESS.

low-level sprinkler irrigation Irrigation by means of fine jets of water issuing from small holes drilled in small-bore rigid plastic pipes laid at ground level. The method gives a larger volume of wetted soil and is less liable to blockage than TRICKLE IRRIGATION. Adjustment of the pressure head, by means of a hand valve, allows the trajectory of the jets to be altered. *See* LAY-FLAT TUBE IRRIGATION.

low water (LW) The lowest water level within the period specified; the lowest level reached by a falling tide.

LW *See* LOW WATER.

lysimeter An installation for the direct measurement of several parts of the hydrological cycle within the soil, including the losses to evaporation, infiltration, surface runoff, the removal of soluble constituents and changes in water storage in the upper horizons of the soil. Lysimeters comprise a large block of undisturbed local soil, preferably covered by natural vegetation, set into the ground and sealed on all four sides and underneath.

The foundation supporting the soil mass enables it to be continually weighed, and, together with observations of the rainfall on the surface and the seepage percolating to the bottom, which is drawn off and measured, the parameters can be measured. Lysimeters are large and expensive installations, and are normally used only at research establishments.

M

macrophytes Plants other than algae, mosses and liverworts.

macropores Rock pores greater than 0.005 mm. *See* MICROPORES, VOIDS.

maelstrom A WHIRLPOOL on the west coast of Norway; any great whirlpool.

magmatic water Water released by molten igneous rocks or magma during the process of cooling and consolidation. *See* JUVENILE WATER.

magnetic method A GEOPHYSICAL PROSPECTING method which detects local changes or anomalies in the Earth's magnetic field. The differing magnetic properties of many rocks allow this method to be used to trace many geological structures such as igneous intrusions, fissured zones forming aquifers, or igneous dikes which form GROUNDWATER DAMS.

main canal *See* CANAL, MAIN IRRIGATION.

main drain A land drain or sewer which leads directly to its point of discharge or OUTFALL.

main pipes (irrigation) The pipes which convey the water from the storage or pumping source to the distribution or discharge points. They may be carried on concrete or brick piers, laid on the surface or buried about 0.6 m (2 ft) below the surface. Main pipes are usually aluminium, fibre cement, plastics, iron or steel, the last two often protected with bitumen layers. *See* RING MAIN.

mainspring A spring the water from which is derived from an underlying stratum, and is thus unaffected by local variations of rainfall. The water is usually potable, but pollution may occur at the outlet.

maintenance The continuous inspection and repair of water supply schemes and river control works.

maintenance of groynes The regular inspection and maintenance of groynes, particularly during the early stages until they are well stabilized. The head or water end is particularly vulnerable to undermining and other damage, and repairs are carried out as soon as possible. *See* AUXILIARY PROTECTION.

maintenance works (land drainage) Works and activities to restore the ravages of nature and to maintain the efficiency of a river system for optimum drainage and plant growth. They may include (1) deepening of channels by removal of silt and shoals, (2) reduction of silting by removal of obstructions, such as fallen trees and weeds, and (3) prevention of damage to banks and bed by fencing and protective works. *See* GROYNES, LEVEE.

main water-table (or **phreatic surface**) The surface or water-table of the ZONE OF SATURATION; it is not applicable to the PERCHED WATER-TABLE.

major dam *See* LARGE DAM.

major divide *See* DIVIDE.

management *See* CATCHMENT MANAGEMENT, GRAZING MANAGEMENT, RIVER ENGINEERING, RIVER IMPROVEMENT, RIVER TRAINING, WATER CONSERVATION.

manhole *See* INSPECTION CHAMBER.

manhole cover A movable cast-iron or steel plate over a manhole or inspection chamber. The plate fits into a cast-iron frame bedded in concrete. In the case of foul drains, the cover is usually formed with a seal to prevent the escape of foul gases.

manned cableway (gauging) A cableway where the operator controls the CURRENT METER from a car which travels on the cable. A special grip enables the car to be moved along the cable by hand.

Manning's formula A formula expressing the relation between average velocity of water flowing in an open channel V, hydraulic mean depth R, and slope of the HYDRAULIC GRADIENT S, containing an empirical constant N, as used in KUTTER'S FORMULA, and varying according to CHANNEL ROUGHNESS:

Metric units:

$$V = \frac{1}{N} R^{0.67} S^{0.50}$$

UK units:

$$V = \frac{1{\cdot}486}{N} R^{0.67} S^{0.50}$$

See CHÉZY'S FORMULA, GAUCKLER-MANNING FORMULA, KUTTER'S FORMULA.

manometer A U-shaped tube for measuring pressure differences in a water pipe. The tube contains a liquid which does not mix with water, such as kerosene, and its two ends are connected to the two points in the pipe between which the difference in pressure is required.

map *See* PLAN.

map, hydrogeological *See* HYDROGEOLOGICAL MAP.

map symbols The letters, numbers or other signs marked on a map to denote rock types or soils, or indicate structural features. The standardized symbols used on a particular map are printed in the margin with the definitions. *See* LITHOLOGICAL MAP.

marigraph A gauge which records the level of the tides at a tidal observation station.

marine borers Animals, such as molluscs and crustaceans usually inhabiting warm waters, which damage timber structures by eating into them; the shipworm and gribble are typical, and in favourable water penetrate almost any wood. The usual remedy is treatment with creosote and similar substances, or the use of concrete as a structural material. *See* SHEATHING.

marine climate The climate prevailing over land areas near the sea. *See* CLIMATE (OCEANIC).

marine surveying (or **marine hydrographic surveying**) The study and mapping of the ocean floor, and the charting of current, etc. *See* HYDROGRAPHY, OCEANOGRAPHY.

maritime plants Plants which grow in salty conditions such as foreshores. They often serve a useful purpose in reducing or preventing scour, and may be cultivated to form a protective covering to soil or rocky surfaces. *See* REVETMENT.

Markov process (or **Markov chain**) A type of stochastic TIME SERIES in which the successive terms in the series are in a fixed relationship to one another. In a Markov chain such as χ_{t-4}, χ_{t-3}, χ_{t-2}, χ_{t-1}, χ_t, χ_{t+1}, χ_{t+2}, χ_{t+3}, χ_{t+4}, the relationship between successive terms is given by

$$\chi_t = A + B\chi_{t-1} + \check{s}$$

where χ_t = the tth term in the Markov chain, χ_{t-1} = the term $[t-1]$th in the Markov chain, A = a constant, B = a serial correlation coefficient, and $\check{s}$ = a random component. The random component is sampled from a statistical distribution selected to give the terms the required size and variation. Markov chains, or similar but more complex mathematical expressions, are commonly used in STOCHASTIC MODELS and NUMERICAL SIMULATION MODELS.

marsh A flat, usually low-lying area of land, often covered with water and supporting a native growth of rushes, reeds, or grasses. The waterlogged condition may indicate that the water-table is at or slightly above ground level. *See* SWAMP, WATERLOGGED.

masonry dam A dam constructed with stone bedded in mortar. The 43.9 m (144 ft) high Vyrnwy Reservoir in North Wales (UK) is one example of this kind of dam.

mass centre In hydrology, usually the centre of gravity of the mass curve.

mass diagram (or **mass curve**) A graph showing cumulative flow quantities, such as the integration of a time–flow curve. Each point on the graph is the sum of all preceding flows. The diagram is of value in water storage studies. *See* DISCHARGE MASS CURVE, PRECIPITATION MASS CURVE, RESIDUAL MASS CURVE.

mass haul diagram A graphical method for estimating if the volumes of 'cuts' and 'fills' in planned earthworks are balanced, the distances excavated material will have to be moved to where is should be placed as fill and, if balance is not possible, the quantities that must be imported or disposed of off site. The basis of the method is to plot the sum of volumes of cut (taken as +) and fill (taken as −) with distance along the length of the earthworks. *See* BALANCED EARTHWORKS.

mass discharge curve *See* DISCHARGE MASS CURVE, MASS DIAGRAM.

massive buttress dam *See* BUTTRESS DAM.

mass movement *See* EARTHQUAKE, LANDSLIDE.

materials surveys The investigation and appraisal of construction materials available at a proposed site. For example, a large dam would require vast quantities of materials, the winning and placing of which would represent a high proportion of the total cost. The surveys would include the availability, at a reasonable cost, of rock for use as rockfill and as coarse aggregate for concrete, sands for concrete, fine aggregate, and also clays and shales suitable for earthfill. The dam design is often adapted to make the maximum use of local material, and this is only possible after careful materials surveys. *See* CONSTRUCTION MATERIALS, DAM SITE INVESTIGATION.

mattress A layer or blanket of brushwood, interwoven or otherwise lashed together, and weighed down with concrete blocks or stones to reduce erosion on the bed and banks of rivers or used with GROYNES. *See* FASCINES, REVETMENT.

maturation pond A large shallow basin for the treatment of wastewater after biological treatment and removal of solids. *See* LAGOON.

mature river A river with a velocity just sufficient to transport sediment delivered by the tributaries; a river with a meandering course, absence of lakes and waterfalls, and which has attained a PROFILE OF EQUILIBRIUM. *See* AGE (of **stream**), OLD RIVER.

maximum capacity (of well) The maximum rate at which water can be withdrawn from a well, expressed in cumecs or cusecs, litres per minute or gallons per minute. *See* ECONOMIC YIELD, TOTAL CAPACITY.

maximum envelope curve *See* ENVELOPE CURVE.

maximum probable flood The heaviest flood that can reasonably be expected to occur along a river or selected stretch of a river. *See* RETURN PERIOD.

maximum probable rainfall The rainfall of a certain amount and duration that can reasonably be expected to occur under present-day climatic conditions in a CATCHMENT AREA.

maximum rainfall intensity The rate of rainfall during any period when the rate is both maximum and uniform. *See* PRECIPITATION (EXCESSIVE), RATE OF RAINFALL.

meadow strip A strip of grassed land or sloping field, usually much larger than a GRASSED WATERWAY and often used as a terrace outlet channel. In addition to yielding a hay crop, the strip acts as a shallow, broad water channel for runoff.

mean annual flood The arithmetic mean or average of a series of annual maximum instantaneous flows.

mean depth The cross-sectional area of water in a stream or channel divided by its surface width.

meander The loop or curve taken by a river as it flows over flattish areas, generally occurring when a river has cut down nearly to BASE LEVEL. *See* FULL MEANDER.

meander belt The valley area enclosed by lines drawn tangentially to the outer sides of loops or bends of a river. The meanders of some large rivers have now been trained between embankments and walls along certain sections, to prevent further encroachment on adjoining land areas. *See* LATE MATURITY.

meander concertina A MEANDER MIGRATION arrested by a natural or artificial obstruction on the river (e.g. a bridge), which results in a crowding-up of the meanders upstream of the obstruction (Figure M.1).

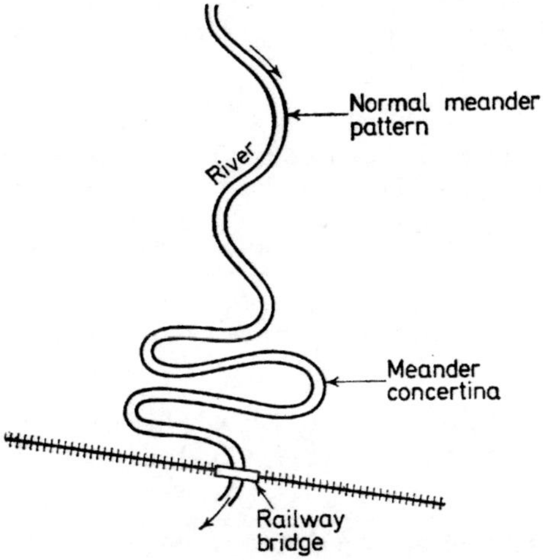

Figure M.1. Meander concertina

meander, full *See* FULL MEANDER.

meander length The length between corresponding points at the outer sides of successive fully formed river meanders.

meander migration The tendency of river meanders to move slowly downstream. A comparison of old and new maps of the same river stretch often shows this downstream shift of the meanders.

meander ratio The ratio expressing relation of MEANDER WIDTH to MEANDER LENGTH.

meander width The width or extent of swing of a fully formed meander measured from midstream to midstream. Also called AMPLITUDE.

mean higher high water (MHHW) The mean level of higher high waters over a long period or within the period considered.

mean high water (MHW) The mean level of high waters over a long period or within the period considered. With semi-diurnal or mixed tides, only higher high water levels are included; in this case, mean higher high water is the same as mean high water.

mean lower low water (MLLW) The mean level of lower low water within the period considered or over a long period.

mean low water (MLW) The mean level of low waters within the period considered or over a long period. With semi-diurnal or mixed tides, only lower low water levels are included; in this case, mean lower low water is the same as mean low water.

mean sea level The level of the surface of the sea if it was stationary with no tides; the average sea level determined by measurements at equal intervals of time over a long period.

mean tidal range The difference in level between MEAN LOW WATER and MEAN HIGH WATER.

mean tide level The level midway between MEAN HIGH WATER and MEAN LOW WATER.

mean velocity At a given section of a stream the mean velocity is obtained by dividing the discharge of the stream by the cross-sectional area at that section. *See* CURRENT METER.

mean velocity on a vertical Mean velocity obtained by slowly moving the current meter at a specified rate along the vertical, or by taking the mean of the velocity at two or more points on the vertical.

mean velocity point The point lying between the bed of channel and water surface, at which the velocity is equal to the MEAN VELOCITY ON A VERTICAL.

mean water (MW) The average level of the surface of still or flowing water. *See* MEAN SEA LEVEL.

measuring section The section used for discharge measurements. *See* GAUGING STATION.

measuring weir A structure or device for measuring rate of flow of water, consisting of a triangular, rectangular or trapezoidal notch in a thin vertical plate, and normally placed at right angles to the flow. The rate of flow is calculated from the depth of water in the notch. *See* CIPOLLETTI WEIR, CRUMP WEIR, RECTANGULAR WEIR, TRIANGULAR NOTCH.

mechanical control *See* ENGINEERING PRACTICES.

medial moraine The middle moraine formed when two glaciers join, consisting of two lateral moraines.

meeting post (or **mitre post**) The vertical timbers at the outer end of a pair of LOCK GATES, mitred so the gates fit tightly when closed. *See* LOCK SILL.

meiyu Literally 'plum rain' in Chinese, referring to the season when heavy rains are required to ripen the plum. This season is about June and July, and the term is applied in the provinces south of the Yangtze. *See* BAI-U.

membrane filtration A simplified routine method of bacteriological analysis of water samples for evidence of pollution by filtering through a membrane capable of retaining bacteria.

Menard pressure permeameter An appliance which measures directly the permeability to water or gas of any rock in a borehole. The field determinations are confirmed by laboratory tests.

mere A pool, pond or lake. In areas where local subsidence occurs, due to shallow mining and other causes, meres are common and are continually extending.

meteoric water The water which falls as rain or is derived from dew, hail or snow. *See* PRECIPITATION.

meteorograph A self-recording instrument or apparatus for recording meteorological phenomena, such as temperature, pressure and humidity; the term is often applied to apparatus used for recording data in upper atmosphere, especially for weather forecasts.

meteorology The scientific study of the atmosphere and its phenomena, mainly pertaining to WEATHER and CLIMATE. *See* HYDROMETEOROLOGY.

meteorology, synoptic A general study or survey of atmospheric processes based on weather conditions prevailing over a widespread area at a given time.

meter A device placed in a pipeline or channel to measure the rate of flow of water. *See* CURRENT METER, FLOW METER.

methane An inflammable, odourless gas, CH_4, which can be derived from sludge digestion and utilized for heat energy.

method of measurement In civil engineering works it is unusual for the total amount of money to be paid to the contractor to be fixed at the start. Normally the amount of work actually done within a large number of categories of work (set out in a Bill of Quantities) is estimated or measured as the work proceeds. The contractor is paid for this at unit rates. As measurement directly fixes the amount paid for the work, it is essential that the procedure or method of measurement is agreed in detail when the contract is placed. Wherever possible it is recommended that a well-established standard method is used; such as the UK Civil Engineering Standard Method of Measurement (CESMM). Otherwise the method of measurement must be set out in the contract.

methylene blue test *See* STABILITY TEST.

MHHW Mean higher high water.

MHW Mean high water.

microbial film *See* BIOLOGICAL FILM.

microclimate The climatic conditions pertaining to a particular place, caused by local modifications in exposure and altitude.

microclimatology The science or study of MICROCLIMATES.

micrometeorology The study of METEOROLOGY pertaining to small areas, such as towns, cities, catchment areas or a river system.

micropores Rock voids smaller than 0.005 mm. If the greater part of the total porosity of a rock consists of micropores, the passage of water through it is inhibited. *See* MACROPORES.

mid-bay bars Banks of sand, silt or gravel formed along the middle of bays. *See* BAY-HEAD BARS.

middle third That portion of the thickness of a GRAVITY DAM which is one third of the total thickness and is centrally located in it. For the stability of a gravity dam, one requirement is that the resultant force must lie within the middle third of the dam (Figure M.2).

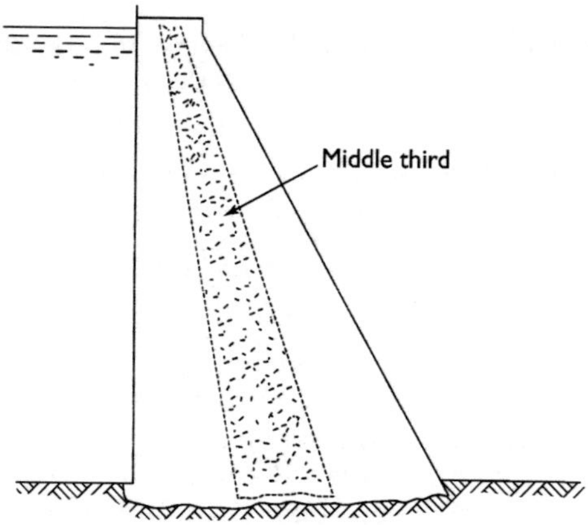

Figure M.2. Middle third

middle thread A line along the course of a stream that would represent the final flowing of its waters immediately before cessation (a legal term).

migration, of meanders *See* MEANDER MIGRATION.

millibar A unit of atmospheric pressure used in meteorology, equalling 10^2 N/m^2, or approximately 1/32 in of mercury.

mills *See* MOULINS.

mill stream *See* LEAT.

mill tail *See* TAIL RACE.

mineral spring A spring with a relatively high content of dissolved mineral matter, and often named after the dominant mineral.

miner's inch A rate of flow of water, originally the discharge from a 1 inch square orifice under a specified head (1 in = 25.4 mm). The unit is used mainly in the western USA, where its volumetric flow value has been fixed by statute in various states.

minimum envelope curve *See* ENVELOPE CURVE.

minimum speed of response The minimum speed at which the angular motion of the rotor of a CURRENT METER remains uniform and continuous. *See* SPIN TEST.

minor dam Sometimes, a dam constructed for small country-town water supply schemes; a dam of smaller size and capacity than a LARGE DAM. A DESIGN RECURRENCE INTERVAL of 500 years is sometimes recommended for minor dams where surcharge is likely to cause loss of life. *See* DESIGN FLOOD.

minor watershed (USA) A subdivision of a watershed which is convenient for investigation, description and detailed planning of water and soil conservation measures; a subdivision of a drainage area of unspecified size. *See* WATERSHED.

'misfit' stream A stream which has lost its headwaters by RIVER CAPTURE and has diminished in size, and is manifestly too small for the valley in which it now flows.

mist *See* FOG.

mitre post *See* MEETING POST.

mitre sill *See* LOCK SILL.

mixed-flow pump A centrifugal pump in which the head is developed partly by centrifugal force and partly by the lift of the vanes, used for pumping large volumes to medium heads.

mixed-flow turbine An inward-flow reaction water turbine in which the water pressure acts both radially and axially on the runner vanes.

mixed liquor A mixture of settled sewage and activated sludge circulating in the aeration tank or channels during activated sludge treatment. *See* LIQUOR.

mixed tides Tides which change, according to the declination of the moon, from diurnal at some periods to semidiurnal at others.

MLLW Mean lower low water.

MLW Mean low water.

model analysis The study of scale models of hydraulic structures, the measurement of the heads and discharges of water, and finally the interpretation of the results to predict the characteristics of the full-size or actual structures. Model analysis is used for dam spillways, river improvement works, harbours and so on. *See* HARBOUR MODELS.

modelling The use of models, in engineering and architecture, as copies or reproductions of a prototype, usually to a very much smaller scale, which can greatly assist in planning and design. The main objectives of modelling include the measurement of technical data for use in design, and aiding an appreciation of the appearance and function of the full-sized works. The two main types of models now commonly in use are PHYSICAL MODELS and NUMERICAL SIMULATION MODELS, the latter being based on CONCEPTUAL MODELS of a parameter or process.

modified Pacific type (coastline) Often, referring to the coastline of Asia which lies behind a chain of islands along the western shores of the Pacific Ocean. *See* PACIFIC-TYPE COASTLINE.

modified velocity The velocity after adjustment for ANGULARITY CORRECTION and DRIFT.

modular flow Flow conditions through a flume or over a weir when the level upstream is independent of the level downstream. *See* DROWNED FLOW.

modular limit The SUBMERGENCE RATIO when the downstream level just begins to affect the flow.

module A contrivance for delivering water at a constant head through an opening 1 in (25.4 mm) high of variable width. *See* MINER'S INCH.

moisture Water diffused in small amounts as vapour or liquid in rocks, soils, or atmosphere. *See* HUMIDITY.

moisture adjustment Modification of observed rainfall in a storm, over a drainage area under study, by use of the ratio of estimated maximum precipitable water to actual precipitable water; it may be used in estimating MAXIMUM PROBABLE RAINFALL. *See* DESIGN FLOOD.

moisture deficiency, field *See* FIELD MOISTURE DEFICIENCY.

moisture equivalent, field *See* FIELD MOISTURE EQUIVALENT.

moisture, field *See* FIELD MOISTURE.

moisture gradient The rate of change in moisture content of a soil with depth.

moisture index (or **Thornthwate index**) An index that indicates the adequacy of precipitation for the requirements of plants.

moisture index, zero The MOISTURE INDEX when rainfall, analysed monthly, is just sufficient to meet maximum transpiration and evaporation over one year.

moisture movement Applied to the increase in length or volume of a material with increased moisture content. *See* BULKING, SWELLING PRESSURE.

moisture regime (or **water regime**, irrigation) The specific state that a crop is said to be subject to when the water content of the soil is permitted to fluctuate only between FIELD CAPACITY and a predetermined state of dryness. The regime is said to be 'dry' where the fluctuation is large, and 'wet' where the fluctuation is small. Maximum plant growth is obtained when the soil is maintained near to field capacity.

moisture status Sometimes, the wetness or dryness of a CATCHMENT AREA, with particular regard to its capacity for absorbing rainfall and preventing this rainfall from becoming runoff.

mole *See* BREAKWATER.

mole drainage (or **moling**) The draining of a stiff clay by forming tubular channels, 75 to 150 mm (3 to 6 in) in diameter, in the soil at a depth of from 0.6 to 0.9 m (2 to 3 ft) by means of a MOLE PLOUGH (Figure M.3). The method is much quicker and cheaper than laying ordinary agricultural drains.

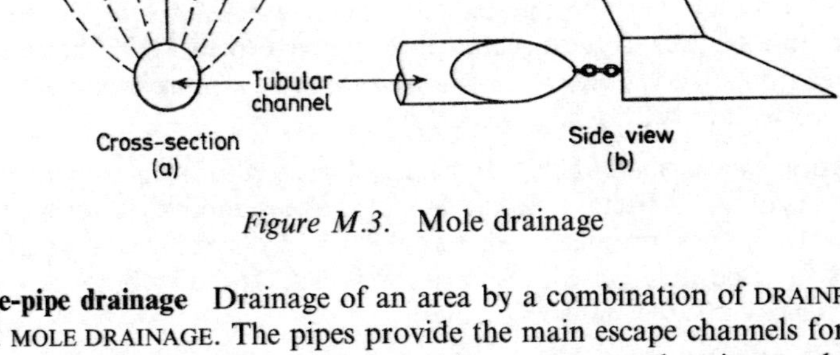

Figure M.3. Mole drainage

mole-pipe drainage Drainage of an area by a combination of DRAINPIPES and MOLE DRAINAGE. The pipes provide the main escape channels for the water; the mole drains provide secondary routes to the pipes and also fracture the soil to increase the rate of percolation. *See* SUBSOIL DRAINAGE.

mole plough (or **mole plow**, USA) An excavating machine or plough consisting of a vertical knife blade with a horizontal bullet-shaped bottom member, from 75 to 150 mm (3 to 6 in) in diameter, which is hauled through the soil for MOLE DRAINAGE. The plough is pulled uphill along the course of the desired drains by a caterpillar tractor. Mole drains can also be used to lay copper or polythene water pipes.

moling *See* MOLE DRAINAGE.

monolith A large hollow rectangular foundation structure, usually of masonry or concrete, with several wells or compartments. It is sunk through water and ground by excavating from within with grab buckets, and sealed around the periphery when solid rock is reached. Hollow monoliths can be designed with sufficient buoyancy to be towed to sites in deep water and then sunk to foundation level. *See* OPEN CAISSON.

monsoon A wind blowing with regularity and persistence during a definite season of the year; especially in the Indian Ocean, the wind blowing from the north-east in winter (dry monsoon) and from the south-west in summer (wet monsoon); a rainy season with winds.

moraine The rock and soil debris which falls on to the surface of a glacier from the valley slopes. It forms two marginal layers or ridges, termed lateral moraines, parallel to the sides of the glacier. In narrow valleys, the glacier surface is often completely covered by slide or scree material. *See* MEDIAL MORAINE.

morass A bog or marsh; an area of soft, wet ground. *See* MARSH, SWAMP.

Moringa oleifera A deciduous tree the seeds of which are a useful coagulant when powdered by grinding, and can be used both at the household level and in small treatment plants to clarify turbid water. *See* COAGULATION.

mosaic A map composed of a number of AERIAL PHOTOGRAPHS which have been enlarged to the same scale and fitted together. It may be either uncontrolled or controlled: an uncontrolled mosaic is cheaper to produce but cannot be used to give accurate measurements because it is not related to any ground survey control; a controlled mosaic is more costly but gives detailed features with greater accuracy.

most probable number (MPN) In bacteriology, a statistical estimate of the numbers of viable bacteria. In the case of the presumptive coliform count, a series of tubes are incubated; the MPN is deduced by examining the tubes for a positive growth response, such as the production of acid and gas, for reference to statistical tables.

moulins (or **mills**) Roughly circular and vertical pits, hollows or cavities formed in the bed of a glacier by the plunging of water through holes in the ice; nearly vertical holes in a glacier formed by running water. *See* POTHOLES.

mound breakwater *See* RUBBLE-MOUND BREAKWATER.

mound, groundwater *See* GROUNDWATER MOUND.

mound spring A spring around which a mound of mineral matter has been precipitated by the spring water or formed by wind-blown sediments and vegetation (Figure M.4).

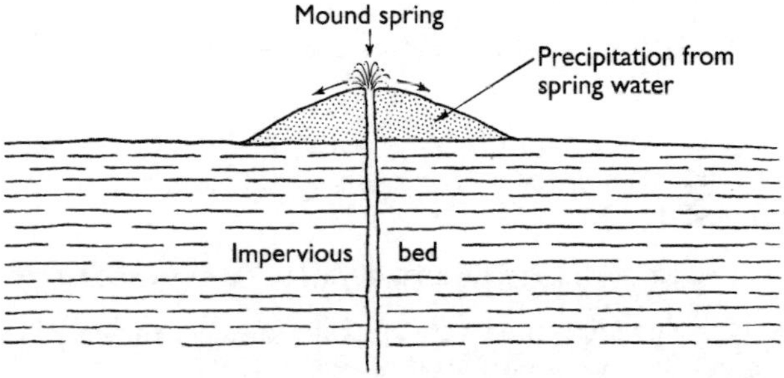

Figure M.4. Mound spring

mountain glacier A glacier occupying a valley between mountains which rise above the SNOW LINE. *See* CIRQUE GLACIERS.

mountain tract The stretch of a stream along its source in the mountains. The water is swiftly flowing and has carved a steep V-shaped depression. In this region, a stream is active in draining the valley slopes and in the erosion and transportation of materials, and even flood waters are easily constrained by the steep gorge section (Figure M.5). *See* PLAIN TRACT, VALLEY TRACT.

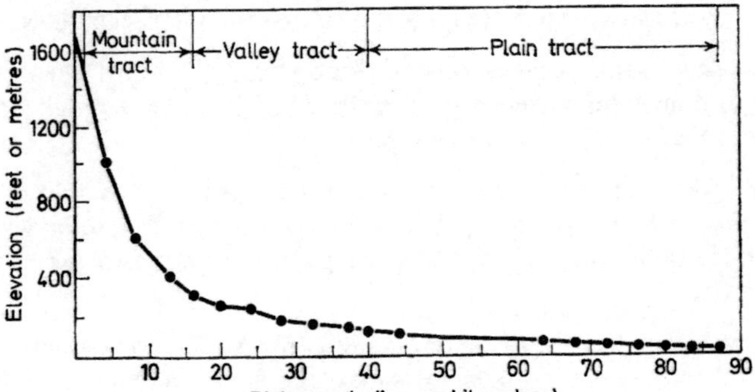

Figure M.5. Mountain tract

movable bridge A bridge which can be moved bodily to permit the passage of ships. *See* BASCULE BRIDGE, SWING BRIDGE, TRAVERSING BRIDGE.

movable dam A barrier which can be removed wholly or in part to allow increased flow during floods. The movable sections may consist of stop logs, gates or any other appliance whereby the area for flow may be increased or controlled. *See* FLASHBOARD.

MPN Most probable number.

muck Dirt, earth or waste material. *See* BACK CUTTING, BACKFILLING, FILLING.

mud A fine-grained, soft earthy mass with a high content of water. It may consist of sand, clay, limestone, or a mixture of these and other materials in any proportion. *See* SLIME.

mud flat (or **estuarine flat**) The area of floor of a bay exposed at low tide.

mudflow The downhill movement of waterlogged sediments and debris, which may result from runoff after heavy rains. *See* LANDSLIDE.

Mulberry harbour A prefabricated harbour used as a bridgehead by the Allies in the Normandy landings during June 1944. The pre-cast concrete sections were made in the UK and then towed to the Normandy coast of France, and there assembled and sunk offshore.

mulching The prevention or reduction of evaporation from soil by spreading a protective layer of small stones, paper, straw, or plant residue on the surface of the soil.

multiple-arch-type dam A lightweight dam consisting of a series of arches supported by parallel buttress walls or piers. The arches transfer the load to the foundations through the buttresses. *See* ARCH DAM, COUNTER-ARCHED REVETMENT, FLAT-SLAB DECK DAM, ROUND-HEAD BUTTRESS DAM.

multiple-dome dam A development of the MULTIPLE-ARCH-TYPE DAM consisting of domes supported by buttresses. The 76 m (250 ft) high Collidge dam in Arizona, USA, is of this type.

multiple purpose reservoir A reservoir constructed to serve two or more purposes, such as water supply, flood control, irrigation, navigation, recreation, pollution control and perhaps power purposes. *See* SERVICE RESERVOIR.

multi-storage system A system of water conservation comprising the use of several reservoirs and associated works. Each unit is planned to operate with the others to achieve maximum economy and efficiency in water use. *See* SIMPLE STORAGE SYSTEM, WATER-RESOURCE SYSTEM.

Muskingum method A semi-empirical method of flood routing, suitable for natural rivers, which was developed during the 1930s in the USA. It uses a series of constants derived from observations taken at the upstream and downstream ends of a channel reach or river reach, during the passage of a known hydrograph. These can then be used in a fairly simple equation to predict the routing of any other, usually larger, hydrograph as it flows down the same reach.

MW Mean water.

N

nail-bournes Temporary streams. *See* BOURNES.

nappe (or **sheet**) The sheet of water flowing over the crest of a wall, dam or weir erected across a stream, having an upper and a lower surface.

narrow-base terrace Similar to a BROAD-BASE TERRACE, except that the base width is normally from 1.2 to 2.4 m (4 to 8 ft).

narrows The narrow channel cut by a stream across a highly inclined or vertical hard bed. Since the resistant rock is roughly vertical, the outcrop does not shift as erosion in the narrows proceeds. The stream is considerably wider in the softer beds on each side of the narrows. *See* RAPIDS, WATERFALL.

natural escape *See* BY-WASH.

natural harbour A sheltered area of sea formed by a favourable form of coastline. *See* ARTIFICIAL HARBOUR, HARBOUR OF REFUGE.

natural levee *See* LEVEE.

natural load The material carried or transported by a stable stream. *See* LOAD.

natural recharge The natural infiltration of surface water or rainfall into underground rocks or aquifers. *See* ARTIFICIAL RECHARGE.

natural reservoirs Lakes which lie at or near the base of mountainous regions and from which rivers flow. They reduce the damage from floods and also increase the navigable season of the river. A good example is the chain of Great Lakes in North America, which exercise a large measure of control over the St Lawrence River.

natural slope The maximum angle at which rock or soil banks will stand without slipping; the steepest slope of the land. *See* ANGLE OF REPOSE.

natural storages Natural features which tend to smooth out the irregularity of stream flow. Natural storages may be underground or on the surface. Marshes, ponds, lakes and any depressions below flood level usually provide temporary flood storage; even a thick cover of vegetation will retard and hold part of any excessive runoff. *See* ARTIFICIAL STORAGES.

natural-strainer well A well which obtains its water from a partly incoherent deposit, and whose SPECIFIC CAPACITY has been increased by the removal of some of the fine particles around the entry section of the well. *See* GRAVEL FILTER WELL.

natural water The water which falls as rain, hail or snow, is derived from dew, or is found in rivers, springs, and surface or underground accumulations. It contains solid, liquid and gaseous substances, which it holds in solution or suspension. *See* FRESH, METEORIC WATER, WATER.

natural well A natural, roughly rounded cavity in the ground which extends downwards into a subsurface body of water. The water does not flow to the surface. *See* KARST TOPOGRAPHY.

navigation Methods of fixing a ship's position and course by nautical astronomy, etc. The term is sometimes applied to a river, the flow of which is controlled by CANALIZATION. *See* SLACK-WATER NAVIGATION.

neap range The mean difference in level between neap high water and neap low water. *See* MEAN TIDAL RANGE, NEAP TIDES.

neap tides The low or smallest tides that occur in the second and fourth quarters of the moon. *See* SPRING TIDES.

needle A heavy timber piece set either vertically or inclined in a stream, to close an opening or to control the flow. *See* FLASHBOARD, NEEDLE WEIR.

needle valve A valve used in regulating the flow to large WATER TURBINES, cone-shaped and terminating in a sharp point.

needle weir A fixed FRAMED DAM built up by heavy vertical timbers, or needles, in contact. The upstream water level may be lowered by removing the timbers as required.

negative artesian head *See* ARTESIAN HEAD (NEGATIVE).

negative confining bed A relatively impervious bed below a water-bearing deposit, where the downward movement of the water is prevented or retarded. *See* PERCHED WATER.

NERC method A method widely used in the UK for the prediction of flood flows from ungauged catchments for any required RETURN PERIOD. It was developed by the UK Natural Environment Research Council (NERC) as part of the work for the FLOOD STUDIES REPORT (1975). The method is primarily intended for use in ungauged catchments in the UK, and uses data shown on specially prepared soil surveys and series of 1:250 000 British Ordnance Survey maps. If any flow records are available they can be incorporated to improve accuracy. Predictions of peak rates of flow can either be made quickly by using a range of regression equations, or a synthetic unit hydrograph method if an approximate full hydrograph is required.

net duty (of water) (or **farm duty**, USA) The quantity of irrigation water delivered to a farm or irrigation holding, usually expressed in acre feet or cubic metres. *See* IRRIGATION REQUIREMENT.

net intake The amount of water which reaches the zone of saturation. It equals the excess of rainfall over runoff, evaporation, and transpiration.

net peak flow Total peak flow minus the corresponding BASE FLOW or sustained flow in a stream or channel.

net supply interval The rainfall period of a storm between INITIAL RAIN and RESIDUAL RAIN.

neutral pressure The HYDROSTATIC PRESSURE in the water occupying the pore space of a soil. *See* PORE WATER PRESSURE.

neutral shoreline A shoreline whose essential features do not depend on either the submergence of a former land mass or the emergence of a former subaqueous surface. *See* COMPOUND COAST.

neutral soil Precisely, a soil with a PH VALUE of 7.0; broadly, a soil with a pH value between 6.6 and 7.3.

névé Snow in a state intermediate between FIRN (loose surface snow) and crystalline glacier ice; the snow and ice that creeps down mountain slopes and forms the source of glaciers.

Nichols terrace A small embankment of earth with a gentle slope on the downhill side, a moderately deep narrow channel and a flat low ridge. *See* TERRACES.

nick point The point in a stream where there is a sudden change in gradient. This occurs at the intersection of old and new STREAM PROFILES.

night soil The contents of external pail latrines, usually removed at night.

nitrates Highly soluble salts which occur naturally through nitrification (the oxidation of ammoniacal nitrogen to nitrates and nitrites by bacteria), and are also constituents of fertilizers. The runoff of leachate from agricultural land can lead to EUTROPHICATION, and also possibly cause blue baby disease, a disorder affecting the amount of oxygen in the blood.

nominal retention level (or **full supply level**) For a reservoir where the surface level is controlled by a weir or other fixed outlet, the lowest crest level at the outlet (Figure R.4). If a reservoir outlet is controlled by gates or another type of movable structure, it is the maximum level at which water can be permanently stored when these are closed.

nomogram (or **alignment chart**) A chart comprising a set of scales for shortening or eliminating calculations. A frequent form consists of three straight lines, each graduated to represent one of the variables in an equation. A set square or straight edge joining values on two of the scales will intersect the required value on the third scale. For example, a simple nomogram may be constructed to represent the values in the equation $discharge = velocity \times sectional\ area$. Similar units are used throughout.

non-artesian well A well in which the water is not under pressure and its level in the well roughly coincides with that of the local water-table. *See* FREE GROUNDWATER.

non-capillary porosity *See* TOTAL POROSITY.

non-carbonate hardness The term which has replaced 'permanent hardness'. *See* HARDNESS OF WATER.

non-contributing area *See* CLOSED DRAINAGE.

non-dimensional unit graph *See* DIMENSIONLESS UNIT GRAPH.

non-effective storage *See* EFFECTIVE STORAGE.

non-flowing artesian well A well which has tapped water with sufficient pressure to cause it to rise in the well, but not sufficient to reach and overflow at the surface.

non-frontal precipitation Rainfall which may occur in an area of low pressure: horizontal convergence and inflow into a DEPRESSION cause a lifting or upward movement of air. *See* FRONTAL PRECIPITATION.

non-irrigable land Land which is too high to allow efficient irrigation by gravity means from existing supply channels. Land less than 150 mm (6 in) below the supply level at the farm boundary is regarded by some authorities as non-irrigable by gravity means. *See* IRRIGABLE LAND.

non-return valve (or **reflux valve**) A valve in a pipe or conduit which allows a flow of fluid in one direction only. Typically, these valves open, or lift, when the water pressure on one side rises above that on the other. This may be caused by the start of pumping, increased river flows or tidal movement. The valve closes, and backflow is prevented, when the pressure difference is reversed. CLACK VALVES (Figure C.6) are a simple type of non-return valve, and illustrate their basic principle. There is a wide range of other types of non-return valve, each to suit the particular requirements of conduit size, rate of flow, time of closure, etc.

non-steady flow A flow in which the velocity changes with respect to both time and location along an open channel.

non-storage routing A FLOOD ROUTING method using empirical relationships, such as the water levels at the inflow and outflow ends of the reach considered. *See* STORAGE ROUTING.

non-tidal river A river beyond the influence of tides. *See* TIDAL CURRENT.

non-uniform flow A flow in which the MEAN VELOCITY varies along the length of the channel.

normal aeration The entire renewal of soil air about once each hour to a depth of 200 mm (7.87 in).

normal cycle (erosion) A CYCLE OF EROSION in which water is the principal erosive agent. *See* EROSION CLASS, GEOLOGICAL EROSION.

normal depletion curve A curve showing normal loss of water from groundwater storage. *See* GROUNDWATER DISCHARGE.

normal depth That depth of water at a certain cross-section in an open channel where the discharge remains steady.

normal distribution (or **Gaussian distribution**) A very important statistical distribution, called 'normal' because it is so frequently observed to fit the variations in natural processes. This result can be predicted theoretically, as it is the distribution most likely to describe the variations in any quantity which in turn is dependent on many other factors. Its probability density function takes the form of the very well-known symmetrical 'bell curve' (Figure N.1). The normal distribution is open-ended so a variable whose occurrence probabilities are described by it may take any value between $+\infty$ and $-\infty$. It is defined by two parameters, the mean (μ) and the standard deviation (σ), a measure of the scatter of the data. Where the true values of these are not known, use is made of $\bar{x}$ and s, the best estimates from the available data. The probability density function $f(X)$ of the normal distribution is as shown. Probabilities that values of a variable X are equalled or exceeded are estimated by the integration of $f(X)$ as below:

$$f(X) = \frac{1}{\sigma\sqrt{2\pi}} e^{[-(X-\mu)2/2_{\sigma 2}]}$$

$$\bar{x} = \frac{1}{n}\sum x_i$$

$$s^2 = \frac{1}{(n-1)}\sum (x_i - \bar{x})^2$$

The probability that value X is equalled or exceeded

$$= \frac{\int_x^\infty f(X)\mathrm{d}x}{\int_{-\infty}^\infty f(X)\mathrm{d}x} = \frac{\text{area A}}{\text{total area}}$$

Once μ and σ, or their estimated values $\bar{x}$ and s, are available, pre-calculated tables for the normal distribution are very widely available to facilitate its application.

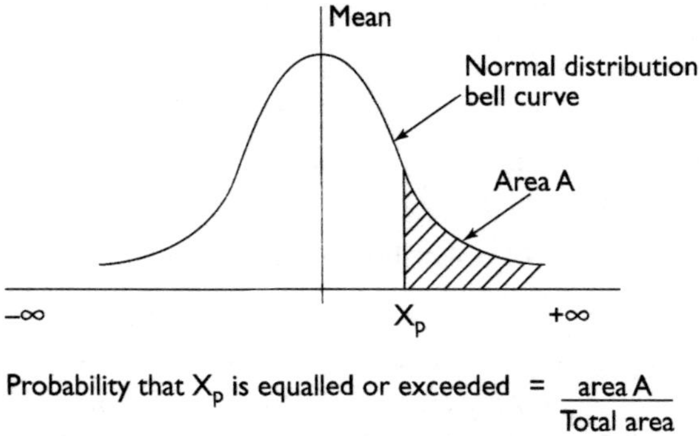

Figure N.1. Normal distribution

normal erosion *See* GEOLOGICAL EROSION.

normal flow The volume of water flowing in a uniform length of stream or channel when the water surface slope and bed slope are parallel.

normal recession curve *See* HYDROGRAPH (RECESSION).

normal soil erosion *See* GEOLOGICAL EROSION.

normal velocity of distribution The distribution of velocities attained in a straight and uniform length of river or channel between two specified points or cross-sections.

normal water level The water level which for 50% of the time during a year or a number of years is equalled or exceeded.

nose A structure closely related to an elongated dome in origin, and consisting of an upwarp of strata that are not closed, i.e. the contours are open in some direction. *See* STRUCTURE CLOSURE.

nose of groyne *See* GROYNES.

notch The opening in a MEASURING WEIR, of rectangular, triangular or trapezoidal shape.

notched weir *See* MEASURING WEIR.

notch, triangular *See* TRIANGULAR NOTCH.

notch, V *See* TRIANGULAR NOTCH.

nullah A normally dry watercourse which becomes full and swiftly flowing after heavy rain (India). *See* FLASHY STREAM.

numerical modelling *See* NUMERICAL SIMULATION MODEL.

numerical simulation model A widely used technique of MODELLING which makes use of mathematical expressions to represent the parameters of a prototype. In turn these are thus based on CONCEPTUAL MODELS. The models are most useful when predicting aspects of a prototype which can be described numerically, such as total crop yield or the volume of water stored in a reservoir. In turn these models give their best results when the parameters being modelled are dominated by only a small number of other parameters, also easily described numerically. Often this type of modelling can be done with only a pocket calculator. However, very large and complicated numerical models to predict the performance of large drainage networks and whole river systems are commonly developed, which need to run on powerful computers.

nun buoy A circular BUOY tapered at the upper and lower ends. *See* WHISTLING BUOY.

nutrient *See* EUTROPHICATION.

***N*-year flood** A flood of a magnitude which, on average, once in a given *N* number of years, is equalled or exceeded. *See* RETURN PERIOD.

O

obliteration of lakes The natural disappearance of ponds and lakes by (1) the enlargement or cutting down of the outlet; (2) evaporation, particularly in arid regions; (3) the filling of the basin with silt, plant growth or both; (4) the lowering of the surrounding water-table, and (5) a combination of two or more of the above causes. *See* PERIODIC TYPE.

observation well A well sunk to assist in groundwater investigations and studies. Checks are maintained on groundwater fluctuations and water quality, etc.

occlusion *See* FRONTAL SURFACE.

ocean A vast expanse of salt water covering the earth's crust.

ocean currents Well-defined flows of water over certain portions of the oceans, caused by differences in density of the water, slope of the sea surface and wind friction. All ocean movements are influenced by the deflective forces caused by the rotation of the earth, etc.

ocean engineering The use of engineering and scientific knowledge of the oceans for the benefit of humankind; the term includes COASTAL ENGINEERING.

oceanography The study of the oceans and their associated phenomena. It covers mapping of the ocean floors and the charting of currents and tides, leading to more accurate shipping routes, better weather forecasting and a better understanding of the mineral wealth on and under the sea bed.

oceanology That branch of science which relates to the ocean. *See* OCEANOGRAPHY.

odour control In water treatment, the removal of odour, normally caused by ALGAL BLOOM, by methods such as ALGAL CONTROL, AERATION, CHLORINATION or ACTIVATED CARBON. In the case of wastewater treatment, objectionable odours may be removed by similar methods or may require masking by chemical aerosols.

offshore terrace A deposit of sand and silt built out into deep water by the combined actions of waves and currents; the deposit is roughly parallel to the shore. *See* BEACHES.

offshore zone The zone beyond the low tide level consisting of deposits of land-derived sediment forming a terrace with a seaward slope. *See* SHORE.

offstream storage A reservoir located some distance away from a stream, usually on river flats. The reservoir is filled either by pumping or by diverting water into it by channel or other means. *See* FARM RESERVOIR, HILLSIDE DAM, RING TANK, TURKEY'S NEST.

ogee An S-shaped curve as used in ogee-jointed concrete pipes, or in elongated form for the downstream face of an overflow dam.

ogee crest The profile of a spillway crest which for a given water depth or head above the crest level gives the maximum discharge. The ogee shape approximates to the underside of the NAPPE of the free discharge over a sharp-crested weir. For a dam with an upstream vertical face, one of the many empirical expressions for this profile is $x^{1.85} = 2.0h^{(1.85-1)}y$, where x and y are co-ordinates with the origin at the highest point on the crest and h is the design head above this crest. Note that the optimum ogee shape is not independent of the design head.

old river A river with a low slope, a sluggish current, a wide floodplain and perhaps many oxbow lakes; a river which meanders widely, eroding during floods and depositing material at other times. *See* MATURE RIVER, YOUTHFUL STREAM.

on-farm storage *See* FARM RESERVOIR.

on-stream storage A reservoir located on a river or watercourse which is filled by the gravity runoff from the stream CATCHMENT AREA.

open caisson A CAISSON which may be cylindrical, open at the bottom and top, and constructed from steel plate, brickwork, mass concrete or concrete blocks. It is sunk in position to form part of the foundation structure. *See* MONOLITH, SHIP CAISSON.

open channel Any channel or conduit within which water flows with a free surface. Typical open channels are streams and canals which are uncovered open channels, while pipes and tunnels when flowing partly full are called covered open channels.

open-end well A well with a pipe form of lining where the water enters near the bottom through a screen or other arrangement in the lining. *See* PERFORATED-CASING WELL.

open pit (or **open well** or **large bore**) A drilled hole or excavated pit at least 0.9 m (3 ft) in diameter or width, which a person can descend to inspect the ground, usually to the water-table or lower if pumping is employed. *See* DAM SITE INVESTIGATION.

open well *See* OPEN PIT.

open wharf *See* WHARF.

operational log *See* DRILLER'S LOG.

operation waste The quantity of water lost from an irrigation district due to the operation of the channel system, including water passed to waste through outfalls or similar structures. *See* IRRIGATION, IRRIGATION REQUIREMENT.

optimum moisture content The moisture content of a soil at which a given amount of compaction produces the highest value for DRY DENSITY.

orifice An aperture or opening in a plate, partition or wall, and generally used for the measurement or control of water. *See* MEASURING WEIR.

orifice, effective area *See* EFFECTIVE AREA OF AN ORIFICE.

orifice meter A device for measuring the flow of water, consisting of an opening in a plate which is fixed within the pipe carrying the water.

original consequent lakes Lakes formed in original hollows or basins in the land surface, such as depressions in glacial till, in lava flows or in sand dunes. The depressions may have been in the ocean floor and preserved when it was elevated above sea level. *See* SINK-HOLE LAKES.

orographic precipitation Rainfall caused by the active cooling of air, as when a current is forced upwards over a range of mountains.

oscillation limits *See* LIMITS OF OSCILLATION.

osmosis The phenomenon of the diffusion of water through a semi-permeable membrane from a lower concentration solute to a higher concentration solute, such as saline water. *See* REVERSE OSMOSIS.

outburst bank The mid-portion of a sea bank or ridge between the uppermost part, or SWASH BANK, and the footing.

outcrop That part of a rock or other deposit which appears at the surface. Outcropping permeable rocks are important as paths to feed underground aquifers. *See* INTAKE AREA OF AQUIFER.

outcrop water Surface water which seeps into and along outcropping porous and fissured rocks; water which passes downwards into tunnels, deep excavations, wells and underground storage.

outfall (1) The site where water is discharged from a conduit into the sea or stream. (2) An irrigation structure used to pass wastewater into a drain.

outlet (river) Any artificial or natural escape channel which branches off a river and delivers water into depressions, drainage areas or into the sea; any artificial or natural escape channel through or over a river bank or levee.

outlet capacity Discharge at the outlet when the main channel is flowing at authorized full supply discharge.

outlet channel Any artificial channel for conveying water from prepared structures such as TERRACES and DIVERSIONS. *See* ESCAPE.

outlet pipe (small dams) A heavy-duty cast-iron or steel pipe placed along the base of a small dam from the upstream to the downstream side (Figure O.1). The pipe flow is controlled by a gate valve, which is normally fixed at the downstream end of the pipe. Strong pipes are used to enable them to carry the heavy earth-moving machines moving over them. The pipes are fitted with square mild steel or reinforced concrete CUTOFF COLLARS (or anti-seep diaphragms).

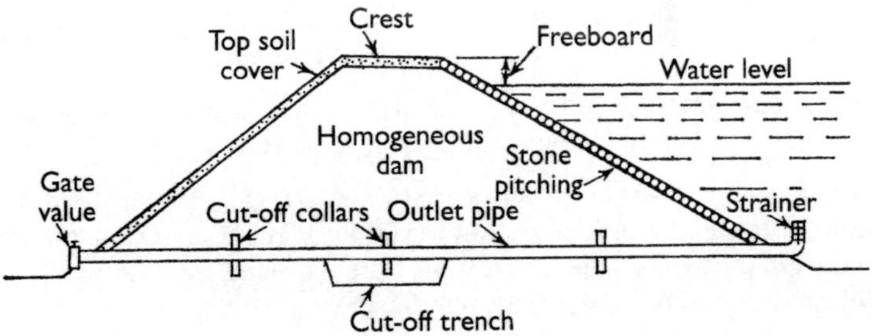

Figure O.1. Outlet pipe (small dams)

outlet works The works that permit water to be released under control from a reservoir or other storage, including the conduits, valves, screens, etc.

outwash fan Accumulations of sand and gravel deposited by streams along the front of a glacier. Streams of melt water often form along the lower part of a glacier, and these transport many subangular fragments which are deposited at the ice-front, where the water velocity is checked. *See* ENGLACIAL MATERIAL, ESKERS, KAMES.

outwash plain *See* VALLEY TRAIN.

overbank flow That part of a flood which overtops the banks of a river. It may cover all or part of the adjacent floodplain and generally flows parallel to the river.

overdevelopment *See* OVERPUMPING.

overfall The crest or part of a dam or weir over which the water flows; the water which overflows.

overfall dam A dam constructed across a stream which allows water to overflow along its entire crest.

overfall weir A low-level weir which permits water to flow over its entire crest.

overflow spillway A structure used on a concrete dam to pass overflow. The dam itself provides the overflow and also the overflow channel. A middle section of the dam forms the overflow, and the water passes down the profile of the dam to the river below. On the downstream base of a large dam, precautions are taken to prevent any scouring by overflow water. *See* APRON.

overflow spring A type of CONTACT SPRING in which the permeable rock dips beneath an impermeable cover. The spring forms an overflow of groundwater at the lower edge of the impervious deposit.

overflow standpipe A STANDPIPE in which the water overflows at the HYDRAULIC GRADIENT or the HYDRAULIC GRADE LINE.

overflow weir A structure which passes overflow from an earth dam reservoir into a channel or spillway leading to the river below the dam, usually constructed at one extremity of the dam and built upwards from solid ground and not from the embankments.

overhead irrigation *See* SPRINKLER IRRIGATION.

overland flow That part of storm water which flows over the ground before it cuts its own channel or reaches a stream. *See* SPILL, SURFACE RUNOFF.

overland runoff *See* SURFACE RUNOFF.

overnight pond A small pond connected to an irrigation system for impounding water, and from which supplies may be drawn when required. *See* PONDAGE, STORAGE.

overpumping (or **overdevelopment**) The pumping out of water from a well at a rate that exceeds replenishment from the outcropping parts of the aquifer. *See* ECONOMIC YIELD, YIELD.

overshot spillway A spillway consisting of a flume formed directly over the dam or embankment. *See* DROP-INLET SPILLWAY.

oversize cobbles (earth dams) Cobbles too large to be placed in earth dams. When a sheepsfoot roller is used for compaction, the maximum diameter should be between 125 and 150 mm (5 and 6 in).

overturn The remixing, which often occurs quite suddenly, of thermally stratified reservoirs. *See* THERMAL STRATIFICATION.

oxbow A bend in a river; the extreme curvatures of a meandering river. *See* FULL MEANDER.

oxbow lake A lake formed in a meander area which has been abandoned by the stream cutting across the narrow neck of the loop (Figure O.2). *See* INGROWN MEANDERS, MEANDER BELT.

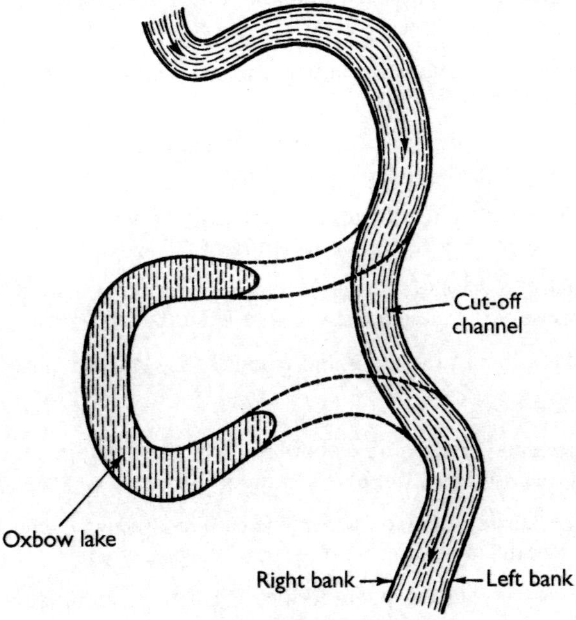

Figure O.2. Oxbow lake

oxidation ditch A closed circuit channel in which screened wastewater is aerated and circulated by a rotor.

oxidation pond A large, relatively shallow, earthen basin used to stabilize organic matter in wastewater by biological oxidation.

ozonization A method which utilizes the powerful oxidizing action of the gas ozone, O_3, to disinfect water supplies and also to produce high-quality water for mineral water factories and for breweries. The ozone may be brought into intimate contact with the water either by blowing it through a distribution system within the water or by means of an injector. *See* CHLORINATION.

P

Pacific-type coastline A coast which runs parallel to the main structural lines of the adjoining land masses. *See* ATLANTIC-TYPE COASTLINE, MODIFIED PACIFIC TYPE.

packaged plants Prefabricated units for water or wastewater treatment, ready for site installation.

packing Hemp or similar material used in a STUFFING BOX to prevent leakage of water from a pump.

paddle A wooden panel used in a lock or sluice or other hydraulic channel to close a water passage. *See* LOCK PADDLE.

paddle hole An opening which allows the entry or discharge of water from a canal LOCK or other hydraulic structure.

padfoot roller (or **sheepsfoot roller**) A roller with projecting teeth, used in the compaction of cohesive soils. *See* SUPERFICIAL COMPACTION.

paludal Pertaining to swamps and marshes. *See* FLUVIAL, LIMNOLOGY.

pan *See* HARDPAN.

pan coefficient The ratio of evaporation from a reservoir, lake, or other body of water to that given by an evaporation pan. *See* EVAPORIMETER.

panel The surface area of water enclosed between the corresponding traces of SEGMENTS in adjacent cross-sections of a stream.

panplain A very broad plain formed by the union of adjacent river valleys. It has a planed-off bedrock floor covered with a thin deposit of alluvium. *See* FLOODPLAIN, PENEPLAIN, VALLEY FLOODPLAIN.

parapet wall (or **storm wall** or **wave wall**) A wall often built on the upstream side of the roadway or pathway formed along the top width of a dam.

Parshall measuring flume (formerly **improved Venturi flume**) A calibrated device or flume to measure flow of water in open conduits, consisting essentially of a contracting upstream section, a throat and an expanding downstream section. The water is intended to flow at Belanger's critical depth (*see* CRITICAL VELOCITY) over a sill placed at the throat.

partial diversion (or **partial cutoff**) A cutoff in a river where total diversion of the current is not desirable. The cut is of sufficient size to take large quantities of flood water, but the river still follows the old course with a reduced flow. A partial diversion may be adopted where property owners on the old course depend on the river for water supply. *See* DIVERSION.

partial duration series A list of hydrologic events, within a specific period, occurring above a selected base and without regard to number. The selected base, in the case of floods, is generally equal to the lowest annual flood to enable at least one flood being included in each year. In contrast, a tabulation or graph in which only the maximum peak for each year is included in the series is termed annual flood peak series.

partially separate system A drainage method in which the rainwater from backyards and the roofs of dwellings flows away with the dwelling sewage, and a different sewer collects the rest of the rainfall. *See* COMBINED SYSTEM, SEPARATE SYSTEM.

Pascal's law A law of fluid pressure (1646) which states that pressure exerted anywhere to an enclosed body of fluid is transmitted undiminished in all directions. This pressure per unit area is uniform throughout and acts at right angles to every part of the surface of the container. *See* HYDROSTATIC PRESS.

passive method A construction method sometimes used to protect buildings erected in PERMAFROST areas. The frozen ground is left intact, and the undersides of buildings are well insulated to prevent heat from the structures thawing the underlying ground. *See* ACTIVE METHOD.

path of percolation *See* LINE OF CREEP.

pathogens Micro-organisms, capable of causing disease, such as coliforms, streptococci and salmonellae.

paving, stone *See* STONE PAVING.

peak flood The point and time of maximum flow during a specific flood. *See* FLOOD.

peak flow, net *See* NET PEAK FLOW.

peak rate of runoff The maximum rate at which water flows from a catchment area during a given storm, usually expressed in cumecs or cusecs. *See* MAXIMUM RAINFALL INTENSITY.

peak unit flow The maximum or peak of a UNIT HYDROGRAPH.

peat A surface covering of more or less decomposed vegetable matter, usually with a high water content. Where peat exists within a reservoir area it may cause acidity and colouring of the water. Remedies include removal of the peat, wholly or in part, or covering the deposit with clean soil from nearby areas.

pegging out *See* STAKING.

pellicular water *See* ADHESIVE WATER.

pellicular zone The ground or depth below the surface to which evaporation effects can penetrate.

Pelton water wheel An impulse type of water wheel which obtains its rotative force from the impulse of a powerful jet of water against buckets spaced at its periphery. The wheel requires a head of at least 275 m (900 ft) for efficient operation. *See* IMPULSE TURBINE.

pendant A small metallic disc or appendage carried by a PENDENT WIRE, and stamped with an observation point number.

pendent wire A wire marking the direction of a section line, and along which the position of observation points is marked by PENDANTS.

peneplain (sometimes **peneplane**) A featureless, roughly flat land area, at or near sea level and formed by long-continued erosion and weathering.

Penman method Two expressions developed by Dr Penman to provide reliable estimates of EVAPORATION from open water and POTENTIAL EVAPORATION from vegetated surfaces, using fundamental physical parameters. Penman's expressions use data on temperature, wind velocity, vapour pressure and solar radiation, and can be used anywhere in the world where these data are available.

Penman–Monteith method One of several techniques developed from the PENMAN METHOD: they give improved results in the estimation of potential evapotranspiration by taking account of soil and plant types.

penning gate A rectangular sluice gate which moves upwards to open, used chiefly in the UK. *See* LIFT GATE.

penstock (1) A pipe which conveys water under pressure to a WATER TURBINE. (2) A sluice gate or floodgate for controlling the flow of water.

percent saturation *See* DEGREE OF SATURATION.

perched stream An INSULATED STREAM, or a stream perched well above the zone of saturation.

perched water A body of subsurface water which is separated from the underlying zone of saturation by a confining bed or impervious stratum.

perched water-table The upper limit or surface of a small body of water above the MAIN WATER-TABLE (Figure P.1). The water is retained in its elevated position by an impervious stratum, and may form a limited source of water supply.

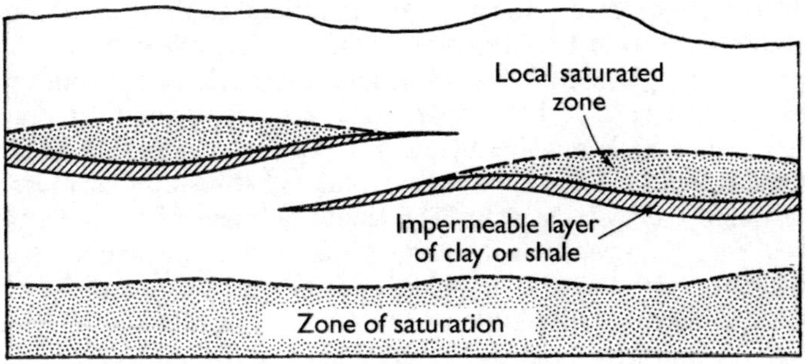

Figure P.1. Perched water-tables

percolating filter *See* BACTERIA BED.

percolation That part of rainfall that enters the soil and subsoil. The water that moves downwards and laterally into streams is termed SHALLOW PERCOLATION, while the water that continues downwards into the zone of saturation or other subsurface accumulation is termed DEEP PERCOLATION. In general, percolation is a maximum in winter and almost nil in summer.

percolation gauge A cylinder or block of earth surrounded by brick, concrete or other material, and so set that water from rainfall may be collected at the base of the cylinder (Figure P.2). Gauges vary in depth from 0.5 to 1.5 m (20 to 60 in) and in area from 0.8 to 4 m^2 (1 to 5 yd^2). They may be left bare or turfed. *See* INFILTROMETER.

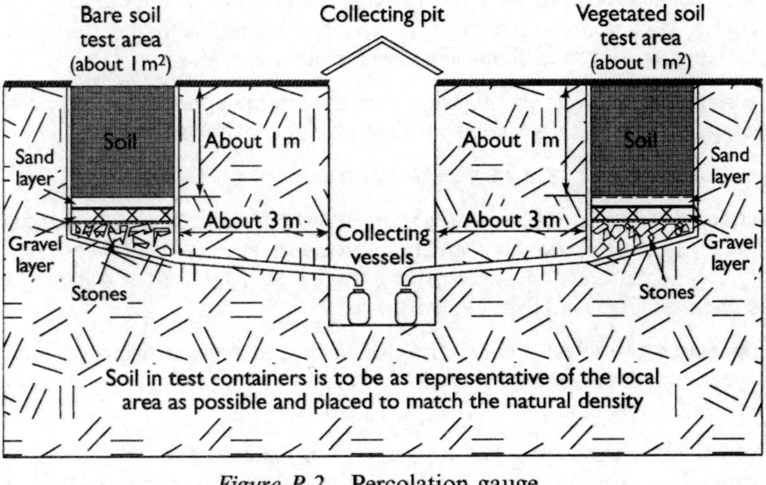

Figure P.2. Percolation gauge

percolation path (dams) The direction or path taken by water percolating beneath a dam from the upstream to the downstream side. It is both a source of leakage from the reservoir and also liable to exert an upward pressure on the base of the dam. When it occurs in permeable foundation rocks a reduction is possible by making the percolation path as long as possible, which lowers the hydraulic gradient between the upstream and downstream faces of the dam. This could be done by a cutoff trench, packed with impervious material to a designed depth and constructed along the length of the foundation near the upstream face. This impervious barrier deflects the percolation downwards and increases its path to the downstream side. The cutoff may consist of a vertical wall of grouted rock or a curtain of sheet piling. Where a dam is to be constructed on porous deposits, a horizontal concrete apron may be formed, extending some distance upstream and downstream from the dam. This construction also increases the length of the percolation path to the downstream side. *See* PIPING, TUNNELLING.

percolation rate The percolation per unit of time.

perennial snowfields Cold regions, such as polar lands and some high mountains, where the snowfall remains throughout the year. *See* SNOW LINE.

perennial spring A SPRING in which the flow of water is unceasing. *See* INTERMITTENT SPRING.

perennial stream A stream that flows continuously throughout the year and often stands lower than the local water-table. *See* EFFLUENT SEEPAGE.

perforated-casing well A well which the water enters through holes in the casing. Commonly applicable to steel- or iron-cased wells, the term also includes wells with different types of casing. *See* OPEN-END WELL.

periglacial Relating to the ground or zone peripheral to a glacier or ice sheet, phenomena associated with the glacial age. *See* PERMAFROST.

perimeter, wetted *See* WETTED PERIMETER.

periodic spring A type of INTERMITTENT SPRING in which the discharge is relatively great at roughly regular and frequent intervals. It often occurs in limestone districts, and its rhythmic discharge may be due to a natural siphon underground. *See* BREATHING WELL.

periodic type (lakes) Some lakes found in arid regions, without outlet, which evaporate to dryness during a portion of the year. *See* LAKE-FLOOR PLAIN, OBLITERATION OF LAKES, PLAYA LAKES.

period of concentration (storm) The storm duration which allows the rainfall on substantially the entire catchment area to have its effect on the flow at the outfall. Sometimes called the RESPONSE TIME. *See* FLOOD HYDROGRAPH.

period of pulsation The average period of a cycle of pulsation at a point in the cross-section of a stream during which the velocity fluctuates between limiting low and high values. *See* MEAN VELOCITY ON A VERTICAL.

permafrost Ground perennially frozen, occurring in regions where the mean annual temperature is below freezing point. It forms a near-surface deposit which is impervious and may be hundreds of feet in thickness. In areas of permafrost certain engineering problems may arise, such as contamination of surface water supplies, disposal of sewage from large camps and foundation work, etc. *See* ACTIVE METHOD.

permafrost table The upper surface or limit of PERMAFROST.

permanent bank protection Permanent measures or works to check erosion along river banks. They may consist of concrete or masonry walls, concrete blocks, or concrete sheet piling, or broken rock tipped from the bank to form a continuous apron. Such works are costly and are used only where important structures, such as bridges or buildings, require protection. They are often used within towns where bank erosion must be arrested. Strong-rooting, quick-growing plants may be viewed as permanent protection. In New Zealand and Australia, poplars and willows are sometimes used because they grow rapidly under a wide range of conditions. *See* BANK PROTECTION, STONE RIPRAP.

permanent hardness (or **non-carbonate hardness**) *See* HARDNESS OF WATER.

permanent stream A PERENNIAL STREAM, or one with an unceasing flow. All streams positioned below the water-table are usually permanent. *See* TEMPORARY STREAM.

permanent wilting point The stage when a soil is so dry that plant roots are unable to extract water at a rate sufficient to maintain turgidity. At this stage, transpiration almost ceases, the plants wilt and the soil is said to be at permanent wilting point. Temporary wilting may occur on a hot day, but in this case the soil moisture may be plentiful. *See* WILTING COEFFICIENT.

permanganate index A rapid measure of oxygen absorbed from potassium permanganate in an acidified sample under standard conditions.

permeability The capacity of a rock, soil, or other substance for transmitting water. Shales and clays are relatively impermeable, while sands, gravels and sandstones, with their much higher pore space and intercommunication, are permeable. Impervious rocks are important in the case of reservoirs and earth dams for ponds, while permeable soils are desirable in the case of roadways and railways. *See* HAZEN'S LAW.

permeability coefficient *See* COEFFICIENT OF PERMEABILITY.

permeability, field coefficient of *See* FIELD COEFFICIENT OF PERMEABILITY.

permeable groynes (or **pile and waling groynes**) Usually a timber construction consisting of a double row of piles strutted together. Walings or horizontal timber planks are fixed to the upstream row of piles and spaced about their own width. Sometimes old steel rails are used as piles. Permeable groynes pointed upstream tend to collect much floating debris, and so lose permeability. The pile and waling type can be built out stage by stage. *See* GROYNES, IMPERMEABLE GROYNE.

permeameter A laboratory instrument for measuring the COEFFICIENT OF PERMEABILITY of a soil sample. The falling-head permeameter is used for impermeable soils such as fine silts and clays, and the constant-head type is used for permeable gravel or sand.

permissible velocity (or **safe velocity**) The maximum velocity of flow of water that can exist in a channel and not cause erosion, depending on the nature of the ground and other factors; the highest velocity at which water may be conveyed safely in a conduit.

persistence A phenomenon that may be observed in long records of natural events where relatively larger and also relatively smaller events are often observed to be grouped together in the sequence. In some cases it is far from clear what causes this, while in others there are apparent physical explanations. For example, one exceptionally heavy fall of rain may be followed by many time-periods of enlarged river flows, as BASE FLOWS remain high because once saturated the underlying rocks drain slowly.

pervious rock A rock that permits the passage of water with relative ease, as opposed to an IMPERVIOUS ROCK. *See* MICROPORES.

pF value For held water in soil, the common logarithm of the suction given in centimetres of water. *See* SOIL MOISTURE SUCTION.

pH value A convenient expression for measuring the ACIDITY or ALKALINITY of a solution, meaning the logarithm of the reciprocal of the hydrogen ion concentration. In distilled water the molecules dissociate $H_2O \rightleftharpoons OH^+ + H^-$ to give a concentration of H^- of 10^{-7}, or a pH value of seven, the neutral point; all values above seven indicate the presence of alkalis, and all below seven indicate acids. A one-unit change in pH value is equivalent to a tenfold change in hydrogen ion concentration. pH values normally lie in the range 0 – 7 – 14. *See* ACID WATER.

photogeology The study and geological interpretation of vertical aerial photographs, of maximum value in open country where rock formations are partially exposed. Even aerial photographs of an area covered by dense vegetation often disclose the main geological features, particularly where colour variations of vegetation can be discerned and correlated to the nature of the soils and subsurface rocks. *See* INFRA-RED PHOTOGRAPHY, MOSAIC.

photogrammetrist A person skilled in preparing maps, both topographical and structure-contour maps, from aerial photographs, and in the study of overlapping photographs in pairs under a stereoscope.

phreatic cycle *See* CYCLE OF FLUCTUATION.

phreatic decline The depression or downward movement of the water-table. *See* DRAWDOWN.

phreatic divide *See* GROUNDWATER DIVIDE.

phreatic fluctuation *See* FLUCTUATION OF THE WATER-TABLE.

phreatic fluctuation, belt of *See* BELT OF PHREATIC FLUCTUATION.

phreatic high The highest level reached by the water-table at a given station and for a given PHREATIC CYCLE. Similarly, the PHREATIC LOW is the lowest level reached by the water-table. Measurement is usually expressed in metres or feet above mean sea level or below a datum on the land surface.

phreatic low *See* PHREATIC HIGH.

phreatic surface *See* MAIN WATER-TABLE.

phreatic water GROUNDWATER; water in the ZONE OF SATURATION.

phreatic water discharge *See* GROUNDWATER DECREMENT.

phreatic wave (or **wave of the water-table**) A swell, undulation or vibration in the water-table that moves laterally away from the zone where a large intake of water occurs in the zone of saturation.

phreatophyte A plant that habitually obtains its water supply either through the capillary fringe or directly from the zone of saturation.

physical geology The science which deals with the activity of natural agents on the earth's surface. *See* ENGINEER–HYDROLOGIST, ENGINEERING GEOLOGY, GEOLOGY.

physical models The small-scale copies or reproductions of engineering and architecture works that are widely used to aid planning and design (Figure P.3). They are of importance in water engineering as a way of studying the performance of full-sized prototypes, or predicting the effects of proposed changes to river channels or flow regimes. Their advantages over numerical models include assisting the visualization of the prototype as a whole; allowing several parameters to be studied in one model; and enabling complex three-dimensional problems to be investigated. However, they can be expensive; they may require considerable skill and take a long time to construct; and they may present complex problems with selecting scales if the overall hydraulic behaviour of the model is to approximate that of the prototype. *See also* MODELLING, CONCEPTUAL MODEL, NUMERICAL SIMULATION MODELS.

Figure P.3. A physical model of a dam and spillway under test

phytometer An appliance to determine transpiration. A sample of soil, in which some plants are rooted, is placed in a vessel and so sealed that water can escape only by transpiration from the plants. *See* EVAPOTRANSPIRATION.

Piedmont glacier A type of glacier formed when two or more valley glaciers unite and spread out on the plains below the mountains, often seen in glacial regions where the SNOW LINE is at a low level.

pier (1) A circular or rectangular column or wall of plain or reinforced concrete or masonry for carrying heavy loads such as a bridge or other structure. (2) A mole or BREAKWATER used as a promenade and landing-stage.

piestic interval The vertical height or distance between two ISOPIESTIC LINES on a map.

piezometer A tube or appliance for measuring PRESSURE HEAD, usually a small pipe tapped into the side of a closed or open conduit and connected to a gauge.

piezometer tubes (dams) Copper or usually plastic tubes inserted in clay fills of dams to measure pore pressure. They are almost standard practice in the UK, where embankment materials give high pore pressures. As a result of experience in several UK earth dams, the British Research Station has developed the BRS apparatus. *See* PORE WATER PRESSURE.

piezometric surface (or **potentiometric surface**) An imaginary surface representing the static levels of the water in an aquifer.

pig (or **go-devil**) A device operated mechanically or hydraulically for scraping and cleaning encrustation from within a section of pipeline.

pile and fascine A method of protecting a steep eroding bank by driving a row of piles at about 3 m (10 ft) intervals, and then packing the space between the piles and the bank with brushwood fascines. The tops of the piles are secured by wire to stakes set well back on the bank. *See* STAKE AND STONE SAUSAGE CONSTRUCTION.

pile and waling groynes *See* PERMEABLE GROYNES.

piles Timber, concrete or steel posts or plates driven into the ground to carry a vertical load or a horizontal load. Steel is usually preferred. Piling is used for foundations and for protective works associated with weirs, aprons, groynes, and general river and reservoir structures.

pillar drains (dams) Pillars consisting of hand-packed stone carried up from the foundations of an earth dam at intervals along its length. They are usually connected at their lower and upper ends by lateral drains. The lower drains are connected to a main drain in which the flow of water can be inspected and measured. *See* VERTICAL SAND DRAIN.

pilot channels Channels or cuts made along the river outside the noses of

GROYNES to establish the main flow in a new line. This is often necessary if the river bed is heavy and difficult to erode. Where machines are available for this work, the excavated material is often dumped along the groynes and also along the bank between the groynes. Willows are often planted in the dumped material to improve the banks and the stability of the groynes.

pilot cutoff (or **pilot cut**) A CUTOFF where erosion is deliberately used to enlarge the cut, only possible where the bed is easily erodible.

pinching of strata The attenuation of strata in a certain direction, finally some beds may 'pinch out' altogether. This structure is important in its effects upon the movement or storage of water within the rocks. A well may not give the expected yield on account of the thinning and perhaps disappearance of the aquifer. *See* STRUCTURAL TRAP.

pipeline A pipe to convey water over long distances, laid on the surface or in a trench which is then filled in or covered over. *See* HYDRAULIC TRANSPORT.

pipeline, irrigation A closed conduit for conveying irrigation water. This method is more expensive than unlined earth channels but has the advantage that the CONVEYANCE LOSS is reduced to a minimum.

pipe outlet (farm dam) A pipe set in the base of a small farm dam to pass COMPENSATION WATER.

pipe outlet (irrigation) A pipe placed in the bank of a farm supply ditch. It is fitted with a slide gate to enable water to be discharged from the ditch on to the land.

pipe sampling The taking of samples of superficial deposits, using a drive pipe.

piping (or **subsurface erosion**) The erosion caused by seepage of water within and around the base of a structure, dam or bank, and which could affect the safety of the construction. It is caused by water seeping through the dam with a velocity sufficiently strong to carry soil particles. A hole or 'pipe' is created at the downstream face and it gradually moves towards the water face. *See* BOIL, TUNNELLING.

piracy (river) *See* RIVER CAPTURE.

piston pump *See* DIRECT-ACTING PUMP, PUMP.

pitched work (or **pitching**) A stone REVETMENT to protect the slopes of a river bank or reservoir from scour. As a protection against heavy waves, the stones are at least 150 mm (6 in) square on the face, with a depth of at least 0.5 m (21 in). *See* BEACHING, REVETMENT.

pitching *See* PITCHED WORK.

pitometer A device based on the PITOT TUBE principle, used for determining the velocity of flow of fluids in closed conduits and pipes.

Pitot tube A device for measuring the pressure and velocity of flowing water. In its simplest form (Figure P.4) it consists of a tube or pipe bent at right angles with open ends. One arm is shorter than the other, and when the device is placed in a current the water rises in the longer arm, thus:

$$H = V^2/2g$$

where H is the height of water in longer arm, V is the velocity of water, and g is the acceleration due to gravity. Other forms consist of two openings (upstream and downstream) and two columns of water. The velocity head is computed from the difference of the two water levels.

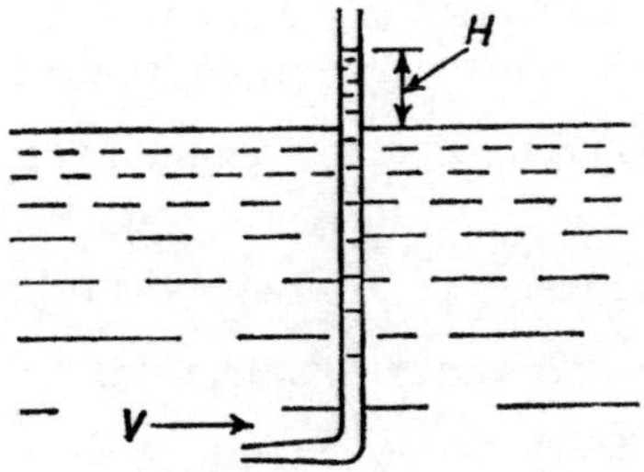

Figure P.4. Pitot tube

pitting The putting down of TRIAL PITS to test and sample shallow alluvial deposits. In soft dry ground, depths of about 15 m (50 ft) may be reached. Pitting may also be used to locate the water-table when at shallow depths, and the pits may be used for water supplies. *See* SHALLOW WELL.

pivot bridge *See* SWING BRIDGE.

pivot point layout The location of observation points in a river, without direct measurement along the DISCHARGE SECTION LINE, by means of a geometrical layout of points on one or both banks.

plains Land forms whose predominating feature is flatness or low inclination; generally, topographic features which cannot properly be described as benches or terraces. Plains of several kinds are formed by rivers. *See* AGGRADED FLOODPLAIN, PANPLAIN.

plain tract (or **alluvial tract**) The stretch of a river approaching its mouth where it flows over a broad floodplain with deposition of fine sediment (Figure M.4). *See* DELTAIC TRACT, VALLEY TRACT.

plan (or **map**) (1) A scale drawing, on a horizontal plane, of part of the earth's surface showing natural and artificial features such as geological exposures, faults, folds, excavations, roads, rivers, dams, drainage, contours, and perhaps irrigation and other relevant features. A mine plan would show mine workings, adits, shafts, etc. (2) A map kept by a water authority showing the positions of all its pipes and fittings. *See* GEOLOGICAL MAP, INDEX MAP, LITHOLOGICAL MAP, WATER PROSPECT MAP.

planation The action of a river when it cuts a flat-bottomed, trough-shaped valley.

plane of saturation The WATER-TABLE. *See* MAIN WATER-TABLE.

planimeter An instrument for measuring an area on a plan, whose perimeter has been traced out by its moving arm; available in several units and scales, it is used in water engineering to compute reservoir area, etc.

plankton The mainly microscopic organisms drifting or floating in the sea or fresh water.

planning (impounding schemes) *See* IMPOUNDING SCHEMES.

planning, waterworks *See* WATERWORKS PLANNING.

plate count (bacteriology) The number of colonies of cultured bacteria grown under standard conditions, the results being used to assess the microbial quality of a water.

playa lakes Broad, shallow sheets of water in desert regions; they generally form and evaporate after a short period, leaving mud flats or PLAYAS.

playas The level plains of ephemeral lakes in desert regions; deposition of sediment occurs during the inflowing of streams after local cloudbursts, and may be followed by a drought of many months' duration. *See* LAKE-FLOOR PLAIN.

plotting position An estimate of the probability of the magnitude of an observed event being equalled or exceeded, as predicted from its position or rank, m, within a set of n observations. Many expressions are available for estimating plotting positions. Figure P.5 shows the relationship between magnitude of flows and probability of exceedance: the plotting position probabilities of seven observations of an annual maximum series of rates of flow are estimated by using the most commonly employed expression, $P = m/(n+1)$:

Observed annual maximum flood (m^3/s)	*Rank in data set*	*Estimated plotting probability (the probability of the flood flow being equalled or exceeded)*
46	1	1/8 = 0.125
40	2	2/8 = 0.25
39	3	3/8 = 0.375
34	4	4/8 = 0.5
33	5	5/8 = 0.625
28	6	6/8 = 0.75
23	7	7/8 = 0.875

The best line drawn through a set of plotting position probabilities, particularly if drawn on special probability graph paper, may indicate the underlying statistical distribution relating events to their probability of occurrence.

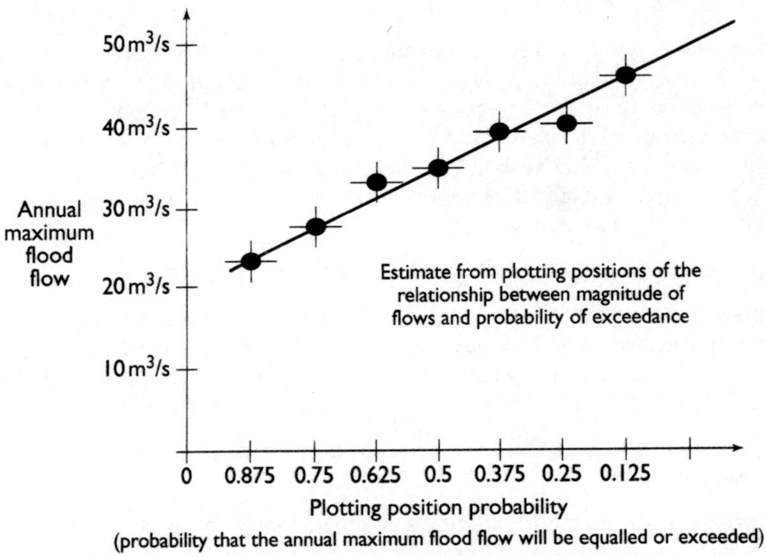

Figure P.5. Plotting positions

plum A large stone of irregular shape placed in mass concrete dams to reduce the cost of the concrete. *See* CYCLOPEAN CONCRETE.

plunger *See* FORCE PUMP, RAM.

plutonic water *See* JUVENILE WATER, MAGMATIC WATER.

pluvial index The index of total rainfall depth which will probably be exceeded or equalled during a certain period in a specified area over a given number of years.

pluviograph (1) A graph indicating stream flow if all precipitation became SURFACE RUNOFF. (2) A graph showing total rainfall.

pluviometer *See* RAIN-GAUGE.

PMP Probable maximum precipitation.

pneumatic caisson A special type of caisson with a pressurized working chamber at its base. The caisson is used on soft foundations and the pressure controls the inflow of soil and water, thus enabling the workmen to carry out the excavation at the bottom of the caisson. Care must be taken to protect these workers from caisson disease.

pneumatic sewage ejector A DISPLACEMENT PUMP, operated by compressed air, for moving sewage.

pneumatic water barrel (or **vacuum tank**) An earlier type of WATER BARREL, used mainly for dewatering a sinking shaft. By means of a hose connection to an air pump at the surface, a partial vacuum is created inside the barrel in the shaft. As a result the water lifts the inlet valve and fills the barrel. The hose is then detached and the barrel hoisted to the surface and discharged. Various types of CENTRIFUGAL PUMPS or SUBMERSIBLE PUMPS are now used.

pocket beaches Small shore areas formed at the heads of coves.

point gauge An appliance for measuring water level, consisting of a sharp point attached to an arrangement which slides on a graduated staff. The point is lowered until it just touches the free surface of the water. *See* HOOK GAUGE, STAGE.

point rainfall A rainfall measurement at a single point or station. *See* RAIN-GAUGE.

point sample (sediment) Taking a sample of SEDIMENT WATER at a single point; a time-integrating sampler or an instantaneous instrument may be used.

poised stream A stream which may be viewed as stable from an engineering aspect, but not possibly from the geologic aspect. *See* REJUVENATION.

polder An area of marshy and low-lying land reclaimed from the sea and cultivated (the Netherlands). *See* LAND ACCRETION.

pollard A tree with its top cut off to allow the stump to produce a more dense growth of limbs and branches. The stump is usually about 1.8 m (6 ft) high, which reduces the tendency to be blown over. 'Pollarding' refers to the practice of pruning and cutting back trees along river banks and other margins.

polyelectrolytes Complex polymeric compounds, forming charged ions in solution, which are used as flocculants.

pond (1) A body of relatively stagnant water, usually smaller than a lake, and often formed artificially by earth ridges or ground excavations. (2) The water retained between two canal locks. *See* FARM POND, REACH.

pondage (or **balance storage** or **regulator storage**) The storage of a volume of water, usually only sufficient for one day, during which outflow varies according to usage and inflow may be fairly constant. The term is often applied to water stored temporarily within low banks at a farm; the water is periodically pumped or diverted into the FARM RESERVOIR.

pondage method A means of SEEPAGE LOSS determination in an earthen channel (Figure P.6). The test section of the channel is isolated with temporary watertight bulkheads and then filled with water. The decrease in volume of water is measured and the seepage is computed with allowances for rainfall and evaporation. *See* INFLOW–OUTFLOW METHOD.

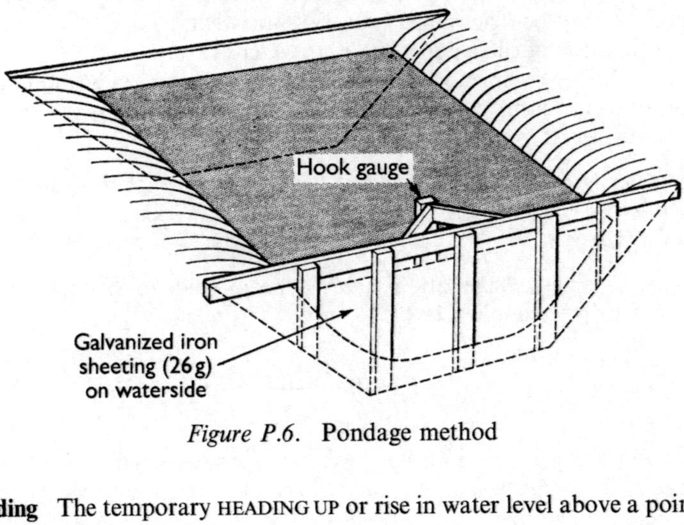

Figure P.6. Pondage method

ponding The temporary HEADING UP or rise in water level above a point in a stream.

pond management The development of sound practices for the protection of ponds from siltation and erosion, with the aim of providing adequate water supplies.

pontoon A BARGE, flat-bottomed craft, or a light framework or float used in military engineering for bridge building and in civil engineering for general duties on construction projects. *See* FLOATING HARBOUR.

pontoon bridge A bridge which is carried on pontoons moored to the river or lake bed. It may be temporary or permanent. Permanent types are sometimes adopted when the foundation material is very soft.

pool spring A spring that discharges water from a deep pool. Many pool springs are related to faulting structures.

poplars Large trees of rapid growth with soft wood, propagated by root suckers or from branch cuttings. Their deep root system is often utilized to bind the soil layers on slopes liable to landslides. *See* TREE PLANTING, WILLOWS.

pore pressure *See* PORE WATER PRESSURE.

pores *See* MACROPORES, MICROPORES, VOIDS.

pore water Water occupying the interstices, voids or pores of rocks or soils. *See* MICROPORES, SATURATION.

pore water pressure (or **pore pressure**) The pressure of water in a saturated soil, which may be determined by a BOURDON PRESSURE GAUGE. During the construction of earth dams, the pore water pressure is often measured during and after completion of work because it gives an indication of the degree of compaction. The compaction process is complete if the pressure is zero. *See* NEUTRAL PRESSURE.

pore water pressure cells Sensitive instruments which measure PORE WATER PRESSURES produced by load fluctuations, such as the rise and fall of a tide.

porosity The percentage ratio of the volume of voids to the total volume of a rock or soil sample. Thus:

$$P = 100\frac{W - D}{W - S}$$

where P is the per cent porosity, W is the saturated weight, D is the dry weight, and S is the weight of saturated sample when suspended in water. As a guide, the porosity of sandy soils is greater than 50%; clays, 45% or higher; sandstones and limestones, up to about 25%; granites and other igneous rocks, less than 1%. *See* EFFECTIVE POROSITY, TOTAL POROSITY.

porosity, effective *See* EFFECTIVE POROSITY.

porous rock A rock which contains pores and transmits water with relative ease. At a dam site or other storage, a porous bed may be sealed off by grouting or by concreting a trench cut along its outcrop. *See* MACROPORES, MICROPORES, VOIDS.

portal The inlet to, or outlet from, a water tunnel, referred to as the upstream or downstream portal, respectively.

Portland cement A substance which binds AGGREGATES into a hard CONCRETE within a few days. It is generally made from chalk and clay by mixing in suitable proportions, burning in kilns, and then grinding to a fine powder. Portland cement is stronger than natural cement and sets more slowly. *See* CONCRETE DAM.

position head The STATIC HEAD of water.

positive artesian head *See* ARTESIAN HEAD (POSITIVE).

positive confining bed A CONFINING BED which prevents the upward movement of groundwater with a positive artesian head. *See* ARTESIAN HEAD (POSITIVE).

post-Pliocene *See* QUATERNARY.

potable water Drinkable water; water used for hygiene and amenity purposes. Usually it must comply with statutory requirements, such as the World Health Organization Guidelines for Drinking Water Quality. *See* INDUSTRIAL WATER.

potamology A treatise on rivers; the branch of hydrology pertaining to streams and rivers. *See* LIMNOLOGY.

potential (groundwater) (1) The sum of the PRESSURE HEAD of groundwater at a given point and the elevation of the point above a fixed plane of reference. (2) The groundwater resources available in a region.

potential energy The energy of water due to its position or height above a given datum. *See* ENERGY HEAD.

potential evaporation The maximum amount of water that can be evaporated by the available winds, solar radiation etc. From open water surfaces the potential evaporation and the actual evaporation are the same, but from soil and vegetated surfaces the actual evaporation is often less – it can never be more – as the available moisture near the surface may be less than there is sufficient energy, etc., to remove.

potential gradient (groundwater) The rate of change in potential in a mass of groundwater. Where no gradient direction is specified, that of maximum gradient is taken.

potential head (or **elevation head**) The height of any particle of water, in metres or feet, above a datum or point.

potential transpiration The TRANSPIRATION that would take place if the water for the vegetation were always adequate. In the UK estimates of potential transpiration have been made for the summer months April to September from the relevant weather data (air temperature, sunshine, air humidity and wind speed). When placed alongside the rainfall for the same places and periods, they indicate the frequency of irrigation need.

potential yield The maximum rate at which water may be extracted from an aquifer throughout the foreseeable future and ignoring the cost of recovery. *See* ECONOMIC YIELD.

potentiometric surface *See* PIEZOMETRIC SURFACE.

potholes Roughly cylindrical holes cut in the solid rock bed of a stream by the swirling of pebbles or sand particles in eddies (Figure P.7). The fragments are rolled at one place, and thus bore downwards a roundish hole which can be up to 6 m (20 ft) in diameter and 3 to 4.5 m (10 to 15 ft) deep. *See* MOULINS.

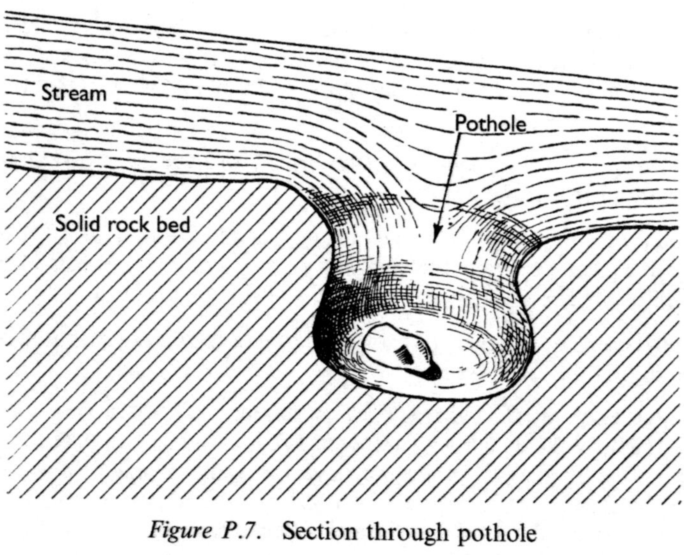

Figure P.7. Section through pothole

power earth auger *See* EARTH AUGER.

power take-off An external splined shaft on a tractor, used for driving water pumps, etc., while the vehicle is idle.

precipitable water The total amount of water produced if all water vapour, in a total vertical column of air of unit horizontal cross-section, were condensed and collected in a receptacle of the same horizontal cross-section.

precipitable water, effective The water vapour in the atmosphere that can be precipitated by efficient storm conditions, expressed as depth of water in millimetres or inches.

precipitation (1) The total amount of water which falls as rain, hail or snow, expressed as millimetres or inches of rainfall over a specified period; moisture deposited as dew. (2) The settling of solids in sewage or other cleaning or separation process; the chemical process whereby the insoluble substance in a solution settles out. *See* RAINFALL, RAIN-GAUGE, RATE OF RAINFALL.

precipitation, excessive A precipitation rate which is greater than certain specified limits selected with regard to the normal rainfall of the area. Excessive storms are those that cause excessive precipitation.

precipitation mass curve A curve showing accumulated rainfall plotted against time. *See* MASS DIAGRAM.

precipitation, trace of A precipitation rate too low to be measured in a rain-gauge, usually below 0.125 mm (0.005 in).

prescribed rights (water) Legal title to the use of water acquired by possession and use over a long period without protest of other parties. *See* WATER RIGHTS.

pressure, effective *See* EFFECTIVE PRESSURE.

pressure filter Generally a cylindrical filter, rather similar to a GRAVITY FILTER, for removing suspended matter by means of a sand bed. It may be vertical or horizontal, depending on the direction of the axis of the cylinder. The drop in pressure between the unfiltered and filtered water pipes is the HEAD available for flow. *See* FILTER.

pressure gauge An instrument which measures fluid pressure, such as the BOURDON PRESSURE GAUGE.

pressure grouting *See* BLANKET GROUTING, CURTAIN GROUTING, GROUTING.

pressure head (or **head**) The water pressure in a system, expressed as either a pressure in N/mm^2 (lbf/in^2) or as metres (feet) head.

pressure, impact *See* IMPACT PRESSURE.

pressure pillow A snow-measuring device consisting of a circular flat bladder about 75 mm (3 in) deep. The pillow is filled with an antifreeze solution, and when snow builds up on it the resulting increased pressure is recorded by an open-end manometer.

pressure-reducing valve A valve designed for regulating or maintaining a reasonably constant pressure. It consists of a weighed piston moving vertically in a cylinder connected by a tube with the main downstream of the valve. It can be made to close at any desired outlet pressure by adjusting the weights on the piston. The valve must be kept in good working order.

pressure-relief valve A safety valve designed to relieve pressure in a pipeline. *See* AIR VALVE.

pressure sounder A device, based on Boyle's law, for measuring depths.

pressure-sustaining valve A valve which tends to close if the upstream pressure falls, and thus a relatively constant upstream pressure is maintained; it is sometimes used on the inlet to service reservoirs to ensure that the pressure does not fall below a fixed minimum.

pressure tunnel A water supply tunnel in which the water is under considerable pressure. It is usually a tunnel driven below some obstruction, such as weak permeable ground or a river bed, lined with reinforced concrete, cast iron or steel, backed with concrete or with broken rock, and grouted.

prestressed concrete Concrete which has been stressed before it is required to sustain a load. The procedure greatly reduces or prevents cracking when in use. Prestressed concrete has been used for water-retaining tanks, and its use appears to be economical for reservoirs if the capacity is greater than 2250 m^3 (0.5 million gal). *See* CONCRETE, REINFORCED CONCRETE.

prestressed gravity dam A type of GRAVITY DAM in which the resistance to overturning, sliding or pivoting is provided by its weight, and the dam is anchored to the foundation rock by a system of stressed steel bars or tendons. The design enables the volume of concrete to be reduced by about 40%, but this saving is partly offset by the cost of steel rods, anchorages, labour, etc.

presumptive coliform test A standard dilution tube method for the determination, by their biological activity, of coliform bacteria present in a sample of water or wastewater, as shown by the production of gas from lactose in the presence of bile salts.

preventative snagging (or **stumping off**, Australia) The removal of trees growing on the bank of a river that are liable to fall into the channel and tear down part of the bank with their roots. They are cut down and made to fall landwards. *See* SNAGGING.

primary treatment Sedimentation (removal of suspended matter), the first major treatment of wastewater, after any preparatory measures such as grit settlement and screening. *See* ADVANCED WATER TREATMENT.

priming (1) The first filling of a reservoir, canal, or other structure with water, which may occur each season, year, or only once during its lifetime. (2) Filling a pump or siphon with water to ensure a full flow at the commencement.

primitive water Sometimes, water trapped in the earth's interior since its formation; INTERSTITIAL WATER of internal origin. *See* MAGMATIC WATER.

principle of Archimedes *See* ARCHIMEDES' PRINCIPLE.

prior rivers A term often used to describe river systems that are prior to the present system but post-date the ANCESTRAL RIVERS.

prismoidal formula A formula to determine the volume of an excavated earth water tank, channel, etc. Thus:

$$\text{volume} = \frac{D}{6}(A_1 + A_2 + 4A_m)$$

where D is the depth of excavation, A_1 and A_2 are the two end areas, and A_m is the area at the midpoint.

probable maximum flood *See* RETURN PERIOD.

probable maximum precipitation (PMP) The theoretical maximum rainfall that can fall. Predictions of PMP are made from consideration of atmospheric physics. PMP as a maximum is not related to a RETURN PERIOD, but comparisons of PMP estimates with those from statistical analysis of long records suggest it approximates to events with a 10 000 year RETURN PERIOD. Estimates of PMP are used to predict the PROBABLE MAXIMUM FLOOD.

probing (1) Investigating underwater conditions of spur pavements, protection of weirs on upstream and downstream sides, and underwater masonry, etc. (2) Investigating ground conditions by forcing down a pointed steel rod up to about 6 m (20 ft) long. It may be used to locate bedrock, shallow water supplies, etc.

Proctor test A standardized test for compacting soils in the laboratory to give comparable results. The density of a soil sample is measured with a known water content, compacted with a standard number of blows in a standard mould. The test is repeated for several moisture contents. By plotting dry density against moisture content (as a percentage of dry weight) the optimum moisture content and optimum dry density are obtained. It is often specified that the soil used as fill material shall be compacted to at least 95% of the optimum (Proctor) density. *See* SUPERFICIAL COMPACTION.

profile Usually a longitudinal section showing ground undulations, near-surface deposits and possibly the planned formation level of a proposed construction, such as a canal or pipeline. *See* STREAM PROFILE, STREAM SECTIONS.

profile of equilibrium A stream with a profile along which all irregularities (waterfalls, etc.) have been removed by degradation, and all basins filled with sediment. *See* DRAINAGE EQUILIBRIUM, GRADED STREAM, REGIME.

profile, stream *See* STREAM PROFILE.

prograding The gradual building out to sea of a coastline by deposition of detritus; often the process is a temporary phase and not long continued. *See* COAST OF EMERGENCE, RETROGRADING.

propeller The curved rotating blade of a centrifugal pump. *See* IMPELLER.

propeller-type current meter A meter incorporating an impeller which is rotated on a horizontal axis by the water flow. *See* CURRENT METER.

prospecting The search for minerals or other occurrences of economic value. Where promising localities have been located, the presence of good water supplies is an important factor in their development. *See* GEOPHYSICAL PROSPECTING, WATER PROSPECT MAP.

protection forest A track of ground covered with trees and woody growth, retained primarily for its beneficial effects on soil and water conservation. *See* SPREADER STRIP.

protective filter *See* FILTER ZONE.

protozoa Small single-cell micro-organisms of the subkingdom Protozoa, including amoebae, ciliates and flagellates.

provinces, groundwater *See* GROUNDWATER PROVINCES.

puddle (or **puddle clay**) A dense clay practically impervious to water, sometimes used to line earthen water channels and in the core walls of earthen dams and embankments; to make watertight by packing with CLAY PUDDLE. *See* HEELING, LINING.

puddling A process by which the structure of a soil is destroyed with a reduction in porosity and permeability, sometimes used in canals, water supply channels and reservoirs to reduce seepage losses. *See* REMOULDING.

pulsometer pump A steam-operated pump in which the only moving part is an automatic ball valve. The valve admits steam alternately to a pair of chambers, so forcing out water which was sucked in by condensation of the steam after the previous stroke. It can deal with dirty water and was widely used for miscellaneous pumping where steam was available. In general, it has been superseded by rotary pumps. *See* SUBMERSIBLE PUMP.

pump A machine or mechanical appliance for imparting energy to water, or for raising water from a lower to a higher level in a pipe. *See* CENTRIFUGAL PUMP, FORCE PUMP, HANDPUMP, LIFT PUMP, PULSOMETER PUMP, ROTARY PUMP, SINKING PUMP, TURBINE PUMP.

pumpage The rate at which water is withdrawn from a well, shaft or other excavation by a pump; the volume of water removed by pumping during a specified period. *See* DRAWDOWN, ECONOMIC YIELD.

pumped storage Any of various storage schemes, such as: (1) A means of storing the energy produced in thermal power stations. Water is pumped up to a high-level reservoir during the night when surplus power is available from steam stations, and released during the peak load period of the day through water turbines to generate power. (2) A form of storage sometimes used where the topography is not suitable for the construction of an impounding dam. The water is pumped from one or more rivers and stored in earthen reservoirs sited a short distance from the river. Sometimes a shallow valley in the area is utilized for storage with an earth dam and the water pumped into it. (3) Increasing the natural drainage to a reservoir by pumping water from an intake in an adjoining catchment area. The total yield is the pumped supply plus the runoff from the reservoir catchment. *See* CATCHWATERS.

pumped-storage hydroelectric station A hydroelectric plant in which surplus low-cost energy generated elsewhere is stored for use in periods of peak demand. The layout consists mainly of low-level and high-level reservoirs, connected by water conduits (usually underground tunnels). During peak load periods, water is drawn from the high-level reservoir through the conduits to operate the generating plant, while during off-peak periods water is pumped back from the low storage to the high storage. The scheme has the advantage of requiring relatively small storage reservoirs, and that hydroelectric generating plant can be started very quickly if demand rises rapidly.

pumping (land drainage) The use of pumps to drain low-lying areas or where the slope for gravity flow is inadequate. The pump may discharge its water into the nearest stream or onto land where gravity flow is possible. In major schemes, the pump may discharge into a central pumping station, which in turn delivers the water from several pumping units into a river, lake or other storage.

pumping level (well) The level from which water has to be pumped after the CONE OF DEPRESSION has been established in the local water-table.

pumping test (1) Yield of water: a test made with a pump in a new well to determine its water-yielding capacity. Quantities and water levels are recorded during the test period. The test pumping rate is usually greater than that at which water will be required, and covers a period sufficiently long to indicate whether the yield can be maintained. *See* ECONOMIC YIELD. (2) Quality of water: water samples during the test are taken to determine, by chemical analyses, the chief constituents and organic purity. Tests may extend over about 14 days and finally a full mineral analysis is often made, which may be used to prescribe treatment and purification processes. *See* WATER QUALITY.

Q

quality (snow) The amount of ice in snow, given as a percentage of the weight of the SNOW SAMPLE.

Quaternary (or **post-Pliocene**) Pertaining to the most recent deposits in the geological column, consisting of gravels, river alluvium, raised beaches, peatbogs, blown sands, morainic deposits, boulder-clays, glacial sands, etc. These materials are handled in most water engineering schemes, and pre-knowledge of their nature and thickness is important during the planning stage. *See* DAM, ENGLACIAL MATERIAL, ICE AGE, RAISED BEACHES, TERRACES.

quick condition Applied to a soil condition in which an upward flow of water has reduced its intergranular pressure and also its bearing capacity. *See* UPLIFT.

quick-return flow Sometimes, that portion of rainfall which enters the upper soil layers and moves laterally into a stream. *See* SHALLOW PERCOLATION, SUBSURFACE RUNOFF.

quicksand An unstable, waterlogged fine sand with no bearing capacity; a sand in which the upward movement of water is so strong that the mass is held in suspension by the water. The most susceptible are the smooth, rounded-grain sands. *See* BOIL, SAND.

quoin post *See* HEEL POST.

R

race A rapid or strong current of water, or the conduit or channel for such a current; a stream from a pond or reservoir to a water wheel which it drives. *See* TAIL RACE.

rack *See* TRASH RACK.

rack and pinion A device consisting of a bar with teeth (a rack) which engages with a toothed wheel (a pinion) to convert rotatory motion to linear, or vice versa.

radial drainage A natural drainage pattern associated with the updoming of a relatively level area with concentric belts of weak and relatively hard beds. The radial drainage was initiated on the domed surface, with streams flowing outwards in nearly every direction. In many cases, the simple dome structure has almost completely disappeared as the streams cut downwards and flow on older rocks. *See* DRAINAGE PATTERN.

radial-flow tank A circular tank with flow from a central inlet to a circumferential weir. The floor usually slopes to a central sump from where the sludge is withdrawn.

radial gate (or **segmental sluice gate** or **Tainter gate**) A gate located on the spillway crest of a dam with a curved upstream face which is horizontally pivoted (Figure R.1). The curved surface is made concentric to the pivot or pin, hence the resultant water pressure passes through the pivot. The design simplifies the hoisting of the gate.

radial well A well where a number of strainer pipes are driven laterally into a water-bearing deposit in radial fashion from a main sump. *See* STRAINER.

radius, hydraulic *See* HYDRAULIC RADIUS.

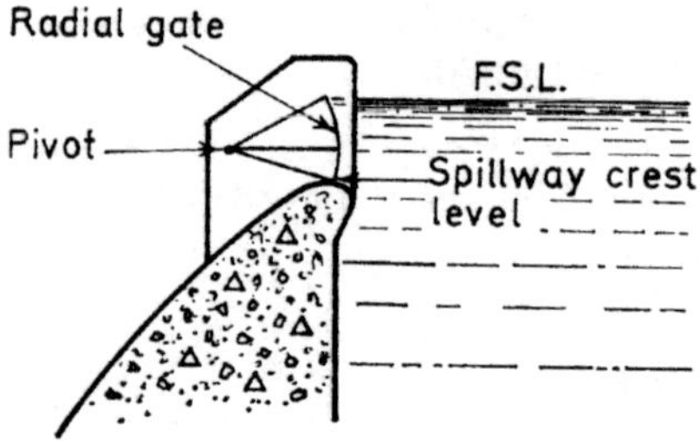

Figure R.1. Radial gate

radius of influence (water well) The radius of the circular base of the CONE OF DEPRESSION formed in the groundwater around a well when pumping is in progress.

raft A flat floating structure, consisting of logs, casks, etc., for conveying persons and materials across a body of water, used in emergencies as a substitute for a boat.

raft bridge A bridge in which rafts are used as supports.

raft foundation A type of foundation often used under heavy structures or in soft, wet ground. It consists of a continuous slab of concrete, generally reinforced, extending under the entire structure. *See* BUOYANT FOUNDATION.

rain Condensed moisture of atmosphere falling as liquid droplets.

rain day A period of 24 hours during which a rainfall rate greater than a specified amount is recorded; the term is used in only some countries.

raindrop erosion *See* SPLASH EROSION.

rainers (or **rain guns**) Large sprinklers with a high output of water and a high working pressure, giving a radius of circle watered from 18 to 55 m (60 to 180 ft). Since only large water drops will reach the outer rim of the circle watered, care is required to prevent soil damage. *See* SPRINKLER IRRIGATION.

rainfall Liquid droplets falling from the atmosphere or clouds during a rainstorm; the volume of rainfall over a specified period. *See* CONVECTIONAL RAIN, DRIZZLE, OROGRAPHIC PRECIPITATION, PRECIPITATION, STORMWATER.

rainfall area The area covered by a storm or period of rainfall which contributes to stream flows, groundwater recharge, etc. *See* AREAL RAINFALL, AREAL REDUCTION FACTOR.

rainfall, average *See* AVERAGE ANNUAL RAINFALL.

rainfall distribution coefficient A storm coefficient obtained by dividing the maximum rainfall at any point within the area of a rain storm by the average rainfall in the area.

rainfall, duration *See* DURATION OF RAINFALL.

rainfall excess The DIRECT RUNOFF, or the amount of rainfall exceeding that taken by infiltration, evaporation, and absorption. Rainfall excess investigations have been aided in recent years by the use of powerful computers and also by rainfall runoff models. *See* RATE OF RAINFALL, SURFACE RUNOFF.

rainfall index An index representing the combined effects of DEPRESSION STORAGE, INFILTRATION and INTERCEPTION; average rainfall above which the volume of rainfall equals the observed runoff.

rainfall intensity *See* RATE OF RAINFALL.

rainfall intensity curve *See* INTENSITY CURVE.

rainfall intensity, maximum *See* MAXIMUM RAINFALL INTENSITY.

rainfall penetration The depth to which a given quantity of rainfall infiltrates: the maximum possible penetration is down to the water-table. *See* INFILTRATION RATE.

rainfall province A region or district throughout which the seasonal concentration and distribution of rainfall are similar.

rainfall rate *See* RATE OF RAINFALL.

rainfall recurrence interval *See* INTENSITY FREQUENCY OF RAINFALL.

rainfall runoff model A model assumed to describe the rainfall runoff process and to make predictions of runoff flows resulting from rainfall. Many of these models have been developed; some are very simple and use only two or three parameters to predict single events from one catchment only. The most complex are of more general application, and use numerous parameters to predict a long series of future runoffs. All but the smallest and most simple rainfall runoff models are usually run on computers.

rain, freezing *See* FREEZING RAIN.

rain-gauge (or **gauge** or **pluviometer** or **udometer**) A device which collects and measures the amount of precipitation, and expresses it as depth of rain to tenths of a millimetre or hundredths of an inch, and includes the equivalent depth of melted snow or hail (Figure R.2). It may consist of a funnel from which the water drips into a graduated cylinder. In the UK rain-gauges are usually 127 mm (5 in) in diameter and set with their rims

305 mm (12 in) above ground level, and placed at sites which are not too exposed nor too sheltered. *See* AUTOGRAPHIC RAIN-GAUGE, BRADFORD GAUGE, SHIELD.

Figure R.2. A common type of rain-gauge – a British Meteorological Office rain-gauge Mk 2

rain-guns *See* RAINERS.

rainmaking *See* CLOUD SEEDING.

rain recorder An instrument which gives a continuous record of precipitation. Trapped water elevates a small float to which a pen is attached for tracing a continuous record on a clock-driven chart. *See* GAUGE.

rain, residual *See* RESIDUAL RAIN.

rainshadow A meteorologically dry area sometimes induced on the leeward side of a line or range of high country. *See* NON-FRONTAL PRECIPITATION.

rain, trace *See* PRECIPITATION (TRACE OF).

rain-wash (1) The action of rain in loosening and washing down rock and soil material to lower levels and finally into reservoirs, rivers, or lakes. (2) A landslide of broken rock and soil caused by heavy rainfall. *See* GULLY DRAINAGE, SHEET EROSION, SOIL EROSION.

rainwater harvesting The collection of rainfall run-off for domestic use, such as from roof catchments, or agricultural use, by means of contour bunds, terraces, or embankments. *See* WATER HARVESTING.

raised beaches Ancient beaches which now stand well above sea level owing to slow relative uplift of the land. Around the Scottish coasts, raised beach terraces occur at elevations ranging up to 30 m (100 ft) above the present sea level, while in Norway and Sweden some beaches lie at over 180 m (600 ft) above the present shore. *See* TERRACES.

raising of dams *See* DAM RAISING.

ram (or **plunger**) The moving cylindrical block in the working chamber of a ram pump (*see* FORCE PUMP).

ramp A short length of drain with a much steeper slope than normal.

ram pump *See* FORCE PUMP.

random numbers Numbers that have an equal probability of occurrence, such as the numbers drawn in a large lottery. Thus in a sequence of random numbers no correlation will be observed between any of the terms. Sequences of random numbers are employed in NUMERICAL SIMULATION MODELLING, and particularly in STOCHASTIC MODELS. In practical applications the random numbers required are usually specified as being drawn from within a fixed range and probability distribution.

random number generator True random numbers are difficult to generate but for certain applications they can be obtained by physical methods, such as throwing dice, drawing counters from a box or measuring natural processes which are known to be random. For NUMERICAL SIMULATION MODELLING pseudo-random numbers are used, which have properties very close to being truly random. These can be generated by a range of mathematical expressions, chosen to produce sequences of numbers drawn from any selected range and distribution. The following expression gives a sequence of pseudo-random numbers n_1, n_2, n_3, uniformly distributed between 0.0 and 1.0:

$$n_i = \text{fractional part of } (\pi + n_{i-1})^5$$

The sequence must be started by choosing arbitrary seed n_0 between 0 and 1.

range (mainly USA) A stretch of grazing land with little subdivision fencing; cover is composed mainly of herbs, native grasses or shrubs. *See* GRAZING MANAGEMENT.

range of tide, great diurnal *See* GREAT DIURNAL RANGE OF TIDE.

rapid flow (or **shooting flow**) The flow in open channel with a velocity greater than the celerity of a gravity wave; flow in open channel in which the FROUDE NUMBER is greater than unity. *See* CELERITY OF WAVE.

rapids The swift and sudden descent of a stream without actual waterfall. They may be formed when a stream passes over a resistant rock with softer beds below; the greater erosion of the latter develops sufficient drop in slope or grade to produce rapids. *See* CASCADE, NARROWS, WATERFALL.

rate of flow *See* DISCHARGE.

rate of infiltration *See* INFILTRATION RATE.

rate of rainfall (or **intensity of rainfall** or **rainfall intensity**) The rate of rainfall expressed as depth of water in millimetres or inches per hour, day or other specified period. *See* MAXIMUM RAINFALL INTENSITY.

rate of runoff The rate at which water flows from a catchment area at any given time, usually expressed in cumecs or cusecs. *See* PEAK RATE OF RUNOFF.

rating (1) In general, the relationship between two mutually dependent quantities. In hydraulics, e.g., the relationship between the water level and the discharge of stream or channel, or between CURRENT METER revolutions and water velocity. (2) The taking of flow measurements to establish a rating. *See* CURRENT METER RATING, DURATION CURVE.

rating curve A curve or tabular representation of a RATING. *See* DURATION CURVE.

rating flume A CONTROL FLUME; a flume containing still water along which CURRENT METERS are drawn at known velocities in order to calibrate them; an open conduit constructed in a channel to ensure good control for flow measurements and values for a RATING CURVE.

rating loop A DURATION CURVE which describes a loop with each rise and fall of the river, caused by higher discharge values for given water levels when the river is rising than when it is falling.

rating table A table showing the relationship between two mutually dependent variables or values, e.g., between discharge of a stream and water level at a gauging station. *See* RATING.

rating tank A tank containing still water through which a CURRENT METER is drawn at a known velocity in order to calibrate the meter.

rational method A method of predicting the rate of discharge from relatively small areas which makes use of both empirical data – particularly of rainfall – and the physical parameters of the CATCHMENT. It was first developed for the design of urban drainage works. The rational method, or techniques developed from it, are widely used for engineering design as they enable discharges to be predicted quickly and with reasonable accuracy for any required RETURN PERIOD. However, although better than most simpler methods, it is not completely rational as good results are only obtained if empirical RUNOFF COEFFICIENTS are used.

rational runoff formula (USA) A formula used to predict the maximum stream flow from a catchment based on the rational method. The basic form is:

$$Q = CAI$$

where Q is the peak rate of discharge for a required RETURN PERIOD, C is a runoff coefficient, A is the area of the catchment and I is the average intensity of rainfall over the whole catchment during the TIME OF CONCENTRATION and with the selected return period. More complex forms of rational runoff formula introduce more realistic information on the pattern of rainfall and additional parameters to describe the catchment, including lengths of watercourses, slopes and types of surface. These increase the time needed for computations but increase the probable accuracy of the predictions.

raw bank A river or other bank in its natural state, or where the protective surface has been eroded or stripped off.

raw-bank system (Australia and New Zealand) A heavy timber construction to provide protection against bank erosion. Tree trunks, placed with butts upstream, are secured by steel cable to vertical bearer logs along the bank and also to posts set well back from the bank. The bearer logs are trenched and layered down-bank, with toes below low water level. Individual trees are wired together with at least two ties per tree. *See* ANCHORED TREES, BATTERED-BANK SYSTEM.

raw water Water in the state in which it is first abstracted from a primary source, such as a river, large lake, aquifer, etc. and before it has received any treatment.

RBC Rotating biological contactor.

reach The distance between two marked or specified stations or cross-sections of a river or stream; the unobstructed stretch of water between two locks. *See* POND.

reaction turbine A WATER TURBINE in which the nozzles or jets are on the moving part and not fixed. *See* IMPULSE TURBINE.

realignment A change in the planned direction of a tunnel, canal, pipeline or other hydraulic excavation. A change in direction (horizontal alignment) often affects its gradient (vertical alignment).

real-time forecasting A method of forecasting that uses observations of parameters such as rates of rainfall, soil moisture and stream flow, which are observed within the catchment while a flood event is in progress. Among the more reliable methods of real-time forecasting of flood flows soon expected at one site on a river is to measure the rate of flow further upstream. Complex methods based on simulation models are now in use: these give early and reliable forecasts, but require the processing of large amounts of data by computer.

receiving waters Streams, lakes and rivers into which sewage and effluents are discharged after treatment. In most cases the quantities and quality of such discharges are regulated by public authorities, whose requirements will dictate the standards to which discharges must be treated. *See* SEWAGE TREATMENT PLANT.

recession curve, normal *See* HYDROGRAPH (RECESSION).

recession hydrograph *See* HYDROGRAPH (RECESSION).

recession of coast The retreat of the coast as a result of wave erosion; it may be very rapid along sandy and low-lying areas. Protective work includes different forms of walls, jetties and bulkheads, with varying results. *See* COAST OF SUBMERGENCE.

recharge of aquifer (or **groundwater increment** or **groundwater recharge**) The introduction of surface water, such as streams and rainfall, into underground storage. Inflow of surface waters must be geologically related to the aquifer to be recharged. This requires knowledge of the local geology, structures, and the trend of the permeable and impermeable deposits. GROUNDWATER DECREMENT, GROUNDWATER INCREMENT, INTAKE AREA OF AQUIFER.

recharge of aquifer, induced (or **induced recharge of aquifer**) Flow of stream water into an aquifer.

recharge of basin *See* BASIN RECHARGE.

recharge of groundwater *See* GROUNDWATER INCREMENT, SEASONAL RECOVERY.

recharge well An INVERTED WELL which conducts surface water into an aquifer at shallow or moderate depth; a type of well often used in arid regions to conserve water. *See* INVERTED CAPACITY.

reciprocating pump Any type of pump which lifts water to a higher level by a series of to-and-fro movements of a piston or plunger. *See* PUMP.

recirculation The return for refiltration of part of the effluent from a trickling filter to maintain a uniform rate through the filter.

recorder *See* RECORDING GAUGE.

recording gauge (or **recorder**) (1) An instrument which automatically records values, quantities or conditions, such as rainfall, flow, stage, pressure or temperature. The values are usually given as a graphic representation on a time basis. (2) A water-level recorder showing the level of water in any conduit, reservoir, or tank over a specific period (usually 7 to 28 days); it is often operated by a float, with levels shown on a chart. *See* GAUGE, WATER STAGE RECORDER.

recovery, cyclic *See* CYCLIC RECOVERY.

recovery, seasonal *See* SEASONAL RECOVERY.

rectangular crib groyne A CRIB GROYNE, rectangular in shape, and filled with stone. The length usually exceeds 6 m (20 ft) and the width between the side walls is about 1 to 1½ times the height. Such groynes can be made with continuous sides and wire or cross-logs used to prevent spreading. Construction is similar to triangular groynes, and wire, rods or spikes are used to bind the logs and walls together. The bank is sloped and grassed. *See* GROYNES, TRIANGULAR CRIB GROYNE.

rectangular drainage A natural drainage pattern where the course of the streams is controlled by fault and joint systems in the superficial deposits, or by folded or tilted strata consisting of alternating strong and weak beds. *See* DRAINAGE PATTERN.

rectangular weir A MEASURING WEIR that has a rectangular notch. *See* NOTCH.

recurrence interval The average number of years within which a given hydrological event, such as flood, is equalled or exceeded. *See* *N*-YEAR FLOOD.

recurrence interval (rainfall) *See* INTENSITY FREQUENCY OF RAINFALL.

recycling The reuse of water in a process or in other processes, or the use of wastewater and sewage sludge in agriculture. Also, the reuse of solid waste components including energy conversion.

reefing float *See* CAPTIVE FLOAT.

reference current meter The observations from CURRENT METERS used in the field for the measurement of river flows are subject to errors from field work, manufacturing inaccuracies and wear and tear. It is therefore good practice to keep one meter as a reference meter. This instrument is used as rarely as possible and then only to calibrate other meters when it is unlikely to be damaged. The reference current meter is itself normally calibrated in laboratory conditions. *See* STANDARD CURRENT METER.

reflected wave The return wave after an incoming wave is reflected from a steep beach, cliff or marine structure.

reflux valve *See* NON-RETURN VALVE.

regelation (of ice) The solidification of ice, previously melted by an increase of pressure, when the pressure is reduced or removed.

regenerated water (or **return water**) Water diverted from a stream or other surface source, for irrigation or other use, which seeps downwards and laterally to return to a stream or other body of water.

regeneration curve A curve showing the rise in water level in a well, plotted against time, after pumping has ceased. *See* CONE OF DEPRESSION.

regime (1) In regime: referring to a canal or stream in which the water flow is such that it neither deposits material on its bed nor picks up material from it. At any particular period, only short lengths of a stream are in regime, while canals after some years tend towards the condition. *See* PROFILE OF EQUILIBRIUM. (2) The prevailing character of a climate or of some climatological factor, such as seasonal distribution of rainfall. Also 'regimen' (USA).

regime, coefficient of *See* COEFFICIENT OF REGIME.

regimen *See* REGIME.

regular velocity distribution The velocity distribution between two defined cross-sections of a channel which does not change for a given flow.

regulated flow A condition where upstream reservoirs regulate the flow in a stream. *See* BASE FLOW.

regulation (stream) The control of the natural flow in an open conduit or stream.

regulator storage *See* PONDAGE.

reinforced concrete CONCRETE in which metal (usually steel rods) is embedded such that the two materials act together in resisting loads or forces. Concrete alone is strong in compression but very weak in tension. Stress calculations are usually made on the assumption that the concrete takes all the compressive forces and the steel takes all the tensile forces. *See* PORTLAND CEMENT.

reinforcing fabric Usually steel mesh, which may be interlaced, woven or welded, and used to secure materials such as vegetation, stone, etc., for river bank protection or for securing and reinforcing certain types of GROYNES and other works. *See* SAUSAGE CONSTRUCTION, STONEMESH GROYNES.

rejuvenated water Sometimes, WATER OF COMPACTION and water released during metamorphism. *See* MAGMATIC WATER.

rejuvenation The condition of a stream where the velocity of flow has increased, as by uplift or tilting of the land, with a renewal of downward erosion of its bed. It often cuts below the level of the plain, parts of which are left as river terraces.

relative humidity *See* HUMIDITY.

relief channel (rivers) A passage, canal or conduit cut between favourable points along the course of a river, used to store water in times of flood and discharged when the level or tide is low. *See* BARRAGE.

relief valve *See* AIR VALVE.

relief well A borehole put down at the toe of an earth dam to lower any high PORE WATER PRESSURE formed by the weight of the dam (Figure R.3).

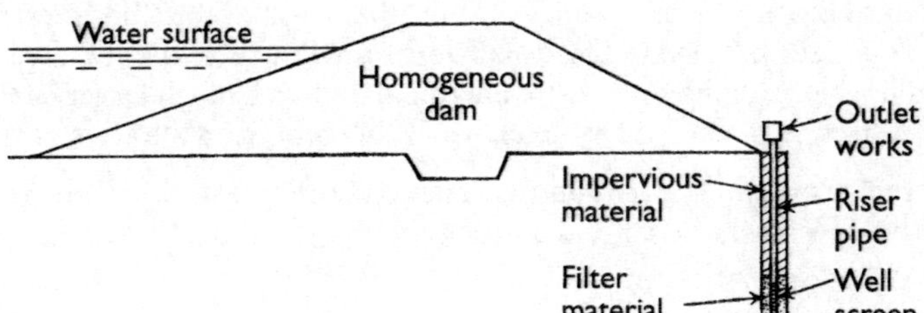

Figure R.3. Relief well

remolding (USA) *See* remoulding.

remote sensing The collecting and recording of data about the earth's surface by cameras, radar or other sensors, operated from aircraft or satellites.

remoulding The derangement of the internal structure of a silt or clay, with the result that it gains compressibility but loses shearing strength. *See* PUDDLING.

removal of support The removal of ore or coal by mining operations under a reservoir or canal. Within a certain depth, such removal of support may cause severe damage to the surface works and may result in a costly legal action.

remove and stockpile topsoil A term used in earthworks contracts to indicate topsoil that is required for reuse to cover embankments, etc., of ordinary fill.

required excavation A contract term used in construction which includes all excavation for the foundations and for structures, but excludes quarries, borrow pits and temporary works.

reservoir An artificial lake into which water drains and is stored for future use; a tank, basin, cistern, millpond or any place in which water has accumulated. A reservoir provides for the storage of irregular sources of water, such as flood and storm water and intermittent streams, and also provides storage capacity during non-irrigation periods. *See* DAM, FARM POND, FARM WATER SUPPLY ENGINEERING, FAULT, IMPOUNDING RESERVOIR, SERVICE RESERVOIR, SILTING, SURFACE WATER SUPPLIES.

reservoir basin The area of low-lying land or valley to be flooded at a dam site. Detailed geological maps are prepared to indicate any faults, zones of weakness, or pervious rocks. The permeability of the rocks may be determined by tests, as by injecting or pumping water down boreholes under pressure and observing the actual losses. *See* DAM SITE INVESTIGATION.

reservoir capacity The volume of water that can be stored in a reservoir up to FULL SUPPLY LEVEL, and usually expressed in cubic metres, acre feet, or millions of gallons. A low stream gradient gives a much larger capacity than a steep gradient for any given height of dam. *See* STORAGE CAPACITY.

reservoir demand The amount of reservoir water required from time to time by consumers. Ideally, the RESERVOIR DRAFT should be equal to the reservoir demand.

reservoir draft The amount of water withdrawn or released from a reservoir periodically to meet the requirements of consumers. *See* STORAGE PERIOD.

reservoir flood routing The process of FLOOD ROUTING for a FLOOD HYDROGRAPH as it passes through a reservoir. Except in the case of long shallow reservoirs, the calculations for reservoir flood routing can be done by relatively simple step-by-step calculations. Apart from the hydrograph of the incoming flood, the only data required are the STAGE–STORAGE RELATIONSHIP for the reservoir and the STAGE–DISCHARGE RELATIONSHIP at its downstream outlet.

reservoir inflow The water which flows naturally or artificially (e.g. by pumping) into a reservoir over a given period.

reservoir inflow hydrograph A graph showing RESERVOIR INFLOW over a given time period.

reservoir, life of *See* LIFE OF RESERVOIR.

reservoir losses Losses to which water held in storage is subject, mainly by seepage through the reservoir floor and embankments and also by evaporation from the water surface. *See* PERCOLATION PATH.

reservoir retention levels The several levels at and above the level of a reservoir's spillway crest which denote the volume available for temporary storage, following flood inflows, and the allowance that must be made over and above that for wave action (Figure R.4). See DAM FREEBOARD, FLOOD SURCHARGE, FULL SUPPLY LEVEL, WAVE FREEBOARD, WAVE SURCHARGE.

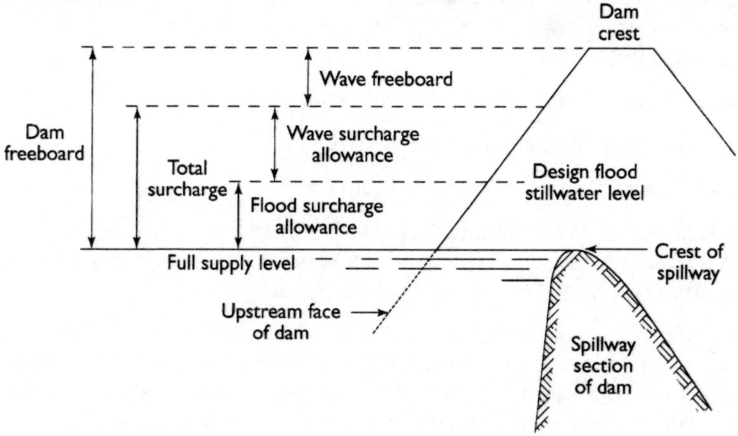

Figure R.4. Reservoir retention levels

reservoir sedimentation The deposition of silt and sediment transported by streams, runoff, etc., into a reservoir. *See* SILTING, LIFE OF RESERVOIR.

residual chlorine Chlorine remaining in water after a specified contact period as free or combined chlorine, available to act as an oxidant. Also 'free residual chlorine' or 'combined residual chlorine', e.g. chloramines.

residual mass curve A curve showing year-to-year residual departures of discharge or rainfall from the arithmetic average accumulated for a given period. *See* MASS DIAGRAM.

residual rain Rainfall near the end of a storm at a rate less than INFILTRATION CAPACITY. *See* INITIAL RAIN, NET SUPPLY INTERVAL.

resistance The friction which flowing water must overcome because of the roughness of the channel or pipe, expressed in metres (feet) head of water. *See* CHANNEL ROUGHNESS, ROUGHNESS COEFFICIENT.

resistance strain gauge *See* ELECTRICAL RESISTANCE STRAIN GAUGE.

resistivity method A GEOPHYSICAL PROSPECTING method in which direct measurements are made of the ratio of voltage to current when a current is forced to flow through the ground to be tested. The conducting property of a rock is controlled by its water content and its salinity. If these values are high then its conductivity is also high and its electrical resistivity is low. The method has been used extensively for water and foundation problems. In certain conditions the SEISMIC METHOD has been preferred.

resoiling The replacement of the original topsoil or of fresh soil along the bed and banks of rivers, canals and reservoirs, on completion of operations to allow the growth of vegetation to reduce erosion and increase stability of slopes.

response time (storm) *See* PERIOD OF CONCENTRATION.

rest water level Sometimes, the STANDING WATER LEVEL.

resurgent water Sometimes, MAGMATIC WATER.

retained water Water retained in a soil or rock by molecular attraction against the force of gravity; the water remaining in a soil or rock after the GRAVITY GROUNDWATER has drained out.

retaining wall A wall constructed to hold back earth or other solid material. A wide range of structural forms are used to resist the pressures applied to retaining walls and they are built from most construction materials including mass concrete, reinforced concrete, masonry and sheet steel piles.

retard A permeable spur built out from the bank into a channel to deflect the main flow of water away from the bank to reduce erosion. *See* JETTY, REVETMENT, TREE RETARDS.

retarding reservoir (or **retarding basin**) A reservoir basin, which at most times is dry and preferably formed by river banks or a natural depression. It enables flood water to be temporarily stored to reduce the rate of flood flow further downstream. Water is usually retained in a retarding basin by a dam at the downstream end, constructed so that it only blocks part of the channel and there is no obstruction at times of low flow.

retention The difference between total precipitation and total runoff on a drainage area. *See* INFILTRATION, INFILTRATION COEFFICIENT.

retention reservoir A reservoir in which water may be stored or released as desired by gated outlets below spillway level.

retention, specific *See* SPECIFIC RETENTION.

retention, surface *See* INITIAL DETENTION.

retrograding Active erosion on a coastline where the shore line is cut back into the land mass. *See* PROGRADING.

retrogression (1) The lowering of bed levels of streams by downward erosion. (2) The lowering of water level at the same discharge. *See* ALLUVIATION.

retrogression, sympathetic *See* SYMPATHETIC RETROGRESSION.

returned sludge Settled activated sludge returned to the aeration tanks.

return period A measure of the most probable time interval between events of a given magnitude being equalled or exceeded. The reciprocal of the return period is the probability of such an event occurring in any one time period. Thus if an event is observed n times in a very long interval of time t, the estimated return period of such events will be t/n time periods, and the estimated probability of this event happening in any one time period will be n/t. The accuracy of the prediction increases with t.

This term is very widely used to express the probable time interval between rare natural events such as major floods, heavy rainfalls or droughts, for which return periods are usually stated in years. However, some care must be taken in the use of this term, as it is often not properly understood that return periods are not the time intervals at which events will regularly occur, but only a measure of the probable time between them based on a very long average.

return water *See* REGENERATED WATER.

reuse of water The repeated use of the same water. The purification of industrial water is performed by chemical treatment or filtration or other methods. The water obtained from rivers is often used several times before it re-enters the river.

reversed filter *See* LOADED FILTER.

reverse flow A flow in a channel or conduit opposite to normal flow.

reverse osmosis An advanced wastewater treatment process to remove dissolved substances, including inorganic salts (Figure R.5). An external force such as pressure applied by a pump is used to reverse the normal osmotic flow through a semi-permeable membrane. The process is also used to remove salinity giving potable water. *See* OSMOSIS.

Figure R.5. Two reverse osmosis treatment plants ready for delivery, each capable of delivering up to 15m^3 a day of desalinated water

revet To reduce or prevent the wearing away of a surface or bank with a REVETMENT.

revetment A lining or wall built along the sides of a stream to protect the bank and minimize erosion; it may be temporary until the bank can be stabilized by solid works or live vegetation. The term is also applied to any covering to prevent scour of rock or soil surfaces such as concrete slabs, grass, maritime plants, faggots, asphalt sheets, etc. *See* APRON, FASCINES, PITCHING, RETARD, RIPRAP, STONE RIPRAP.

Reynolds number (or **Reynolds criterion**) A dimensionless number symbolized by Re, named after Professor Osborn Reynolds and applied to fluid flow in a tube. The Re value is the same for all fluids at the CRITICAL VELOCITY.

$$Re = \frac{\text{mean velocity of flow} \times \text{pipe diameter}}{\text{kinematic viscosity}}$$

The number is sometimes used in MODEL ANALYSIS, and should be the same in both model and corresponding structure. *See* FROUDE NUMBER.

rias *See* DROWNED VALLEYS.

ria-type coast A coastal area with long parallel estuaries (or rias) running far inland, as in south-western Ireland, developed where the river valleys and ridges are at right angles to the coastline. *See* DALMATIAN-TYPE COAST, FIORD COAST.

ridge terrace An embankment of earth to reduce erosion by diverting runoff water across the slope of the ground, instead of allowing it to flow directly down the slope. It usually consists of a long, low ridge of earth with a gentle slope along the sides, and a shallow channel along the crest. *See* BENCH TERRACE, BROAD-BASE TERRACE, NICHOLS TERRACE.

ridging for irrigation Controlling the flow of irrigation water in fields by constructing small ridges, borders or embankments.

right bank The right-hand bank of a river facing downstream (Figure A.4).

rigid joint A joint in a pipe which allows no movement. The type includes welded, flanged and turned-and-bored. *See* SEMI-RIGID JOINT.

rill erosion Soil erosion by rills or streamlets; the rills can be levelled out by normal cultivation. *See* GULLY EROSION.

ring dam An earthen dam, usually of circular shape, built on level ground from material excavated from within the storage basin to impound water for farm supplies and irrigation (Figure R.6). These small dams may have storage ratios varying between 2:1 and 6:1. *See* HILLSIDE DAM, STORAGE/EXCAVATION RATIO, TURKEY'S NEST TANK.

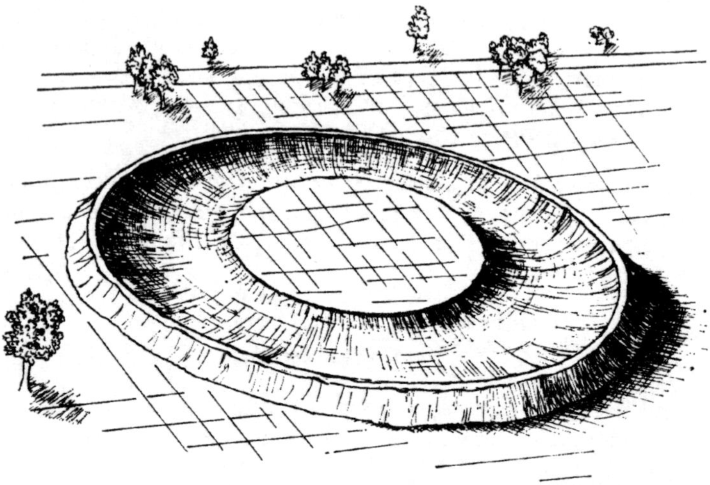

Figure R.6. Ring dam

ring levee An artificial, irregular ring-shaped LEVEE sometimes used to protect an area of land enclosed by a river and an effluent which later rejoins the main river. The ring levee is built to protect the island so formed.

ring main A main irrigation pipe installed in the form of a complete circle. It enables a greater area to be irrigated than in the case of a single line. Friction losses are reduced, since the water has two conduits to reach each distribution point. The pipe column can thus be of smaller diameter, although its total length is greater. A ring main may also be used in a water supply system. A complete circuit is the aim, in which case a supply is available in either or both directions. It is often connected to all or most of the TRUNK MAINS in the area.

ring tank *See* RING DAM.

riparian Pertaining to, or situated on, the banks of a river or body of water, or the rights or ownerships along the banks of rivers, lakes or canals; a dweller on a river bank. A riparian owner has rights over part at least of one bank.

riparian land Usually the land situated along the banks of a natural body of water.

riparian rights The rights of a riparian owner to the use of a stream or other water which borders or flows through his land. *See* WATER RIGHTS.

Ripple method A graphical method of estimating from a record of inflows the volume of reservoir storage required to sustain a given yield (Figure R.7). Graphs are plotted of both the cumulative reservoir inflows and the required cumulative yield; the required storage volume is read off as the maximum separation between these two lines. The method is now little used except for preliminary design, as unless very long records of inflow data are available it gives only approximate predictions, and no indication of the probability of failure.

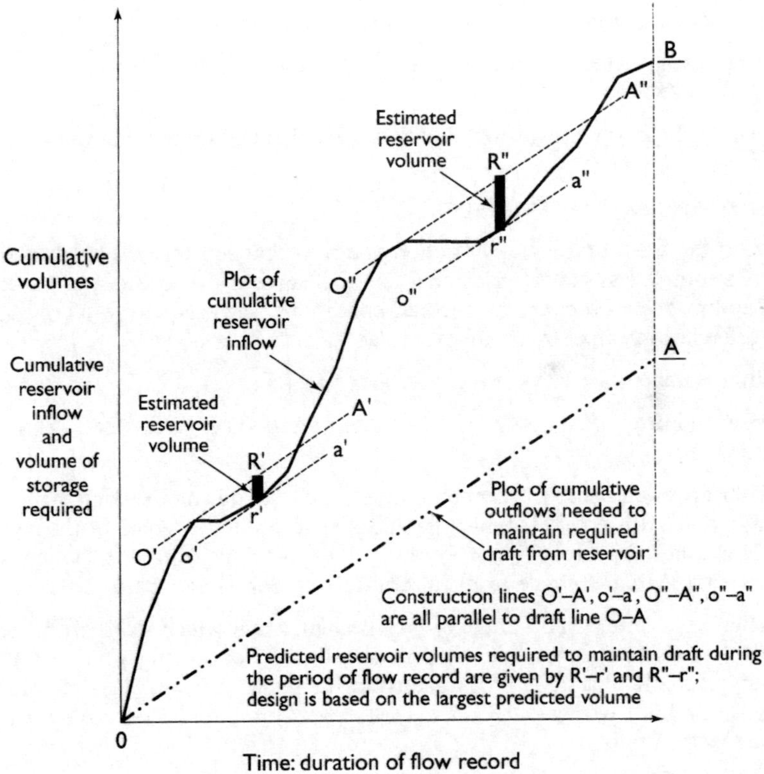

Figure R.7. Ripple method

riprap (or **enrockment**) A covering or backing of broken stone or brush and stone on the face of a dam or the bed or banks of a river as a protection against erosion, often used along irrigation channels or river improvement works. *See* REVETMENT, STONE RIPRAP.

rising tide *See* FLOOD TIDE.

river A stream of water, larger than a brook or rivulet, flowing in a natural open channel towards the ocean, a lake, or another river. In addition to the part it plays in draining land areas, a river may supply large quantities of water for industry and towns. From a legal or administrative point, the term 'river' may include any lake or artificial storage along the river. *See* ALLUVIAL RIVER, BRAIDED COURSE, CONSEQUENT STREAMS, CREEK, INCISED RIVER, OLD RIVER, SUBSEQUENT STREAM, TIDAL RIVER.

river bank erosion *See* EROSION, STREAM BANK EROSION.

river basin The land area which a river and its tributaries drain. *See* CATCHMENT AREA.

river bed That relatively flat part of a river channel where water normally flows.

river breathing *See* BREATHING.

river capture (or **piracy**) The action of a river that cuts into the headwater of another, less active, river and captures some of its tributaries. River capture occurs because of unequal conditions, such as nature of rocks, difference in slope or size of rivers. *See* 'MISFIT' STREAM.

river channel *See* BANK, CHANNEL, RIVER BED.

river control *See* RIVER ENGINEERING, RIVER IMPROVEMENT, RIVER TRAINING.

river diversion tunnel A tunnel to divert the river during the construction of a dam. It is excavated into the hillside and around one end of the new dam, and has a capacity to take flood flows. The upstream or inlet end is covered with a concrete or metal grid to stop debris and trash.

river engineering The branch of civil engineering which deals with the control of rivers and their improvement and flood mitigation. A full appreciation of a river's characteristics demands a knowledge of local geology and hydrogeology. *See* LAND ACCRETION, RIVER IMPROVEMENT, RIVER TRAINING.

river flats (or **flats**) Deposits of silt, gravel, etc., bordering a river along its lower reaches, and sometimes subject to flooding to a greater or lesser degree. *See* ALLUVIAL FLAT.

river floodplain swamps Swamps which form in the wider and flatter reaches of rivers, usually but not always in their lower reaches. Commonly they are caused where LEVEES have formed and are frequently overtopped, but then prevent drainage back to the river channel.

river forecasting Estimating or forecasting river discharge and water levels, based on HYDROLOGY and METEOROLOGY. *See* HYDROMETEOROLOGY, RIVER ENGINEERING.

river gauge, river gauging *See* CLOUD VELOCITY GAUGING, COLOUR VELOCITY GAUGING, DILUTION METHODS, FLOAT GAUGING, GAUGE, SALT VELOCITY METHOD, STREAM GAUGING.

river improvement Mainly, engineering works and practices to improve the flow conditions of a river, and to control erosion and siltation. The works may include deepening, widening, straightening or diverting the channel; prevention or reduction of damage by flooding; control of erosion of banks and bed; removing obstructions from bed and banks to improve flow; construction of groynes, weirs, levees, revetments, etc., for river training; construction of new outlets for the river, canalization, etc.; and the planting of trees, shrubs, or grasses for protection of bed, banks, or associated works. *See* RIVER ENGINEERING.

river loads *See* LOAD.

river piracy *See* RIVER CAPTURE.

river profile *See* STREAM PROFILE.

river stage The height of the water surface relative to a fixed datum. *See* HOOK GAUGE, STAGE.

river terraces Natural benches or terraces which border many rivers, usually narrow but often of considerable length along the river. Several terraces may be present on one or both sides; they usually represent the remnants of FLOODPLAINS.

river training Mainly, engineering works to improve the tidal and non-tidal sections of rivers for navigation and other purposes. Improvements may include dredging to remove shoals and to secure deeper water, and restricting the flow to a selected channel by construction of training walls through an estuary, or by the use of low groynes to concentrate the flow in the main stream. In general, efforts are made to concentrate the scouring effect of the current and to keep the channel clear and deep. *See* FLOOD CONTROL.

rivulet A creek, brook or small stream.

roaded catchment (Figure R.8) *See* ARTIFICIAL CATCHMENT.

roadside erosion control *See* HIGHWAY EROSION CONTROL.

rock and soil symbols *See* MAP SYMBOLS.

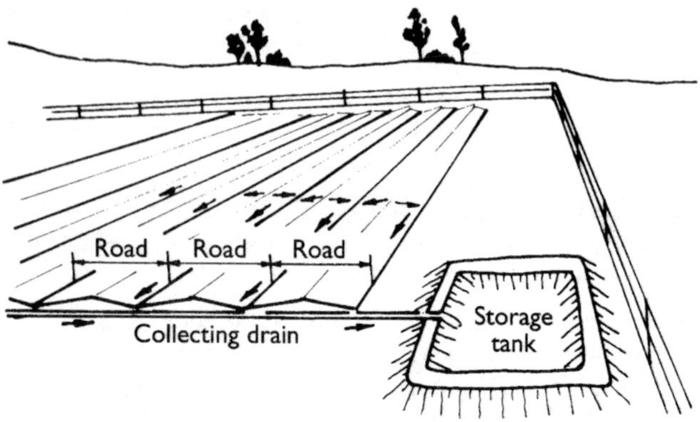

Figure R.8. Roaded catchment

rock excavation An earth dam contract term which includes all solid bedrock which has to be loosened by blasting, and cannot be broken up by heavy pneumatic equipment, and includes boulders over 0.765 m^3 (1 yd^3).

rockfill dam A fill dam composed of more than 50% of pervious rockfill material. The type has many modifications. The watertightness is provided by an impervious membrane or facing which is usually placed on the upstream face of the dam and is connected to a CUTOFF WALL.

rockhead Another name for BEDROCK.

rodding Cleaning out drains with DRAIN RODS which are inserted through a manhole or cleaning eye. *See* INSPECTION CHAMBER.

rod float *See* FLOAT ROD.

roller A machine used to compact soils and rockfill in earth dams and embankments. It consists basically of a drum that can have either a smooth surface or projected teeth. It can be self-propelled or towed. *See* PADFOOT ROLLER, SMOOTH-DRUM ROLLER, VIBRATING ROLLER.

roller gate A CREST GATE, in the form of a hollow cylinder, for regulating the flow at dam spillways. It is carried at each side on large toothed wheels which mesh with steeply inclined racks up which the gate moves when being opened. *See* SECTOR GATE, SLIDING GATE.

rolling-up-curtain weir A type of FRAMED DAM in which the frame remains upright. It is built up with horizontal planks, connected by chains, and drawn up when being opened. The planks increase in thickness with their depth below water.

rollway An overflow spillway or the overflow portion of a dam.

root The part of a dam construction which becomes rooted or embedded in the ground where it joins the hillside. *See* DAM FOUNDATION.

root concentration An expression of the root concentration in a soil. The soil is usually divided into zones 150 or 300 mm (6 or 12 in) thick, and the percentage of the total root system given for each zone.

root of groyne The end of the groyne buried or rooted in the bank of a river. *See* GROYNES.

root zone That soil horizon or zone in which 90% of plant roots are located.

rope and washer pump A VLOM pump made of basic materials, comprising a rope knotted to hold a series of rubber washers (made from old car tyres), rotated by a pulley (made from an old car tyre rim), on a wooden axle (Figure R.9). The rubber washers are guided into a plastic or bamboo

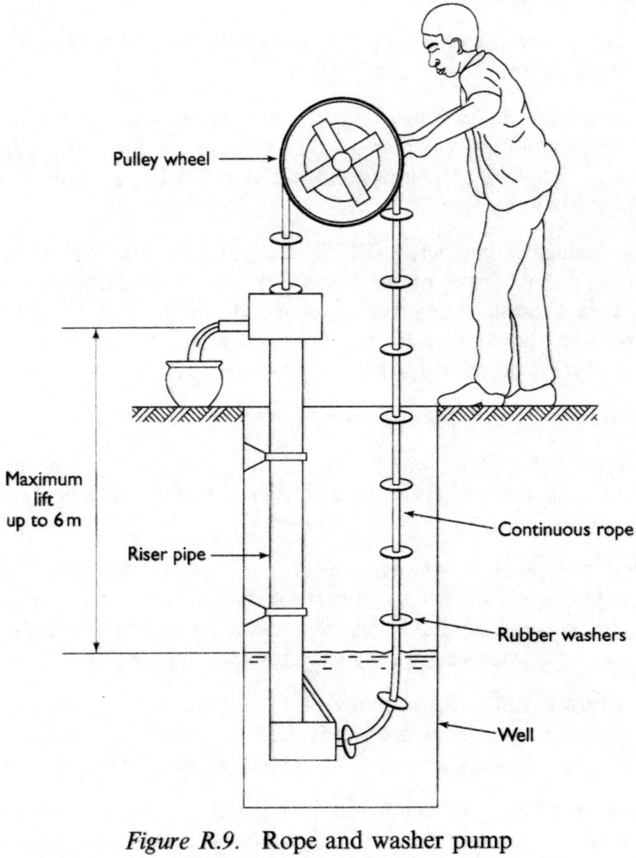

Figure R.9. Rope and washer pump

riser pipe to discharge into a delivery bucket. Sources can be wells, boreholes, ponds and streams. *See* HANDPUMP.

roster system A system in which irrigators receive their supply of water on fixed dates during the season. The dates are shown on cards supplied to the irrigators. *See* ROTATION IRRIGATION.

rotary drill A fast and continuous method of drilling a deep well. The drill bit remains in the hole and additional drill stem is added as the hole deepens. A mixture of clay and water is forced through the bit. The cuttings are carried upward to the surface, and into a tank where they settle out, and the mud is pumped back into the well. When the drilling is complete, a perforated casing is inserted in unconsolidated aquifers to support the walls and allow water to enter the well. The drill is operated by electricity or compressed air. *See* CABLE TOOL.

rotary meter An instrument for measuring the velocity of flow in a river or any open channel. *See* CURRENT METER.

rotary pump A pump consisting of rotating elements and capable of raising large volumes of water at low pressure. It does not depend on centrifugal force to lift the water and it may operate at almost any speed. *See* PUMP, TURBINE PUMP.

rotating biological contactor (RBC) A unit for treating wastewater, comprising a slowly rotating shaft on which are mounted a series of closely spaced discs or mesh drums. The lower sections of the discs (or drums) are submerged in the wastewater so that biological film is carried on the wetted surfaces and then oxidized on rotation out of the wastewater.

rotational slide *See* FELLENIUS'S CIRCULAR ARC METHOD.

rotation irrigation A scheme by which a number of irrigators, receiving water from the same lateral, agree to take the entire flow in turn for a limited period. *See* CONTINUOUS DELIVERY.

rotavirus A type of virus, named after its wheel-like appearance under a microscope, the numerous strains of which are often spread by contaminated water, and are a major cause of diarrhoea that kills several million children each year through unprotected supplies.

roughing filters Filters installed upstream of fine filters and screens for the removal of major solid particles, particularly from highly turbid waters. Normally they consist of beds of stones, large gravel and coarse sand.

roughness coefficient (or **coefficient of rugosity**) A value indicating the channel roughness, used in certain formulas when calculating the velocity or flow of water in a stream or channel. *See* BAZIN'S FORMULA, KUTTER'S FORMULA, MANNING'S FORMULA.

roughness of channel *See* CHANNEL ROUGHNESS.

round-head buttress dam A form of concrete BUTTRESS DAM, consisting of a radial upstream water-supporting face which transmits the water pressure to a buttress below. *See* MULTIPLE-ARCH-TYPE DAM.

round worm *See* HOOKWORMS.

routing flood *See* FLOOD ROUTING.

rower pump A low-lift irrigation HANDPUMP, comprising a plastic tube with a rubber diaphragm foot-valve and a piston operated by a shaft with a handle (Figure R.10). The shaft is inclined at about 45° and operated with an energy-efficient rowing action.

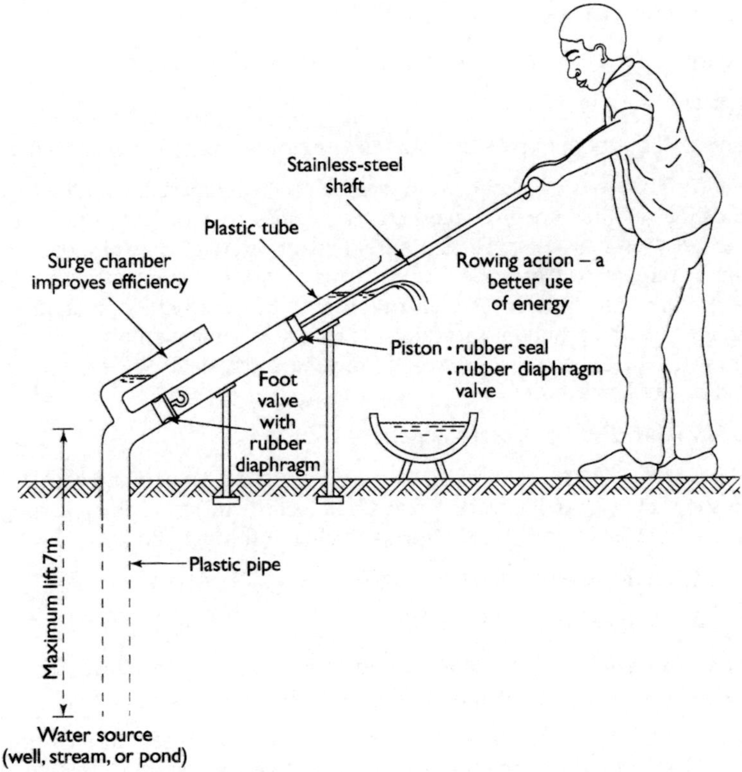

Figure R.10. Rower pump

rubble Any mixture of concrete or brick fragments or small stones of irregular size and shape.

rubble concrete Concrete in which large blocks of stone are placed with a minimum of 150 mm (6 in) of space between them, so that the concrete may be properly rammed and no voids are left. The stones are roughly squared and arranged in courses, so that they break joint. This form of concrete is often used for the HEARTING of masonry dams.

rubble drain (or **blind drain** or **spall drain** or **stone drain**) A drainage trench or channel filled with rubble or stones of suitable size through which the water flows.

rubble-mound breakwater (or **mound breakwater**) A common type of BREAKWATER composed of stones weighing up to 5 t deposited at a nearly flat slope between tide marks, and at a steeper slope below low water. The underwater part is often stabilized by injecting fluid bitumen and sand at 200 °C (392 °F). *See* BLOCKWORK.

rugosity *See* CHANNEL ROUGHNESS, ROUGHNESS COEFFICIENT.

rummel *See* DRY WELL.

runner The rotating part of a WATER TURBINE or other hydraulic wheel.

runoff That part of precipitation which flows from a catchment area and finds its way into streams, lakes, etc. It includes both DIRECT RUNOFF and GROUNDWATER RUNOFF. The amount of runoff is affected by (1) condition and nature of soil or rock; (2) intensity of precipitation; (3) slope; (4) vegetation and (5) wind. A high runoff may be expected if the rainfall is sudden and the ground is saturated or frozen. By gauging the streams that drain a catchment area its runoff may be calculated. *See* CATCHMENT YIELD, RATE OF RUNOFF.

runoff coefficient *See* IMPERMEABILITY FACTOR.

runoff cycle That section of the HYDROLOGIC CYCLE between precipitation or snow melt over land areas and direct return to atmosphere through EVAPO-TRANSPIRATION or subsequent discharge through streams.

runoff, depth *See* DEPTH OF RUNOFF.

runoff, overland *See* SURFACE RUNOFF.

runoff percentage The total amount of water which flows from a catchment area, given as a percentage of total rainfall on the area.

runoff plots Small and relatively normal plots of land used for measuring DIRECT RUNOFF and perhaps other determinations, such as soil erosion and removal of soluble materials.

run-up (or **wave run-up**) The maximum height attained by a wave running up a dam face, above the height of steady water level, including a FLOOD SURCHARGE when present.

S

saddlebag groynes GROYNES consisting of shingle or other material covered over with wire mesh which has ample overlap at the outer end and sides. Rounded stones are placed over the mesh which is then turned back and fastened. The mesh is thus weighed down by a sausage along each side and a mattress at the end. *See* MATTRESS, SAUSAGE CONSTRUCTION, STONEMESH MATTRESS.

safe velocity *See* PERMISSIBLE VELOCITY.

safe yield (aquifer) *See* ECONOMIC YIELD.

sag pipe *See* INVERTED SIPHON.

salinity (groundwater) The content of totally dissolved solids (TDS) in the water, measured by the electrical conductivity method. For sea water the content is about 3.5% or 35 000 mg/L. In some areas salinity is a hazard, as it may rise to levels higher than can be tolerated for irrigation. *See* BRACKISH, pH VALUE.

Salmonella A genus of entero-pathogenic bacteria, found in polluted waters, responsible for typhoid fevers and food poisoning.

salmon ladder A provision made in some sluices, weirs or other barriers to permit the free passage of fish. It consists of a series of small pools, arranged in steps, down which a constant stream of water flows. *See* FISHWAY.

saltation (streams) *See* LOAD.

salt balance The difference between the total solids removed by drainage water from land annually and the total dissolved solids brought in by irrigation water annually. *See* CRITICAL CONCENTRATION.

salt index A value or formula for determining the suitability of a water supply for irrigation.

salting The accumulation of excess salt in the soil with harmful effect on crops and vegetation, possibly caused by the rise of saline groundwater to near-surface level. The soil may be cured by drainage and then LEACHING.

salt load The load of salt carried by a stream, usually expressed in tonnes or tons. *See* SALINITY.

salt slug The introduction of a solution of highly concentrated salt into a stream of relatively low or no salinity.

salt velocity method (or **Allen's method**) A method to determine water velocity by introducing concentrated salt brine at a station and recording the time it takes to travel a known distance downstream. The passage of the brine at the end points of the known distance is detected by the change in electric conductivity of the flowing water. *See* CHEMICAL GAUGING.

sampling station The site or cross-section selected on a stream or open channel at which the dilution or cloud is measured, sampled or observed. *See* DILUTION METHODS, GAUGING STATION.

sand A loose accumulation of particles, mainly quartz, with a grain size between 0.02 and 2.0 mm. Sands and sandstones are important deposits as water reservoirs, allowing the migration of water downwards and laterally. *See* MICROPORES, POROSITY, QUICKSAND, SILT.

sand blanket A covering layer of sand, about 150 mm (6 in) thick, to prevent the cracking of bentonite or clay in reservoirs, canals and water channels when they are empty.

sand boil A BOIL (1) that has developed downstream of an earth dam or embankment with sandy foundations. Sand boils are best prevented by installing a GRADED FILTER, but a temporary remedy is to surround the boil with a wall of sandbags to develop a back pressure and so decrease the hydraulic gradient and stop the loss of sand.

sand catcher (or **sand-grain meter**) An instrument through which the water flows, allowing any sand to settle out and its quantity to be measured. *See* SAND TRAP.

sand drain *See* VERTICAL SAND DRAIN.

sand filter A filter for purifying domestic water. It has a layer of coarse stone at the bottom, grading upwards to fine quartz grains at the top. The water flows slowly downwards and the upper surface is renewed periodically to remove bacteria. A similar type of filter is used for treating sewage effluent, but the sand is coarser and air has access to the sand. *See* GRADED FILTER, SLOW SAND FILTER, UPFLOW SAND FILTER.

sand-grain meter *See* SAND CATCHER.

sand pump *See* BAILER.

sand pump dredger *See* SUCTION DREDGER.

sand reef *See* BARRIER BEACH.

sand screen (boreholes) A tube type of screen for boreholes, generally non-ferrous, with narrow openings for the entry of water. It may be used with gravel packing. *See* PERFORATED-CASING WELL, TUBE WELL.

sand trap A basin-shaped enlargement in a stream, ditch or channel, where the silt and sand carried by the water is deposited due to the drop in velocity; means may be included to remove such material. *See* DESILTING BASIN, DESILTING STRIP.

saturated air Air containing the maximum possible amount of water vapour at the given temperature; air at a RELATIVE HUMIDITY of 100%. *See* VAPOUR PRESSURE.

saturated soil A condition when all the interstices of a soil are filled with water.

saturated surface *See* WATER-TABLE.

saturation The condition when water (or other substance) has filled all the voids of a rock or soil and it can assimilate or contain no more. *See* HUMIDITY, ZONE OF SATURATION.

saturation deficit A measure of the 'dryness' of the air; expressed by:

$$\frac{100 - RH}{100}p$$

where RH is the RELATIVE HUMIDITY and p is the SATURATION VAPOUR PRESSURE.

saturation gradient Slope of the SATURATION LINE.

saturation line (1) A line indicating the uppermost limit of flow of water through an earthen dam, ridge and subsoil, and marked on a cross-section of the structure or mass. (2) The WATER-TABLE.

saturation vapour pressure The pressure that would be observed at any given temperature inside a closed vessel that contained both the gaseous and liquid phases of only one substance (such as water) where there is a state of equilibrium between the rates of evaporation and condensation.

saturation zone *See* ZONE OF SATURATION.

saucer A large flat float tied to a ship to elevate and float it past a shallow area. *See* CAMEL.

sausage construction A type of STONEMESH CONSTRUCTION used for river bank protection. It consists of tubes of wire mesh filled with stone, preferably rounded. The tubes are laid parallel down the bank with the lower ends closed, and then filled from the top end. The bank is usually battered down to a regular slope before the tubes are filled. Where possible the lower ends of the tubes are extended about 0.9 m (3 ft) beyond the foot of the bank across the bed of the stream. This prevents erosion and undermining along the base of the bank. The structure is sometimes reinforced by a growth of roots and branches. *See* CRIBWORK, UNDERWATER APRON.

sausage dam A dam composed of rock fragments which have been wrapped into cylindrical bundles with wire and laid in a vertical or horizontal position to form a barrier across the stream or channel.

schistosomiasis (or **bilharziasis**) The disease caused by the parasite of the genus *Schistosoma* (or bilharzia). The parasite has a life cycle from human, to water, to snail, to water and back to human. The two types of infection are urinary or intestinal, with different snail hosts. It can be a debilitating, even fatal, disease.

schmutzdekke *See* ZOOGLOEAL LAYER.

scour The wearing away of the coast and sea bed, or the banks and bed of a river, or of terraces, diversion channels or dams, mainly by waves and water action. *See* EROSION, SHORE EROSION.

scour, bed *See* BED SCOUR.

scouring organisms Invertebrate animals, such as worms and larvae, which graze on the microbial film of biological filters.

scouring sluice (or **scour pipe** or **washout valve**) A sluice, pipe or opening, in the lower part of a dam, controlled by a gate, through which accumulated silt, sand, gravel or other material may be occasionally ejected. It may also be used to lower the water level in the reservoir in an emergency.

scour pipe *See* SCOURING SLUICE.

scour protection Appliances and practices to prevent scour of rock or soil surfaces. *See* REVETMENT, VEGETATIVE PRACTICES.

scour valve A valve or outlet provided at low points in pipelines through which suspended solids or silt-carrying waters are being delivered.

scow *See* HOPPER BARGE.

scraper An earth-moving machine consisting basically of a large scoop mounted on rubber-tyred wheels. The bottom incorporates a cutting edge which, when lowered, is dragged through the soil. When full the machine moves to the discharge point where the material is dumped through the bottom of the vehicle in an even manner. Used in the construction of earth dams.

screened well A cased well in which the water enters through one or more screens and not through holes in the casing. *See* OPEN-END WELL, PERFORATED-CASING WELL.

screens (intake works) Usually flat or round metal bars built into a frame, to keep out trash and debris, fitted at the entrance end of the structure for abstracting water from a river or other body of water. Facilities are provided for cleaning the screens. Fine screens may also be fitted in the INTAKE WORKS adjacent to the pump house. *See* BAND SCREEN.

scumboard A board which dips below the surface of a fluid so as to arrest the movement of scum.

sea A continuous volume of salt water which may form part of an OCEAN, e.g. the Tasman Sea, North Sea, etc.

sea cliff A steep or vertical rock face formed by wave action. The height depends on the elevation of the land on which the sea advances, and the slope depends on the nature of the cliff rocks and rate of cutting. *See* WAVE-CUT TERRACE.

sea defence works Mainly structures and practices (1) to protect low-lying land against flooding by building walls or embankments, (2) to prevent or minimize erosion of the foreshore, (3) to limit the movement of sand and shingle by erecting groynes, and (4) to protect sand dunes by the planting of trees, grass, and shrubs. *See* SEA WALLS.

seas *See* FORCED WAVES.

seasonal recovery (groundwater) The replenishment of water in the zone of saturation during and following a wet season with rise in the water-table level.

sea walls Protective walls built along the coast to arrest the force and erosive action of the waves, consisting of a low bank of sand, earth or other material; agricultural land is often protected by banks on account of their relative low cost. They have a sloping seaward face covered with a pitching of stone blocks or other hard material. The wall may consist of concrete-filled bags, secured with steel spikes and surmounted by a parapet wall, the top of which is well above the level of the highest known tide. Walls of stone, concrete or reinforced concrete, with a stepped or gently sloping seaward face are sometimes used, and also timber stakes and continuous timber barriers with shingle filling between them. The sea wall may be constructed in conjunction with a system of groynes.

sea water intrusion The seepage of sea water into and under a zone of fresh water (Figure S.1). Saline water has a slightly higher density than fresh water, which leads to the interface between salt and fresh water in AQUIFERS adjacent to the coast being usually inclined and not vertical. As saline water is denser than fresh this inclination will be away from the land, as the heavier salt water will advance under the fresh water zone. The profile of the interface between fresh and salt water is governed by the local geology and the seepage flow of fresh water coming from the land, but is naturally relatively static. However, this equilibrium can be disturbed if water supplies are pumped from aquifers adjacent to the coast, and in extreme cases salt water will be drawn in as far as the points of abstraction and their yield will become too saline for further use.

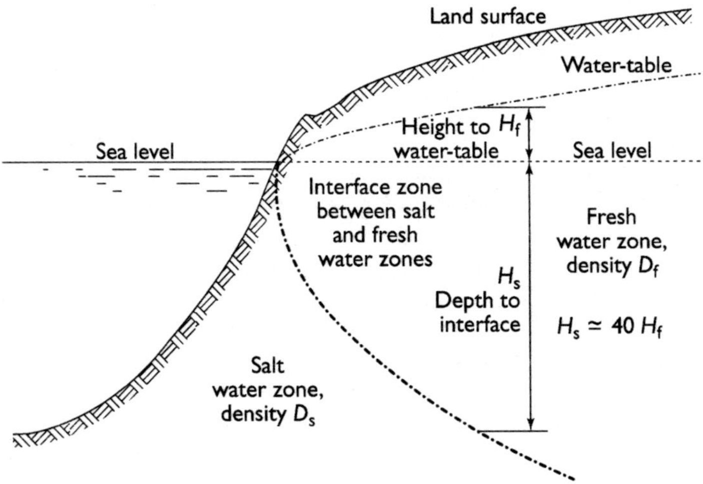

Figure S.1. Sea water intrusion

secondary main Generally, a medium-diameter pipeline forming a link between the TRUNK MAIN and SERVICE MAIN. It may be used to supply the larger consumers direct. *See* DUPLICATE MAINS.

secondary treatment Wastewater treatment by biological methods following primary treatment by sedimentation. *See* ADVANCED WATER TREATMENT.

second-feet (USA) Rate of water flow in cubic feet per second or CUSECS; 1 cusec = 0.028317 CUMECS.

second-foot (USA) Rate of water flow equal to one CUSEC, or one cubic foot per second.

second-foot day Rate of water flow equal to one cubic foot per second for 24 hours. Equivalent to 2447 m^3 (86 400 ft^3) of water.

section The representation, on a suitable scale, of geologic, hydrogeologic or other features on a vertical or inclined plane. *See* CROSS-SECTION, STREAM PROFILE, STREAM SECTIONS.

sectional tank A water tank made from pressed-steel sections or units, commonly 1.2 m (4 ft) square with external flanges for bolting together. Usually 4.9 m (16 ft) is the maximum depth, and in this case tie bars are used within the tank.

sector gate (or **sector regulator**) A type of ROLLER GATE in which the roller is not cylindrical but a sector of a circle. Usually the gate opens when moved upwards.

sediment (or **silt**) Rock particles transported by, suspended in or deposited by a stream or other body of water, usually expressed as grams per litre of water by weight, or cubic centimetres per litre by volume. *See* LOAD.

sedimentation tank *See* SETTLING TANK.

sediment charge Ratio of volume or weight of silt to volume or weight of water passing a given cross-section over a specified period.

sediment concentration The ratio of weight or volume of sediment in water to total weight or volume of the mixture, normally expressed in parts per million (ppm) or grams per cubic metre (g/m^3) for low values of concentration and in percentage for high values.

sediment discharge The sediment transported in a given period for a given stage of a river, expressed in weight or volume units.

sediment discharge curve A graph showing the relation between SEDIMENT DISCHARGE and stages of a river.

sediment hydrograph A graphic presentation of variation in SEDIMENT DISCHARGE or SEDIMENT CONCENTRATION with respect to time.

sediment runoff The sediment discharge in a specified period of a day, week, month, or year and termed daily runoff, weekly runoff, etc.

sediment spatial concentration The weight of sediment contained in a unit volume of stream or channel flow.

sediment transport competency *See* TRANSPORT COMPETENCY.

sediment transport concentration The amount of sediment transported compared with the amount of sediment and water carried by a stream past a given cross-section. *See* LOAD.

sediment water Water mixed with particles of mud, clay, silt or other material and obtained from, or existing in, a stream, canal, lake or other body of water. *See* SLIME, WASH LOAD.

seep Broadly, water which oozes out of the ground over a limited area, indicating the outcropping of the water-table; a small outflow of water in a cutting or foundation.

seepage The percolation of water, generally outwards through the soil or rocks of an earthen channel or storage; the slow passage of water through a rock or soil mass. Seepage varies with the nature of the material, its condition, its position and other factors. *See* EFFLUENT SEEPAGE, INFLUENT SEEPAGE.

seepage force *See* CAPILLARY PRESSURE.

seepage loss The volume of water lost by seepage through the banks or bed of a canal, earthen channel or storage, generally expressed in loss in depth of water (mm or in) per 24 h. The loss varies with the nature and permeability of the ground and is difficult to measure accurately. *See* INFLOW–OUTFLOW METHOD, PONDAGE METHOD, PERMEAMETER, SEEPAGE METER.

seepage meter An appliance for determining seepage loss in earthen channels, consisting essentially of a watertight bell, a plastic bag and a plastic connecting hose. It is used mainly for detecting localized losses rather than losses over a considerable length of channel. *See* PONDAGE METHOD.

seepage spring A spring which occurs in sand or gravel in a valley or cutting located below the water-table, or along the upper surface of an impervious deposit. Seepage springs often occur along the line separating pervious and impervious rocks. They do not always flow continuously, but dry up in periods of drought or low rainfall. *See* FISSURE SPRING, VALLEY SPRING.

seep-off *See* SUBSOIL FLOW.

segment A specified portion of discharge section of a stream or channel, bounded by two consecutive verticals, the free surface above and the bed below.

segmental sluice gate *See* RADIAL GATE.

seiches Phenomena in which some types of earthquakes, with a long period between fairly large ground displacements, may initiate a relatively slow rocking motion, with appreciable magnitude, in the whole mass of water in a large reservoir or lake. The movement could cause considerable damage by the spillage of water over the top of the dam or over the banks of a lake. Also, it is considered that heavy winds or even sudden changes in atmospheric pressure may cause irregular variations or oscillations of the water level.

seismic method A GEOPHYSICAL PROSPECTING method based on the speeds of transmission of shock waves through the ground to be tested. A shock wave is initiated by firing an explosive charge at a known point (shot) and recording the travel times for selected waves to arrive at receivers. The wave speed varies from about 600 m/s (2000 ft/s) for loose sediments up to about 6000 m/s (20 000 ft/s) for granite. Although limited in its application, the method has been used to investigate the base of drift deposits and drift-filled channels. In this respect, relative depths are more reliable than absolute depths.

selected filling The procedure of placing the finest and most adhesive of the bank-forming materials next to the core wall.

self-acting movable flood dam A small timber dam *AB* (Figure S.2) pivoted to a back stay *CD*, the point *C* being placed at a distance of two-thirds *AB* from the top. Hence when the water level is below *A* the CENTRE OF PRESSURE falls below *C*, and the dam is stable. If, however, the water flows over *A*, the centre of pressure rises above *C*, and the dam tips over. This occurs immediately during a flood. Each section has to be replaced when the flood has abated.

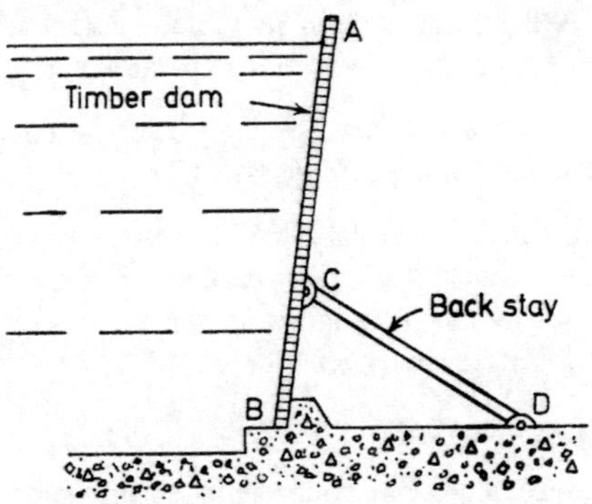

Figure S.2. Self-acting movable flood dam

self-cleaning gradient The slope of a pipe of a particular diameter at which the flow will carry away any solids normally contained in the water. Approximate gradients are 1 in 40 for a 100 mm (4 in) pipe, 1 in 60 for a 150 mm (6 in) pipe and 1 in 90 for a 230 mm (9 in) pipe. The gradient is often stipulated under bylaws affecting drains and sewers. It should not be too steep nor too flat.

self-docking dock A type of FLOATING DOCK in which each section can be unbolted and lifted up on the others and repaired as required.

self-propelled hopper A self-propelled HOPPER BARGE.

semi-arid Relating to a land or climate where the rainfall is greater than under arid conditions but still the amount limits the growth of many crops. Irrigation or dry-land farming methods are commonly required for crop production.

semi-diurnal tides Tides with two high waters and two low waters in each lunar day, that is, tides requiring roughly one half of a lunar day to complete their cycle. In general, the semi-diurnal tide is the most common. *See* DURATION OF TIDE.

semi-humid Referring to a climate or land which is somewhat intermediate between ARID and HUMID, with a tendency towards the humid condition.

semi-rigid joint A joint between individual sections of pipe which permits limited movement. The traditional caulked lead spigot and socket joint is of this type, although many more modern designs permit more movement and are easier to install. *See* FLEXIBLE JOINT.

separate system A drainage system in which sewage and rainwater are carried in separate sewers or drains. *See* COMBINED SYSTEM.

septic tank A tank in which sewage is subject to bacterial action and reduced to harmless substances. *See* CESSPIT.

septicity A condition due to lack of dissolved oxygen and oxidized nitrogen in sewage caused by hot weather or high sulphate content, leading to slime growths of anaerobic organisms and foul odours, particularly in long flat sewers.

S/E ratio *See* STORAGE/EXCAVATION RATIO.

serial correlation coefficient The coefficient relating successive terms, such as in a TIME SERIES, which gives a measure of how one term in the sequence tends to be dominated by the one before it. Natural river flows often show persistence, and these coefficients are widely used to reproduce this in NUMERICAL SIMULATION MODELS. In a series of terms such as χ_{t-4}, χ_{t-3}, χ_{t-2}, χ_{t-1}, χ_t, χ_{t+1}, χ_{t+2}, χ_{t+3}, χ_{t+4}, if successive terms are related by $\chi_t = \rho\chi_{t-1}$ an estimate of the serial correlation coefficient ρ is obtained from

$$\rho = \{\text{expected value } [(\chi_t - \mu)(\chi_{t-1} - \mu)]\}/\sigma^2$$

where μ and σ are the mean and variance of all the terms in the series.

service basin *See* SERVICE RESERVOIR.

service main Generally a small pipe from 75 mm (3 in) to 150 mm (6 in) diameter for conveying water to consumers in the local area or street in which they are laid. *See* SECONDARY MAIN, TRUNK MAIN.

service reservoir (or **distribution reservoir** or **clear-water reservoir**) A reservoir which forms part of a water distribution system. It is designed to meet the hourly and daily demands of the consumers, and also emergencies such as firefighting, the breakdown of pumping equipment and breaks in the mains supply pipeline.

set (1) The direction of flow of tide, current, etc. (2) A timber frame for supporting a tunnel.

settlement The gradual downward movement of an embankment or dam by gravitational compaction of soil and earth fill. Settlement of dam embankments will take place during and after construction. An allowance ranging from 20 to 40 mm per metre (0.25 to 0.5 in per foot) height of dam is usually made during construction so that the surface will eventually settle to the required level. Allowance for settlement is also made in the design of any structures built on the dam.

settling basin *See* DESILTING BASIN.

settling tank A tank in which settleable solids are removed from water or wastewater at velocities low enough for the solids to be deposited by gravity. *See* CLARIFIER.

sewage The discharge from sanitary appliances attached to domestic, industrial or other buildings. It flows down a drain connected to a public sewer or sometimes to a SEPTIC TANK.

sewage disposal The treatment or disposal of sewage by any method or process such as bacterial action, filtration, chemical precipitation and also SEWAGE FARMS.

sewage farm A term little used now and in the past often mistakenly used in place of SEWAGE TREATMENT PLANT. It refers to a farm that takes sewage, usually after treatment and as sludge from settlement tanks; more solid sewage is used as manure and any liquid effluent is used to irrigate the land. A farm with sandy soil is preferred for the use of sewage in this way, since porous sands provide both some filter action as well as under-drainage. Care is needed: if pathogens are present they may not all be removed by both treatment and farming. See BROAD IRRIGATION.

sewage gas (or **sludge gas**) A gas composed of two-thirds methane and one-third carbon dioxide, collected from tanks where sewage sludge undergoes DIGESTION.

sewage treatment plant A facility for dealing with domestic sewage and commercial effluents, all except the smallest quantities of which need to be collected and in nearly all cases given some form of treatment. The main objectives of the treatment of domestic sewage are to reduce the levels of SUSPENDED LOAD or suspended sediments, BOD and COD to levels acceptable for discharge into RECEIVING WATERS. Treatment must also be given to remove or reduce the levels of other substances, particularly those that may be present in industrial wastes. A very large number of methods are now in use for treating sewage and effluents: in any specific case the factors affecting the selection of the required process will include the volume to be treated, its contents and their concentration, the regularity or otherwise of the discharge, the space available for plant and equipment, and the requirements set for ultimate discharge.

sewer A large pipe, conduit or channel usually covered over for carrying off sewage or water to the sewage disposal works. *See* SEPARATE SYSTEM.

sewerage An authority's planned system of sewers through which sewage is carried off to a disposal works. *See* COLLECTING SYSTEM.

sewer pill A wooden skeleton-framed ball which scrapes and cleans the sewer walls as it floats along; its diameter is slightly less than that of the sewer.

s.g. Specific gravity.

shadoof An ancient Egyptian device used to lift water from the Nile.

shaft-sinking pump A pump used for keeping a shaft dry during sinking operations. It is long, narrow and designed to deal with about 4.5 m^3/min (1000 gal/min) from the greatest estimated depth at which water will be encountered. It is often suspended from the surface, with facilities for raising and lowering when SHOT-FIRING at the shaft bottom. Sinking pumps are usually of the electrically driven centrifugal type, and allow for additional stages to be fitted as the depth increases. The pump may be suspended by a single-drum, worm-driven capstan engine with a very low speed. *See* SUBMERSIBLE PUMP.

shale A compacted clay or mud, containing 70–80% argillaceous minerals.

Shale is an important deposit in groundwater engineering: when intact it will not allow the passage of water, or only at a very slow rate. *See* CONFINING BED, PERCHED WATER-TABLE.

shallow manhole An inspection pit or MANHOLE in which the cross-sectional area is the same all the way up.

shallow percolation That part of precipitation that percolates downwards and laterally into streams. *See* DEEP PERCOLATION.

shallow well A well that taps water above the first impermeable rock layer, limited to a depth of less than 30.5 m (100 ft); it is usually lined with stone, brick or concrete, and the upper length made watertight. Periodic tests are desirable to detect possible contamination of water. Shallow wells are liable to seasonal fluctuations in yield and may dry up completely during a period of drought. *See* DEEP WELL, TUBE WELL.

sharp-crested weir A weir for measurement of discharge, consisting of a notch cut in a thin plate, having a sharp edge on the upstream side of the crest. *See* MEASURING WEIR.

sheathing A covering of sheet metal to protect underwater timber against MARINE BORERS.

sheepsfoot roller *See* PADFOOT ROLLER.

sheet *See* NAPPE.

sheet erosion The wearing away of soil layers by RAIN-WASH, especially on sloping ground; it is a serious problem where the root-bound surface layers of a sandy soil have been removed. *See* ACCELERATED EROSION.

sheet flood A flood which has covered an alluvial fan or cone with a sheet of water. Also, a broad, shallow sheet of running water which appears and rises rapidly, generally after a cloudburst, and soon subsides and disappears – characteristic of some semi-arid regions.

sheet ice A sheet of ice formed near the banks of a stream, lake or pond and extending slowly towards the centre, usually occurring where the velocity is sluggish.

sheet piles Closely set and interlocking strips of wood, steel, concrete or other material, driven vertically to form a continuous lining as bank protection, control seepage and the movement of sub-soils, to form a COFFERDAM, or to keep loose earth or water out of an excavation; they may also be used to stabilize permanent structures. *See* TIMBER SHEET PILING.

shelf ice A thick mass of freshwater ice attached to, and extending from, the land. It may be afloat or aground and fed in part by glaciers, and is often found along a continental glacier or in large bays. *See* ICE CAPS.

shell pump *See* BAILER.

shield A RAIN-GAUGE attachment which shields the gauge from wind and eddies, giving a more accurate rainfall figure.

shigella A bacterium present in many polluted waters which causes bacillary dysentery.

shingle A beach or river deposit consisting of coarse gravel and pebbles of a size ranging from about 25 to 200 mm (1 to 8 in). *See* GRAVEL.

shingle bank A deposit or ridge of shingle formed along rivers and coasts where the water velocity is checked, sometimes found at river bends and along the lower reaches where the flow is too sluggish to carry the material further. *See* COMPETENCE.

shingle-carrying rivers Rivers in which the material transported is mainly shingle and stone, derived perhaps from the erosion of steep slopes or landslides or faulting of beds. Transportation of shingle is active during floods and the material piles into heaps as the velocity diminishes. This causes the river to wander and erode its banks with widening of the channel.

ship caisson (or **sliding caisson**) A CAISSON for closing the entrance to a lock, dry dock or wet dock, consisting of a floating steel box fitting into a recess in the dock wall when the entrance is open, or on to grooves in the walls of the entrance when it is closed. *See* GATE CHAMBER, STEP.

shoal A shallow; a sandbank exposed at low tide or water; a submerged bar across a river bed.

shooting flow *See* RAPID FLOW.

shore The zone or area between the highest high water line and the lowest low water line. *See* BACKSHORE, FORESHORE.

shore current A flow of water parallel to the shore line, formed by waves running inwards at an angle. The shore current carries material, in a zigzag fashion, across the shore. *See* LONGSHORE DRIFT.

shore drift The coarse material covering the bottom along the shore in shallow water or where disturbance reaches the bottom. The material may include that delivered to the sea by streams or that produced by wave action. *See* LITTORAL DRIFT.

shore erosion The wearing away of the land belt adjacent to oceans and around lakes by the action of water, waves and wind. *See* UNDERTOW, WAVE EROSION.

shore platform A wave-cut platform formed at the base of a headland and exposed at low tide, covered with sand and pebbles. *See* BEACH RIDGE.

shore protection Safeguarding a shore from the scouring effect of wave action and currents by the construction of BREAKWATERS, GROYNES and REVETMENTS. *See* WAVE EROSION.

shot-firing Increasing the supply of water in boreholes put down in sandstone, chalk or limestone by detonating small charges of gelignite or dynamite to shatter and enlarge the fissures. Water increases of up to about 300% are possible if the explosives can be sited to give maximum results. *See* ACIDIFICATION.

shower A fall of rain or hail of short duration from convection clouds. *See* CONVECTIONAL RAIN.

shrinkage The decrease in volume of soil or fill material through the reduction of voids by mechanical compaction, superimposed loads or natural settlement. *See* SUPERFICIAL COMPACTION.

shuttering (or **formwork**) Temporary works to support concrete during its placement and hardening. Shuttering may be made of timber, sheet steel or similar material.

side-channel spillway *See* LATERAL-FLOW SPILLWAY.

side contraction *See* CONTRACTED WEIR, END CONTRACTION.

side-entrance manhole A deep manhole in which the access shaft is located to one side of the INSPECTION CHAMBER which is reached by a passage from the bottom of the shaft.

side pond A storage pond at the side of, and connected to, a lock chamber. It reduces considerably the lockage loss.

side slopes *See* BATTERS.

Siebe diving helmet A diving helmet devised by Augustus Siebe in the 19th century and used by divers to the present day. A closed helmet is sealed to the watertight diving suit which is inflated by air to counterbalance the water pressure on the diver.

sight rail A horizontal board fixed at a specific height above a required level, used with BONING RODS for checking the levels set out for drains and trenches.

significant wave height The wave height, trough to crest, that is exceeded by only a small stated percentage of waves.

sill The horizontal overflow section of an irrigation check, or dam spillway, or measurement structure. Also, an underwater structure erected across a river to control or retain the level of the river bed. *See* LOCK SILL, NAPPE.

silt In hydrology, particles carried by, suspended in or deposited by a river or other body of water. *See* SEDIMENT. In geology, sediment composed mainly of fine sand and clay of dimensions between 0.02 and 0.002 mm, which becomes unstable in the presence of water and 'quick' when saturated. *See* SAND.

silt basin *See* DESILTING BASIN.

silt bench A bed, bank, terrace or flat of silt material bordering a river or other body of water. It may form banks or flats along the lower reaches of a river or in the channel proper where flow is sluggish. *See* RIVER FLATS, TERRACES.

silt box A removable iron box fitted at the bottom of a GULLY for collecting silt and grit. It is removed as required for discharge of silt. *See* GULLY SUCKER.

silt-carrying rivers Those in which the material carried is mainly silt and sand. The rivers tend to have deep, well-defined beds, and meandering channels with wide floodplains. *See* SHINGLE-CARRYING RIVERS.

silt ejector *See* HYDRAULIC EJECTOR.

silt factor A coefficient which denotes the size of silt carried as LOAD by a stream or other body of water.

silt grade The average size of silt particles; sediment of silt size. *See* SILT.

silting The deposition of silt in a reservoir, river, sea bed, lake or overflow area. Where it becomes excessive and obstructs flow of water, or of navigation, the material is removed by dredging or by groynes. The silting up of a reservoir is caused by inflowing streams depositing their sediment there. If the quantity is considerable, the water storage capacity of the reservoir may be seriously impaired in a small number of years. The remedies include washing the silt out through passages in the dam. Silt traps are sometimes constructed on the streams feeding the reservoir or a bypass for flood waters around the reservoir. *See* SCOURING SLUICE.

silt jetties Long, narrow banks of silt on the sides of streams in digitate deltas.

silt sampler *See* BOTTLE SILT SAMPLER.

silt tank A small artificial basin, located on the upstream side of a farm or other reservoir, to retard the velocity of the incoming water and induce the deposition of silt and debris. The silt tank is connected to the main storage by concrete pipe or other conduit. *See* DESILTING, DESILTING STRIP.

simple storage system A water conservation scheme, comprising a single reservoir and operated independently of any other water resources scheme. The system includes the reservoir, the catchment and all necessary works, pumps, structures and irrigation layouts, etc. *See* MULTI-STORAGE SYSTEM.

simulation modelling *See* NUMERICAL SIMULATION MODEL.

single-acting Referring to a RECIPROCATING PUMP in which water is delivered only every second stroke of the piston or plunger. *See* PUMP.

single-purpose reservoir A reservoir constructed to supply one specific demand, which may be water conservation, hydroelectric power or flood protection.

single-stage pump A CENTRIFUGAL PUMP with a water lift of 30 m (100 ft). Normally, a six-stage pump would have a lift of 180 m (600 ft).

sink A water lodgement or trap at a pumping station. *See* SUMP.

sinker A plummet or weight attached to a CURRENT METER when taking velocities in streams. The sinker is usually of streamline shape.

sinker groyne A type of GROYNE developed in New Zealand and used mainly to retard the underscour of groynes. A STONEMESH GABION, concrete block or other heavy weight is placed at the outer end of the groyne, and secured to a HOLDFAST on the stream bank by wire cables to which obstructions, such as tree branches, are fixed. *See* TREE RETARDS.

sink-hole lakes Small lakes formed in cavities or SINKS which have become clogged with debris, usually found in areas where limestone underlies the surface soil. *See* CRUSTAL-MOVEMENT LAKES.

sinking pump *See* SHAFT-SINKING PUMP.

sinks Rough holes and cavities in the ground which join underground passages, usually occurring in areas with limestone near the surface, portions of which have been dissolved by seepage of water. *See* KARST TOPOGRAPHY, SWALLOW HOLES.

sinuosity The meanders and loops in the course of a river flowing over relatively flat valley areas; the ratio between the length along the sinuous course of a river between two points and the straight-line distance between the two points.

sinuous flow *See* TURBULENT FLOW.

siphon A closed pipe and valves to conduct water from one level to a lower level on the opposite side of an intervening ridge of a height not exceeding about 7.6 m (25 ft). The difference in level between the inlet and outlet ends of the pipe column must be sufficient to give a motive head which overcomes the frictional resistance of the pipe and valves. The siphon was often used at an earlier period where power was not available. The term is sometimes used in irrigation to describe INVERTED SIPHONS.

siphon spillway (or **automatic siphon spillway**) A spillway which operates on the siphon principle, used in concrete dams where flood discharges are not very large and where space is limited.

site The spot, locality, or area of land or water inspected, investigated or selected for a dam, waterworks or wells. *See* INDEX MAP.

site investigations The collection of basic information on the geological structure and physical characteristics of the soils and rocks at the site of new water engineering projects or works, usually performed during the early planning stage. A project involving water supplies or storage would require information on the pervious and impervious rocks, on faults and on the water-table. The putting down of trial pits or boreholes is often necessary. *See* DAM SITE INVESTIGATION.

site laboratory A laboratory located at a dam or other works in which tests can be performed to control fill placing, compaction and so on. As a minimum it can perform moisture content, Proctor and *in situ* density, plastic and liquid limit, and shear strength tests. During the initial period the laboratory is often used to establish standard procedures.

site plan A plan, prepared to a suitable scale, showing the location of a proposed dam or other project. It may show the land to be requisitioned, with ground, water-table and perhaps rockhead contours. Fault lines, buried channels and springs are usually shown. Each site plan is drawn to disclose information relevant to the project. *See* GEOLOGICAL MAP, INDEX MAP.

skimming Obtaining relatively clean surface water by shallow overflow and holding back sand, silt and debris carried as bottom load. Also, the levelling out of soil or surface irregularities.

skin friction *See* SURFACE DRAG.

slack water Water with no current or a very slight current; water about turn of tide, especially low tide.

slack-water navigation (or **still-water navigation**) Navigation in water channels with a very sluggish current. In a river the water may be made deeper with weirs, and the river thus divided into reaches which are separated or connected by LOCKS. *See* SLUGGISH RIVER.

sleet A fall of rain mingled with hail or snow, or of snow melting while falling.

sliced blockwork BLOCKWORK for the construction of BREAKWATERS in which the stone or concrete blocks form sloping, nearly vertical courses. Each block is lowered by crane and slides into position much more easily than in COURSED BLOCKWORK. *See* TITAN CRANE.

slide *See* LANDSLIDE.

sliding caisson *See* SHIP CAISSON.

sliding gate A type of spillway CREST GATE with a vertical lift. It is difficult to raise against the water pressure, and consequently its use is usually restricted to small spillways. *See* ROLLER GATE.

sliding-panel weir A frame weir consisting essentially of wooden panels which slide between a pair of grooved uprights.

slime Water containing fine particles which do not settle readily to the bottom; a fine oozy clay or mud. In the USA slime is defined as a suspension containing 50% or more of 200 mesh material or material below 0.0625 mm (1/400 in).

slip (1) *See* LANDSLIDE. (2) A stone or concrete structure which slopes down to the water's edge; a sloping plane on the bank of a river, used for shipbuilding; a contrivance for hauling vessels out of the water for repairs, etc.

slip circle The assumed circular arc mode of failure in a clay or dam embankment. *See* SLIP SURFACE OF FAILURE.

slip dock A dock with a sloping bottom from which water is excluded by a gate.

slip surface of failure A form of SLIP in a dam, embankment or other cutting when the ground is composed of homogeneous clay or similar material. The surface of failure closely follows the arc of a circle, which usually intersects the toe of the bank. Fixing the approximate position of the centre of rotation of the movement is important. *See* FELLENIUS'S CIRCULAR ARC METHOD.

slipway A landing or shipbuilding SLIP.

slope The gradient or inclination of a line or surface, expressed (1) in degrees from the horizontal plane, (2) as one vertical linear unit to a number of horizontal linear units, or (3) as the difference in elevation between two points divided by their horizontal distance apart. *See* BATTERS, GRADIENT.

slope gauge (1) An inclined staff or plank graduated according to slope to give a more accurate depth reading than a vertical staff. (2) A gauge to determine water surface slope, fixed above and below the DISCHARGE SECTION LINE of a stream. *See* STAFF GAUGE.

slope of stream *See* STREAM GRADIENT.

sloping of banks Bringing a river bank to a flatter slope and planting protective vegetation. This may suffice to check erosion if the flow is moderate and floods rare. The sloping bank is planted with strong rooting grasses and perhaps with shrubs or bushy trees along the lower part. The soil is often covered with brush until the grass has established itself. The lower part often needs additional protection against base erosion and caving of banks. *See* BANK PROTECTION.

slotted lining (boreholes) A perforated or slotted lining sometimes used in boreholes in consolidated rocks where falls may occur. The perforations range from 13 to 75 mm (½ to 3 in) in diameter, and formed at 75 to 300 mm (3 to 12 in) centres. The slots are from 150 to 300 mm (6 to 12 in) in length and from 6 to 20 mm (¼ to ¾ in) in width. Similar slotted casings are used in loose ground with GRAVEL PACKING.

slough (1) A marsh, fen, swamp or quagmire. (2) A secondary river with a sluggish flow (USA). (3) The breaking off or sliding of a sloping bank.

sloughing (1) The casting off of large amounts of suspended slime and solids which are detached from the media of trickling filters by scouring organisms. The material is discharged with the effluent. This process occurs with the onset of warmer weather after winter. (2) The sliding of overlying material due to soil saturation.

slow sand filter A filter to remove particulate matter from a water supply and form a barrier against bacterial pollution, usually a rectangular open basin, up to about 0.9 ha (2 acres) in area and built below finished ground level (Figure S.3). The vertical or sloping filter walls may consist of brickwork, masonry, and clay puddle or concrete. The floor is paved with concrete or brick on clay puddle on which the under-drains are laid. About 0.6 m (2 ft) of successively finer gravel is laid over the drains, followed by 0.6 to 0.9 m (2 to 3 ft) of graded sand. The use of slow sand filters is declining, largely on account of high cost. *See* FILTER.

Figure S.3. A slow sand filter serving an African village of about 300 inhabitants. This type comprises a horizontal flow roughing filter in the foreground with the main vertical flow fine sand filter bed and a treated water tank beyond

sludge cake Dewatered solids from wastewater forming a semi-solid state.

sludge conditioning *See* CONDITIONING (SLUDGE).

sludge-drying bed Sand beds or lagoons with under-drains where liquid sludge is discharged for dewatering by drainage and evaporation.

sludge gas *See* SEWAGE GAS.

sludger *See* BAILER.

sludge sample The broken material produced during drilling and sometimes used as a sample of the rock passed through. *See* CORE.

sluggish river A river with little flow or current, in which peaks of flood form more slowly because of low slope, or because of retarded or reduced current by heavy withdrawal or storage in upper reaches of a river system. A very sluggish river may have only fine silt as its bed load. *See* OLD RIVER.

sluice box (irrigation) A structure made of steel, wood or other material to enable the irrigator to control the flow of water from distribution ditches and into each bay. Also called bay outlet or ditch outlet.

sluice gate A water-gate or floodgate; a channel for passage of water fitted with a vertically sliding gate for controlling the flow; an opening in a structure for passing water at high velocity and for discharge of silt and debris. *See* SCOURING SLUICE.

sluice valve A valve in which a bronze spindle operates a solid tapered slide with gunmetal faces. It is not a regulating valve, and when so used it may be seriously damaged. *See* VALVE.

sluicing A term sometimes used for HYDRAULICKING or HYDRAULIC EXCAVATION. *See* HYDRAULIC MINING.

slushing *See* HYDRAULIC FILL.

small-area floods The flooding of a relatively small area by high-intensity storms with a duration of up to 24 h. *See* LARGE-AREA FLOODS.

smoke test A test carried out with smoke in a new pipeline to locate any leakages, or to assess the efficiency of drains.

smooth-drum roller (or **flat roller**) A roller with a smooth drum used to compact rock in a rockfill dam, and to compact a filter zone in an earth dam. It may be tractor-towed or self-propelled.

snag A stump or trunk of a tree which forms an obstruction in a river.

snagging The removal of obstructions from the channel of a stream, particularly fallen trees. The clearing of a river channel lessens the risk of overflows and minor floods, and is often one of the first steps in improving a river. For removing snags, tractors are often used, particularly those with winches attached. *See* PREVENTATIVE SNAGGING.

snifter valve A valve designed to admit a small quantity of air into a pipeline to keep an air vessel charged with air.

snore A pump suction end which is submerged in the sump and through which water passes to the pump at the upper end of the pipe. It may be a cast-iron cylinder, with an oval or flat-bottomed STRAINER. The water entering the pump is thus relatively free from dirt and debris. *See* FLOATING STRAINER.

snout The projecting end of a glacier, usually in an area below the SNOW LINE.

snow Atmospheric vapour frozen into ice crystals of needle-like or feathery structure and falling to earth in white flakes or spread on it as a white mantle; the flakes may consist of a number of single crystals matted together.

snow course A line permanently marked on the SNOW COVER, along which snow samples are taken at suitable distances and periods to determine depth and SNOW DENSITY. The measurements enable forecasts to be made of the subsequent runoff. *See* SNOW SURVEY.

snow cover (or **snow mantle**) A layer of snow and ice which covers an area for a considerable period.

snow density The water content of snow, expressed as a percentage by volume. In sampling a snow cover, it is the ratio of the scale reading in millimetres (or inches) of water to the depth in millimetres (or inches) of the snow core. *See* SNOW SAMPLER.

snowdrift A bank of snow formed in depressions and sheltered places by the action of wind, or by snowfalls in strong wind.

snow fence A barrier or fence to impound snow in areas where melting in place contributes to soil moisture; a fence of slat and wire or other construction to intercept and hold drifting snow and thus safeguard engineering works, railways and roads.

snowfield A large tract covered with snow, usually in mountainous or polar regions. *See* SNOW LINE.

snow-gauge An instrument for measuring the amount of snow that falls during one storm or over a specified period at one place. *See* SNOW SAMPLER.

snow hedge A fence or barrier of bushes, shrubs or low trees planted to intercept drifting snow.

snow line The lower limit or line in altitude of perpetual snow on mountains, varying in height according to climate. In equatorial regions, the line may be as high as 5500 m (18 000 ft) above sea level, while in polar regions it is below sea level.

snow mantle *See* SNOW COVER.

snow melt The formation of water when snow melts; the water so formed, which may seep downwards, become part of surface runoff or evaporate. *See* EFFECTIVE SNOW MELT.

snow sample A core extracted by means of a snow sampler from a point on the SNOW COURSE to determine snow thickness and density.

snow sampler A device consisting of light jointed tubes for taking snow samples. A spring scale gives a direct reading of the depth of water contained in a given sample of snow. *See* SNOW-GAUGE.

snow stake A vertical stick, prop or post, generally permanent, which indicates the depth of snow in areas with a thick snow cover.

snow survey The sampling of the snow cover at selected points to determine snow depth, density and water content. The water stored as snow on a catchment area and the subsequent runoff may be estimated from the survey data.

snubber A fitting used for the protection of pressure gauges against hydraulic shock and sudden changes of pressure. *See* HYDRAULIC PRESSURE SNUBBER.

soakaway *See* DRY WELL.

soda ash Sodium carbonate, used for pH correction of aggressive waters.

sod strips Strips or bands of sod or turf left in a natural water channel to prevent erosion; narrow ridges or bands of grass or other close-rooted crop spread across a channel to retard the flow of water. *See* MATTRESS.

soffit The uppermost part of the internal surface of a culvert, drain, sewer or channel (Figure S.4). *See* INVERT.

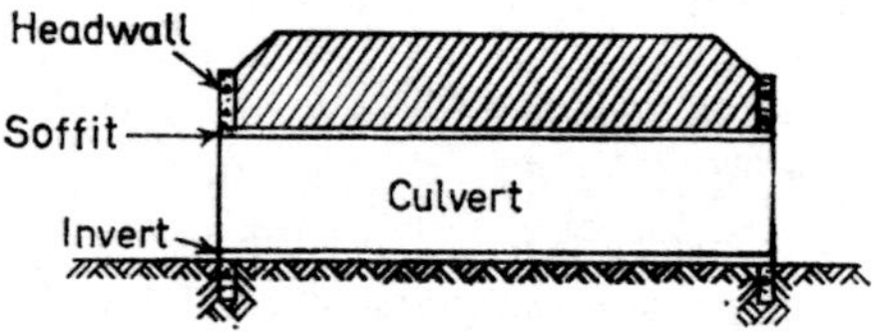

Figure S.4. Soffit

soft water Relatively pure water which lathers readily when mixed with soap. *See* HARDNESS OF WATER.

soil A natural material developed by the chemical and physical disintegration of rocks. It forms the upper layer of earth on which rain falls and plants grow. It is composed of mineral particles, some organic matter and water, and may range from fine clay or mud to gravel or boulders. To the engineer, soil is the superficial material that is not hard or solid rock. Soils may be classified into certain well-defined types, depending mainly on the size and nature of the particles, as shown in the table.

Grade	*Dominant grain size (mm)*
Gravel	More than 2
Sand	2 to 0.02
Silt	0.02 to 0.002
Clay	Less than 0.002

soil conservation The management of land and soil, including drainage or irrigation where necessary, to obtain the maximum profitable production of crops and stock on a sustained-yield basis. *See* DRAINAGE, IRRIGATION, SOIL EROSION, SUBSURFACE DRAINAGE CHECK, WATER PROSPECT MAP.

soil conservation survey A report and a map or aerial photograph showing the physical land features and associated data in sufficient detail for planning soil conservation and land usage measures.

soil conserving crops Grasses and other crops that prevent or retard erosion, and maintain or replenish soil organic matter. These crops, when cut for hay, yield considerable aftermath or after-grass.

soil discharge The loss of water from the ZONE OF SATURATION through evaporation directly from the soil or rocks. It occurs only in areas where the water-table is very close to the land surface. *See* EVAPORATION DISCHARGE.

soil erosion The wearing away and removal of the soil layers mainly by the action of wind, rain and running water. The process may remove the entire mantle of soil, making the land valueless for cultivation; it sometimes occurs in areas of over-cultivation and ill-planned deforestation. Soil erosion often creates problems for engineers concerned with floods and silting in reservoir basins, rivers and canals. The remedies include control of over-grazing, afforestation on the steeper slopes, contour trenching or ridging, construction of barriers, and weirs or detention dams, etc. *See* GULLY DRAINAGE, SHEET EROSION.

soil evaporation The evaporation of water from the surface and near surface soil layers.

soil mechanics The systematic study of the nature, composition, classification, consolidation and related properties of soils. This information determines their suitability for dams, canals, reservoirs and other water engineering works. *See* DAM FOUNDATION, DAM SITE INVESTIGATION, SITE INVESTIGATIONS.

soil moisture (or **soil water**) Water occurring in the voids of the soil mantle, gradually discharged by evaporation or by transpiration of plants. *See* BELT OF SOIL WATER.

soil moisture deficit The quantity of water, expressed in millimetres (or inches), required to restore a soil to FIELD CAPACITY. In general, plants growing in heavy (clayey) soils can withstand higher deficits than plants growing in light (sandy or coarse-grained) soils. Therefore, plants in sandy soils need higher rainfall or more frequent irrigation than plants in clayey soils. *See* PERMANENT WILTING POINT.

soil moisture suction The pressure (below atmospheric pressure) which causes water to adhere to or be retained in soil or rock above the zone of saturation. *See* CAPILLARY WATER, ZONE OF AERATION.

soil probe A pointed steel rod sometimes used to measure the depth of water penetration in soil irrigation. *See* PROBING.

soil-saving dam A ridge or barrier of earth constructed across a stream or gully to impound runoff and silt, used to minimize erosion and promote the deposition of eroded soil material. The term is often restricted to dams which are larger and of a more permanent nature than CHECKDAMS.

soil-saving dike (USA) An earth ridge formed along the lower end of an irrigation field, often with an adjustable outlet, used mainly to conserve soil or sediment within the area.

soil structure The arrangement or grouping of the primary soil particles into soil aggregates, which may be coarse, medium or fine. Soil structure affects the rates of absorption and movement of water.

soil water *See* SOIL MOISTURE.

solar water disinfection A method that can be used to disinfect water contained in transparent bottles or plastic bags at household level, and also in continuous-flow systems, comprising solar collectors and heat exchangers, for institutions such as hospitals and schools. Careful monitoring and control are essential.

solid map A geological map showing only those formations that are solid, and usually referred to by engineers as BEDROCK, and none of the unconsolidated deposits or drift that overlay them. *See* GEOLOGICAL MAP.

solids-handling pump Commonly, a centrifugal pump for dealing with water or liquids containing sand, ore tailings or fine coal; it is designed to resist abrasion, and rubber linings are often used, as they last longer than iron or steel. *See* PULSOMETER PUMP.

solid waste Unwanted material discarded in the form of refuse, scrap and sludge, which is either landfilled, incinerated or recycled.

solid wharf *See* WHARF.

soling Large stones in PITCHED WORK.

solution cavities Irregular underground passages and pits formed in certain soluble rocks, such as limestone, by the seepage and flow of water. In some cases, percolating solutions have filled the cavities with minerals. *See* SINKS.

solution valley A valley formed by the chemical wearing away of soluble rocks, often broadly U-shaped in cross-section.

sorption The taking up of dissolved solids, either on the surface of a solid or organism (ADSORPTION) or into the interior (ABSORPTION).

sorting Sometimes, the progressive reduction in size of particles forming the LOAD of a stream as it flows towards the sea or other body of water.

sounder, echo *See* ECHO SOUNDER.

sounding Determining the depth of water in a river, ocean or reservoir at a particular spot by sending out a sonic or ultrasonic sound impulse and measuring the interval between the transmission and the reception of an echo. For accuracy, many factors must be known, such as the type of water, etc. The term is sometimes used in measuring the depth to bedrock, or thickness of soil, by forcing a steel rod into the soft ground.

sounding lead Sometimes, a sounding weight. See SOUNDING LINE.

sounding line (or **lead line**) A strong cord up to about 180 m (100 fathoms) in length, knotted at fathom intervals, and used for taking SOUNDINGS in hydrography. A weight is attached to take the line downwards to the bed of a river, lake, reservoir or sea.

sounding line corrections The adjustments made to sounding line measurements to allow for departures of the line from the vertical. *See* AIRLINE CORRECTION.

sounding rod (or **sounding stick**) A graduated rod or stick weighed with lead at one end and used for determining the depth of relatively shallow water, such as streams and canals.

sounding weight (or **lead**) *See* SOUNDING LINE.

sounding wire *See* LOG LINE.

sowing with grass seed A dam contract term which includes sowing, watering, fertilizing and maintenance during the construction period. Measurement is in square metres or square yards.

spall drain *See* RUBBLE DRAIN.

spatial concentration of sediment Sometimes applied to the sediment contained in a unit volume of flow. *See* SEDIMENT TRANSPORT CONCENTRATION.

specification A document giving a minute technical description, or enumeration of particulars, materials, methods of construction and workmanship of any item to be purchased or any construction work to be carried out under the contract. For large civil engineering works the five documents that taken together define the contract are normally the Conditions of contract, Specification, Bill of Quantities, Drawings and Contractors' tender.

specific capacity (of well) The rate at which water can be pumped from a well per unit of DRAWDOWN; it is only useful where the drawdown varies approximately as the yield from a well. It may be expressed in litres per minute per metre (drawdown) or gallons per minute per foot (drawdown).

specific discharge The discharge per unit area of catchment. The term is often used when comparing floods.

specific energy A value of flow energy in a conduit or open channel, expressed as energy head above base of channel. For open channels if friction and other losses are neglected this is normally written

$$d + V^2/2g = \text{a constant} = \text{specific energy}$$

where d is the depth, V is the velocity of flow and g is the acceleration due to gravity. *See* BERNOULLI'S THEOREM.

specific gauge reading Water level at a specific station for a given discharge.

specific gravity (s.g.) The ratio of the weight of a given volume of soil, rock or other substance to the weight of an equal volume of water at a given temperature.

specific retention The ratio of the volume of water that a soil will retain against the force of gravity to its own volume; it is determined by first allowing the soil to become thoroughly saturated and then noting the change of weight as drainage proceeds. *See* EFFECTIVE POROSITY, SOIL MOISTURE SUCTION.

specific yield The amount of water which a unit volume of soil or rock will yield after being saturated and then allowed to drain under specified conditions, expressed as percentage of volume. *See* EFFECTIVE POROSITY.

speedy moisture tester An instrument for the rapid determination of moisture in a soil sample. It uses calcium carbide, and a pressure gauge is calibrated to give direct values of moisture percentage.

spill Overflow of water in tidal estuaries during the rise of tide or of riverbanks during high floods.

spillover That part of OROGRAPHIC PRECIPITATION which reaches the ground on the lee side of a mountain or barrier, having been carried forward by horizontal wind.

spillway (dams) A structure which passes floodwater through, over or around a dam (Figure S.5 and S.6). Provision is usually made to prevent scour and the removal of rock at the downstream face of the dam by spillway discharges. This may consist of a concrete apron at the toe or other constructions. In the design of spillways, rainfall records, particularly of past heavy storms, are important. *See* APRON, BY-WASH, EMERGENCY SPILLWAY, FLASHBOARD, FUSE PLUG SPILLWAY, OVERSHOT SPILLWAY, ROLLWAY.

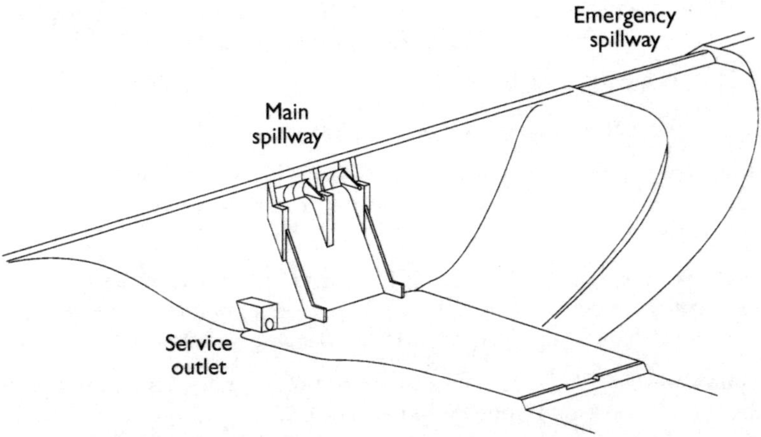

Figure S.5. Main and emergency spillways

Figure S.6. A large spillway passing floodwater past a dam

spillway capacity The capacity of a dam spillway to pass flood water. For example, spillways for large farm dams should be capable of carrying the 1-in-100-year flood. *See* RETURN PERIOD.

spillway crest level The upper level of the crest structure which regulates water flow in a spillway (Figure R.1). *See* FLASHBOARD.

spillway gate *See* CREST GATE.

spillway overflow *See* OVERFLOW SPILLWAY.

spin test A check of the condition of a CUP-TYPE CURRENT METER by blowing into the cups and spinning the rotor. *See* MINIMUM SPEED OF RESPONSE.

spit A ridge of land or deposit of sand or shingle projecting from headlands and terminating in the open sea; a narrow and submerged ridge across the mouth of a bay. *See* BAY-MOUTH BARS, HOOKED SPITS.

splash erosion The loosening and erosion of soil particles and soil layers by the action of raindrops. *See* RILL EROSION.

sporadic permafrost Scattered islands of PERMAFROST in ground which is predominantly unfrozen.

spout-delivery pump A pump which cannot deliver water above its own height and is similar to the DIAPHRAGM PUMP.

spray irrigation *See* SPRINKLER IRRIGATION.

spraylines A method of irrigation often used in fields that can be watered in rectangular strips from 9 to 15 m (30 to 50 ft) in width. The spacing of the spraylines is related to the throw of the water jets, which depends on the water pressure available. Spraylines consist of galvanized steel, aluminium or plastic tubing of 25 to 50 mm (1 to 2 in) internal diameter, with special connections to facilitate movement. The tubing is fitted with nozzles spaced at about 0.6 m (24 in) and designed to give the desired output and spray pattern.

spreader (1) Any appliance which distributes water uniformly in or from a channel. (2) A machine fitted with wide blades for spreading soil, subsoil or rock excavated from a pond, drainage ditch or other cut.

spreader strip A plot of land, of variable width, planted with an erosion-resistant crop, to regulate and fan out the runoff from cultivated land on the rise side. The strip is planted across the general slope, and helps to distribute runoff and control erosion. *See* BORDER DIKES.

spring A natural outflow of groundwater at the surface. Springs often serve as useful sources of water. *See* ARTESIAN SPRING, BOURNES, CONTACT SPRING, FAULT SPRING, FISSURE SPRING, THERMAL SPRING, TUBULAR SPRING, VALLEY SPRING.

spring range The mean difference in level between spring high water and spring low water. *See* MEAN TIDAL RANGE.

spring tides The highest ordinary tides that occur at, or soon after, the new moon and the full moon. *See* EQUINOCTIAL TIDES, NEAP TIDES.

sprinkler irrigation The spreading of water over soil by means of sprays from pipes or pipe projections above ground level. Rotating sprinkler heads are often used with one or more nozzles which discharge the jets. The sprinklers may each water a circle of from 7.6 to 13.6 m (25 to 45 ft) radius, and be spaced from 6 to 10.6 m (20 to 35 ft) intervals along the lateral line according to the planting pattern and other local factors. *See* SPRAYLINES.

spud (or **anchorage spud**) A steel post under a DREDGER which provides anchorage during dredging operations. The post is lowered by ropes or a toothed rack until it is secured in the river or sea bed.

spur (or **wing dam**, USA) (1) An artificial obstruction formed from the bank of a channel and projecting into it for training the flow of water and for bank protection; a similar structure for protecting a shore from erosion. A spur may be constructed of stone, earth, timber or brushwood; the head is reinforced to resist scour. (2) A minor topographic feature, such as a small hill extending from a large dominant range of mountains.

spur dike *See* SPUR.

spur terrace A short ridge of earth for impounding or diverting runoff.

squall A sudden and violent gust or successive gusts of wind, especially with snow, sleet or rain.

square check irrigation A modification of CHECK IRRIGATION in which rectangular or square basins are used to confine irrigation water.

square-mile foot A volume of water one foot deep covering one square mile, equivalent to 789 440 m^3. *See* ACRE FOOT.

stability (1) The condition established in a channel carrying water when ACCRETION balances SCOUR. (2) The resistance of a river bank or valley slope to sliding or deformation, depending largely on the shearing strength, the water content of the material and the degree of slope. *See* BATTERS, CAVING.

stability number A number pertaining to the stability of a soil embankment. Thus:

$$\text{stability number} = HW/C$$

where H is the critical height of sloped bank, W is the effective unit weight of soil and C is the cohesion of soil.

stability test (or **methylene blue test**) A test of the stability of sewage effluent after biological treatment in which methylene blue is added to a sample. The time taken to decolorize the blue dye gives a measure of the stability of the effluent.

stabilization pond *See* OXIDATION POND.

stabilized grade That grade or slope of a stream or channel along which neither erosion nor silting occurs. *See* PROFILE OF EQUILIBRIUM, REGIME.

stack (UK) A high detached rock rising out of the sea and near the coast. *See* CHIMNEY.

staff gauge A vertical staff, plank, rod, metal plate, wall of a pier, etc., with a graduated scale by which depth of water in a river, dock or canal may be read. At stations where a STAGE DISCHARGE RELATION is available, staff gauge readings provide an estimate of river flow. *See* INCLINED GAUGE, SLOPE GAUGE, SOUNDING ROD.

stage (or **gauge height**) Elevation of water surface, usually in a stream, relative to a fixed datum or plane of reference. *See* BANKFUL STAGE, BED-BUILDING STAGE, FLOOD STAGE, RIVER STAGE, WARNING STAGE.

stage discharge relation The relation between water level and discharge of a river or conduit at a specific station for a given condition of flow. *See* DISCHARGE CURVE.

stage profile *See* STREAM PROFILE, SURFACE PROFILE.

stage recorder *See* WATER STAGE RECORDER.

stage relation *See* GAUGE CORRELATION.

stage storage relationship The relationship between the STAGE of the water surface in a reservoir and the volume held in storage.

stake and stone sausage construction A river protection method in which the bank is sloped, and on the face vertical willow stakes are placed. Over these, long willow stakes are fixed horizontally near the top of the bank. Stone sausages (*see* SAUSAGE CONSTRUCTION) are then laid at about 3 m (10 ft) centres down the bank slope, and anchored by wires to posts fixed well back from the bank. *See* FASCINES.

staking (or **pegging out**) Fixing the location or line of drains, channels, terraces, ponds and similar structures, and the placing of metal or wood stakes as markers; staking a drainage ditch includes marking the line and also the depth or slope of the ditch.

standard current meter A calibrated current meter used for checking and comparing the rating of other current meters. *See* REFERENCE CURRENT METER.

standard flood Flood data used in certain equations in which it is assumed that the duration of the storm is equal to the PERIOD OF CONCENTRATION, the storm is stationary and its area just covers the catchment, the rainfall is of constant intensity over the catchment and throughout the storm, and the catchment has a constant slope and runoff coefficient.

standardization An agreement between authorities, manufacturers and engineers on certain qualities, tests, dimensions and tolerances of a certain practice, product, part or machine, and to adopt same in engineering practice. It offers many advantages such as interchangeability of machine parts and a reduction in stocks of spares.

standard mix CONCRETE mixed in the volumetric proportions of one part cement, two parts sand, and four parts coarse material.

standard project flood A flood representing a synthesis of flood occurrences in a given area. Its frequency is unspecified, but it forms a sort of datum or standard against which other floods are compared. *See* DESIGN FLOOD.

standard solution A solution of known concentration as used in volumetric analysis.

standard system of levels The system of levels used at gauging stations and other sites, and related to a recognized datum.

standing water level (or **rest water level**) The water-table when left undisturbed for a period; in general, the level at which water stands in wells when no pumping has taken place. *See* FREE WATER LEVEL.

standing wave flume A flume containing a constriction which makes the flow change from SUB-CRITICAL FLOW to HYPER-CRITICAL FLOW and in which the measurement of one water level, the upstream one, enables a calculation of discharge to be made.

standpipe (1) A pipe projecting vertically upwards from a water main under pressure. It is open at the top and the water level will indicate the pressure in the water main. *See* SURGE PIPE, WATER HAMMER. (2) A circular reservoir of reinforced concrete or steel constructed above ground level. (3) A pipe connected to a water main; it has a tap or hose attachment to supply local water needs (Figure S.7).

Figure S.7. A village standpipe

stank (1) To seal off or make watertight. (2) A small timber COFFERDAM sealed with clay. The term was formerly applied to a pool, ditch or tank.

starling A series of timber piles driven into a river bed on the upstream side of a bridge pier, and also downstream in tidal rivers, to protect the structure from floating ice blocks and debris.

static delivery head The vertical height, usually in metres or feet, between the pump centre and the outlet of the delivery pipe.

static head The head produced by static pressure of water, equal to the pressure divided by unit weight of water. The difference in water levels between two locations, such as the impounding reservoir and the service reservoir; in a pump, the sum of the static suction head and static delivery head.

static level *See* HYDROSTATIC LEVEL.

static suction head The vertical height, usually in metres or feet, from the free water level in the sump to the centre of the pump. *See* SUCTION HEAD.

stationary dredge *See* DREDGER.

staunching piece (or **staunching bead**) The vertical gap left in a concrete dam between successive bays. The concreting of the gap is carried out after the shrinkage of the adjoining bays.

staunching rod A strong rubber rod used at a CREST GATE to form a watertight joint; the rubber is compressed between the gate and the dam structure.

steady flow A flow in which the flow rate does not change with time.

steening (or **steining**) The lining of a well with bricks or stones, which may be laid with mortar but are usually dry. *See* DRY WELL, WELL SINKER.

step The installation of a LOCK GATE into the vertical position; a gate is usually towed to the site in a horizontal position, and stepping is slow and awkward.

step irons (or **foot irons**) Staples of galvanized malleable cast-iron, built into the brick or concrete wall of a MANHOLE to provide access to a deep sewer.

sterilization The process of making liquids, instruments, surfaces, vessels, etc., completely free of living micro-organisms. This level of decontamination is required for surgical instruments and for some work in laboratories, but is not necessary for general water treatment, where DISINFECTION is sufficient.

Stevenson's formula A British formula for calculating the height of waves. It applies only to unobstructed deep water, and for distances of up to 400 miles (640 km).

$$H = 1.5\sqrt{L}$$

where H is the height of the wave in feet, and L is the FETCH in nautical miles. *See* FORCED WAVES, HAWKSLEY'S FORMULA, WAVE HEIGHT, WAVELENGTH, WAVE PRESSURE.

stiff-fissured clay A clay which is firm when dry at depth, but is intersected by numerous cracks along which water can seep readily.

stilling basin *See* STILLING POOL.

stilling pool (or **stilling basin** or **water cushion**) A water basin formed at the foot of a dam spillway or waterfall or rapids to reduce the velocity and scouring action of the descending water. *See* APRON.

stilling well A well in which the water enters through small inlets in a chamber or pipe; these dampen the waves or surges while allowing the water level within the well to rise or fall steadily with the movements of the main body of water. *See* STRAINER WELL.

still-water navigation *See* SLACK-WATER NAVIGATION.

stirrup wire The wire connecting the lower and upper sheets of wire mesh in MATTRESSES and other river protection works.

stochastic model A NUMERICAL SIMULATION MODEL in which the terms are generated by a STOCHASTIC PROCESS. Such models are commonly used in the modelling of TIME SERIES representing hydrological events, as they enable the probabilities of a range of events to be more easily predicted than is possible with DETERMINISTIC MODELS.

stochastic process A sequence of terms in which successive terms are generated from a combination of a random component and also constants and/or deterministic relationships, such as correlation coefficients. The random part of each term cannot take any value, but is restricted to one sampled from a suitable statistical distribution with a mean, variance, etc., chosen to give the terms the required magnitude and variations. A typical example of a stochastic process is a MARKOV PROCESS or Markov chain.

stockade groynes A modified form of the CRIB GROYNE with shorter timbers. A box structure is formed, with open top and bottom, by saplings strung vertically on steel cables and filled with stone or shingle. The construction is stiffened by cross ties and braces. *See* CRIBWORK.

stock pond *See* FARM POND, TANK (EARTH).

stockwater development The development in rural areas of new and improved schemes of water supply for stock. It includes investigations covering water sources, both surface and underground, and the design of associated delivery conduits and storages. *See* SURFACE WATER SUPPLIES.

stone, angular *See* ANGULAR STONE.

stone apron *See* UNDERWATER APRON.

stone drain *See* RUBBLE DRAIN.

stone-filled work All shore, river, canal, and dam works in which stone (angular or rounded) forms an essential construction material. *See* BEACHING, GROYNES, RIPRAP, SAUSAGE CONSTRUCTION.

stone, loose *See* LOOSE STONE.

stonemesh A form of river bank protection consisting of stones 50 to 150 mm (2 to 6 in) in size enclosed in a wrapping of wire mesh. As the wire will gradually rust and fail, the construction is viewed as semi-permanent. It should, however, hold for a few years until the establishment of grasses or willows to bind the stone together. *See* STONE RIPRAP.

stonemesh groynes These groynes are used mainly to divert the river flow away from an eroding bank; they are used extensively in New Zealand in the smaller rivers, particularly where rounded stone is plentiful and where timber suitable for cribwork is not available. They consist essentially of rounded stones held in a wrapping of wire mesh, and are often built up with rectangular gabions laid in courses; the construction tapers as the work proceeds to the top. *See* GABION.

stonemesh mattress (or **stonemesh apron**) A type of STONEMESH CONSTRUCTION used where protection of beds or banks of rivers from erosion is of special importance. It may be used around bridge piers, at the roots and heads of groynes, and below weirs, etc. It consists essentially of a layer of stones, some 300 mm (12 in) thick, with a surrounding layer of mesh to form a mattress or apron. Should scour commence to undermine the construction, the weight of the stones causes the apron to sag and adapt itself to the new surface, while the stones are held in position and cannot be rolled away. *See* STONE RIPRAP, UNDERWATER APRON.

stone paving A hard floor or paving consisting of roughly flat stones, hand-placed on the downstream side of a river structure to minimize scour of the bed. *See* APRON.

stone pitching A facing of stones, up to 0.45 m (18 in) or more in thickness, above the BEACHING on the WATER FACE of an earth dam as a protection against erosion and wave action.

stone riprap The use of loose, angular stones to check river bank erosion, used particularly in areas where good hard stone can be obtained cheaply. The stones should be durable and each of a weight sufficient to resist transportation by the strongest current likely in that river. If the water is deep and the eroding bank nearly vertical, it may be cheaper to use extra stone instead of sloping the bank. Greater protection is provided by placing a layer of gravel or small stones against the bank and then building up the large stones over it. *See* BEACHING, SLOPING OF BANKS.

stoning Preventing scour along a river bed by means of a layer of broken stone. *See* REVETMENT, RIPRAP.

stop *See* BAY OUTLET.

stopbank (New Zealand) *See* LEVEE.

stop-log *See* FLASHBOARD.

stop-plank *See* FLASHBOARD.

stop valve A valve to stop or start the flow of water, usually in a pipe. *See* DISCHARGE VALVE.

storage The impounding of water in underground or surface reservoirs, tanks, canals, rivers or basins, from which supplies may be drawn as required. *See* ARTESIAN BASIN, BANK STORAGE, CHANNEL STORAGE, CONSERVATION STORAGE, DEPRESSION STORAGE, GROUNDWATER STORAGE, PONDAGE, RESERVOIR, VALLEY STORAGE.

storage balance *See* PONDAGE.

storage canal *See* CANAL (MAIN IRRIGATION).

storage capacity The maximum volume of water which a reservoir can hold without overflowing, expressed in cubic metres, cubic yards, millions of gallons, or acre feet. The required capacity of a FARM RESERVOIR depends on the approximate water requirements for domestic consumption, stock and irrigation, and includes estimated evaporation and seepage losses. Where the BORROW PIT is inside the natural reservoir basin, the total capacity equals the natural storage plus the volume of the borrow pit. *See* PONDAGE, STORAGE/EXCAVATION RATIO.

storage curve *See* CAPACITY CURVE.

storage curve, groundwater *See* GROUNDWATER STORAGE CURVE.

storage cycle Total time or period at the end of which the level or volume of water in a reservoir is the same as at commencement – a few hours, 24 hours, one year or a longer period.

storage, economic *See* ECONOMIC STORAGE.

storage/excavation ratio (or **storage ratio** or **S/E ratio** or **water-to-earth ratio**) The ratio of volume of water stored to the volume of earth excavated for the bank or dam. Thus a storage ratio of 10:1 means that the volume of water is ten times the volume of earth excavated. *See* DAM SITE INVESTIGATION, STORAGE CAPACITY.

storage of aquifer The quantity of water released from storage in an aquifer with a given lowering of head.

storage period The period without water inflow to the reservoir. Over such a period the consumer requirements must be met wholly from the water held in storage.

storage, pumped *See* PUMPED STORAGE, PUMPED-STORAGE HYDRO STATION.

storage ratio *See* STORAGE/EXCAVATION RATIO.

storage regulator *See* PONDAGE.

storage reservoir A reservoir for storing water as part of a flood protection scheme or for water supplies. *See* SERVICE RESERVOIR.

storage routing A method of FLOOD ROUTING by determining the actual volume of inflow to, outflow from and storage in the reservoir.

storage, surface *See* INITIAL DETENTION.

storm A violent disturbance of the atmosphere with heavy rain, or snow or hail, and often thunder and lightning; a violent gale with rain or snow. Storms vary greatly in duration, in intensity and in movement; some are part of seasonal conditions and last for months, while others are isolated phenomena; some move rapidly while others remain almost stationary for days and cause heavy local precipitation and floods. *See* HAILSTORM, MONSOON, PRECIPITATION (EXCESSIVE), THUNDERSTORM.

storm distribution pattern A plan, chart or layout showing variation in depth of rainfall over an area for a specific storm.

storm flow Excess water flowing after heavy rain, at or near the surface, into streams or rivers.

storm flow, subsurface *See* SUBSURFACE RUNOFF.

storm lane (or **storm belt**) The belt or tract in which storms are frequent.

storm pavement A gently sloping paved bank to a BREAKWATER.

storm rainfall Rainfall associated with a given storm; it may affect many thousands of square kilometres (or square miles) and vary in duration from a few minutes to several days.

storm rainfall profile The relationship between rainfall intensity and time during the most intense and heavy falls of rain, which in general are produced by the same atmospheric conditions (Figure S.8). This makes the forms of the most critical storm profiles relatively constant: typically they are bell-shaped, which for engineering design can be assumed fixed for any selected probability of occurrence. Storm profiles are normally stated in a dimensionless form so that they can be corrected for local rainfall. Different patterns are expected from summer and winter storms, and the profile to be used depends on design requirements.

storm runoff The runoff from a CATCHMENT or drainage area after a storm or snow melt. The term usually refers to conditions where increased runoff flow reaches a measurable amount soon after the occurrence of rain or thaw, such as where all, or a large part, of the catchment is covered by one rain storm. It is often expressed as a depth uniformly covering the given drainage area.

storm sewer A SEWER which normally has no flow but carries water discharged after heavy rain.

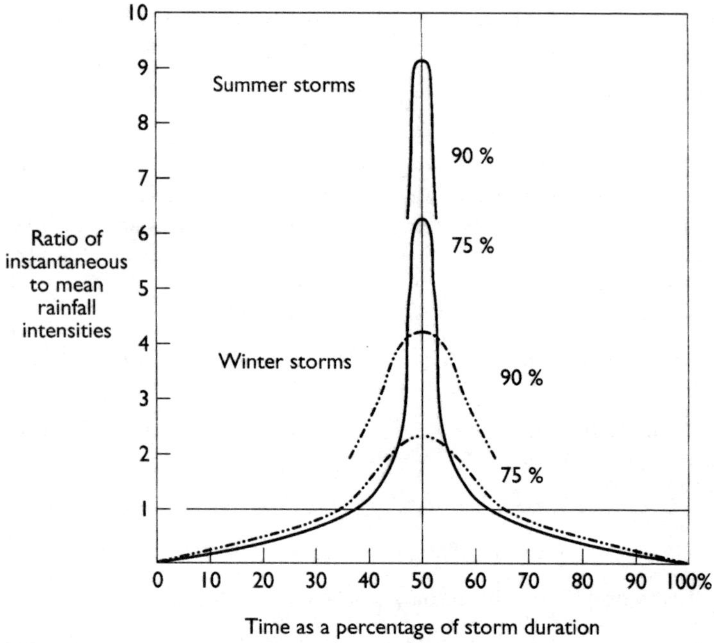

Figure S.8. Storm rainfall profiles

storm wall *See* PARAPET WALL.

stormwater (1) The water which drains off a catchment area during and after a heavy fall of rain or snow. (2) Often, in a combined system of drainage, a stated excess of drainage water above the dry-weather flow at which the sewage is permitted to overflow into a stream after passing through STORMWATER TANKS. In general, before the sewage overflows into STORM SEWERS the total flow must be three times the average dry-weather flow. *See* COMBINED SYSTEM, PARTIALLY SEPARATE SYSTEM.

stormwater tanks Tanks that hold excess stormwater. The period of retention ensures that sediment and solid material settle from the stormwater before it passes into a stream. *See* SILT BOX.

storm wave The overflow of low coasts, not normally flooded, by TIDAL CURRENTS and sometimes strong wind. *See* TIDAL WAVE.

strainer A perforated pipe, cylinder of wire-gauge, or similar tube. It allows the passage of water but not large particles, and is used in RADIAL WELLS, STRAINER WELLS and PUMPS. *See* FLOATING STRAINER.

strainer well A well in which a strainer is positioned in the borehole where it intersects the aquifer to exclude solid particles above a certain size. *See* STILLING WELL, WELL SCREEN.

stranded caisson *See* BOX CAISSON.

stratification *See* THERMAL STRATIFICATION.

stratigraphic bores Boreholes put down primarily to establish geological sequence in a given area. In hydrogeology, establishing the position, structure, and continuity of the pervious and impervious deposits and faults is a major objective. *See* STRUCTURE BORES.

stratum spring *See* CONTACT SPRING.

stream (or **brook** or **rill** or **rivulet**) A general term for a flow of water in a natural surface channel; a flow of water in an open or closed conduit; a jet of water issuing from an orifice or notch. *See* BROOK, POTAMOLOGY, SUBSEQUENT STREAM, SUBTERRANEAN STREAM, WATERCOURSE.

stream at grade *See* GRADED STREAM.

stream bank *See* BANK, LEFT BANK, RIGHT BANK, RIPARIAN.

stream bank erosion The removal of soil and rocks along the banks of a stream by scour and caving of overlying deposits; some bank erosion also occurs by runoff from adjoining land, especially with steep slopes.

stream bank erosion control The use of mechanical or vegetative methods to reduce or control the removal of stream banks by water action. *See* JETTY, RETARD, REVETMENT, RIPRAP.

stream bed erosion The detachment and removal of material from the bed of a stream by the movement of water, pebbles and boulders. *See* LOAD, SOD STRIPS.

stream flow depletion The annual amount of water flowing into a basin or valley, minus the amount that flows out of a basin or valley; it is usually less than CONSUMPTIVE USE of water, which in addition includes rainfall and water from underground sources.

stream flow routing A process of determining hydrograph at a point downstream from an upstream hydrograph aided by local knowledge of inflow and storage considerations. *See* FLOOD ROUTING.

stream gauging Measuring the area of cross-section in a stream channel and the velocity of water flow to determine discharge; measurements are usually taken over a period to determine the trend or changing amounts of discharge. *See* GAUGING STATION.

stream gradient (or **slope of stream**) The general slope of the water surface or bed of a stream. *See* SURFACE PROFILE.

streamline A line following the direction of flow of a stream at all points along its course at a given time. *See* STREAM PROFILE.

streamline flow (or **laminar flow**) A flow in which each particle moves in a direction parallel to all other particles.

stream loads *See* LOAD.

stream profile (or **thalweg**) A curve indicating the elevation at key points along the course of a stream from source to mouth; a line following the deepest part of the bed of a stream. The stream is assumed to be straightened out on plan. In general, a river has the steepest slope nearest its source or head, and least slope at its mouth, but in between there may be many irregularities.

streams (fault control) Streams controlled by fault lines. The fault may form a scarp which acts as a confining bank, or form a zone of weak resistance to erosion which is sought by streams for their channels. In these cases the stream flows along relatively straight and definite lines, and does not meander as normal. Fault control of streams shows up clearly in aerial photographs. *See* DRAINAGE PATTERN.

stream sections (1) *See* CROSS-SECTION. (2) Sections along hillside streams showing the rock exposures in the channel. Geological mapping in some regions consists largely of drawing sections along streams and connecting up identical beds.

strike (or **level course**) A line on a rock stratum at right angles to the full dip or slope of the stratum; a line drawn on a map to represent an imaginary line on the rock or ground of a definite level at all points.

stripping A water engineering contract item applied to topsoil removal which is not required for reuse, and the stripping of borrow pits and quarry areas of unsuitable surface materials overlying usable materials. The unit rate includes all operations in getting the material to the disposal area. Measurement is in cubic metres (1.31 cubic yards).

structural analysis (or **tectonic analysis**) An investigation of the geological features shown on a map, and the making of deductions as to the rock succession, the rock structures, and the forces active during the period; the analysis forms part of all hydrogeologic investigations.

structural trap A geological structure in which the porous and non-porous rocks are so arranged as to be favourable for the accumulation of water. *See* FAULT TRAP.

structure (1) Any artificial construction. (2) Major geological features. (3) In soils the spacing and pattern of particles.

.**structure bores** Exploratory BOREHOLES drilled into the ground to determine geological structure. Often linked to groundwater exploration.

structure closure (or **closure**) The elevation in metres, feet or other units of the highest point of a flexure above the lowest 'closed' contour. Closure, nose, trap, fault, and related structures are important to the water-supply geologist. *See* NOSE, STRUCTURAL TRAP, FAULT TRAP.

structure contours Contours drawn on the upper surface of a rockhead, aquifer or other deposit and used to depict its form and depth. The information for drawing structure contours is obtained from maps, boreholes, wells, springs and perhaps geophysical surveys.

stuffing box A cylindrical recess compactly filled with hemp or similar packing material to prevent leakage of water from a pump. *See* GLAND.

stumping off *See* PREVENTATIVE SNAGGING.

sub-artesian water ARTESIAN WATER which the pressure causes to rise in a well but not to overflow at the surface.

sub-critical flow A flow in an open channel in which the velocity is less than the CRITICAL VELOCITY and the depth of the flow is greater than the CRITICAL DEPTH.

sub-glacial stream A stream formed by melting ice near the termination or SNOUT of a glacier. The water seeps down through crevasses or cracks to the bottom of the glacier.

sub-humid A climate or region where the moisture content is below that under humid conditions but normally still adequate for the production of many crops without irrigation. *See* SEMI-ARID.

sub-irrigation Irrigation in which measures are taken to control drainage discharge and to elevate the water-table locally to or nearly to the root zone. *See* SUBSURFACE DRAINAGE CHECK.

submarine blasting Blasting or detonation of explosive charges underwater to break rock during dredging, or in wet shafts or hydraulic excavations.

submarine core drill A drill capable of taking cores and samples of the deposits beneath the sea bed to assist in planning dredging operations, etc.

submarine river crossing A pipeline laid across a river below bed level.

submeander Small loops or windings formed within the banks of a stream, resulting from relatively low flows after flood has eased.

submerged float *See* SUBSURFACE FLOAT.

submerged intake A type of INTAKE WORKS often used to obtain a supply of water from a lake. A dredged channel is formed along the bed of the lake in which a pipe is laid with flexible joints. Loose material is placed round the pipe with hard material on top. The intake end has a bellmouth opening covered with wire mesh or screen. A variation consists of a tunnel excavated under the bed of the lake to the intake point. *See* EXPOSED INTAKE.

submerged orifice An opening or orifice which is wholly below the TAIL WATER level of the structure.

submerged weir A weir in which the TAIL WATER level is equal to or higher than the crest level. *See* FREE-FLOW WEIR.

submergence ratio (or **drowning ratio**) The ratio of level of TAIL WATER to level of HEADWATER with reference to a structure crest when both water levels are higher than the crest. The distance downstream and upstream from the crest at which the water elevations are measured should be such that the structure does not influence the levels. *See* DOWNSTREAM TOTAL HEAD.

submersible pump Usually an electric centrifugal pump which can operate when wholly submerged in water. The motor is totally enclosed and fully protected, and its position can be lowered in a shaft, well or borehole as the water is withdrawn. Corrosion-resistant bronze and stainless steel materials are used with sealed motor stator windings. Control of the pump can be automatic and remote. It is often used for dewatering shafts, mine workings and other inundated works, and also for pumping large volumes of water for irrigation from deep wells. Submersible pumps have capacities up to 450 m^3/h (100 000 gal/h) and heads up to 300 m (1000 ft). *See* SHAFT-SINKING PUMP, TURBINE PUMP.

subnormal pressure (water) Pressure converse to that in ARTESIAN WATER; it is related to CONFINING BEDS in which the water is pressing downward.

subpermafrost water Water underlying the permafrost. *See* SUPERPERMAFROST WATER.

subsequent stream A stream whose course is determined by outcropping belts of weak or soft strata. *See* CONSEQUENT STREAMS, SUPERPOSED STREAM.

subsoil Broadly, the decomposed and incoherent material below the soil layers.

subsoil agricultural drain *See* AGRICULTURAL DRAIN, SUBSOIL DRAINAGE.

subsoil drainage The removal of subsoil water by construction of open intercepting ditches and the laying of drainpipes. Concrete or earthenware pipes, placed at a depth of about 0.9 m (3 ft), are often used. The pipes are butt-ended, 150 mm (6 in) or more in diameter, and laid with 6 mm (¼ in) open joints. The distance between subsoil drains is a maximum in sandy soils and a minimum in clay. The drains are covered with stones or other suitable material. The term also applies to the removal of surplus water in sandy or gravelly subsoil by natural drainage to streams or lower levels. *See* CATCHWATER DRAIN.

subsoil flow (or **subsurface flow** or **seep-off**) Water which infiltrates upper soil layers and seeps laterally into streams or emerges as spring water; rate of flow or discharge of subsurface water. *See* SHALLOW PERCOLATION.

subsurface drainage check The control of drainage discharge in an area to maintain the water-table at a level suitable for SUB-IRRIGATION. It may consist of a gate or other structure placed in a deep drainage ditch or covered drain.

subsurface erosion *See* PIPING.

subsurface float (or **double float**) A body which moves with the stream at a known depth, attached by a line to a SURFACE FLOAT which indicates its position and movement.

subsurface flow *See* SUBSOIL FLOW.

subsurface geology The geology of the rocks occurring below the surface; the study and interpretation of geologic data provided by boreholes and wells or by geophysical methods. At dams and other works, the subsurface geology is explored in some detail by trial bores or other means, and maps and sections prepared. *See* SURFACE GEOLOGY.

subsurface ice (or **interstitial ice**) Ice occurring below the surface.

subsurface irrigation (1) The use of below-surface porous tiles or similar materials for irrigation. (2) A method of disposing of settled wastewater by discharging it into trenches of open-jointed tiles surrounded by a medium such as clinker.

subsurface runoff (or **subsurface storm flow** or **interflow**) That portion of precipitation which forms a subsurface flow and is discharged into a lake or stream. *See* SHALLOW PERCOLATION.

subsurface storm flow *See* SUBSURFACE RUNOFF.

subsurface water All water occurring below the earth's surface, in the liquid, solid or gaseous state. *See* GROUNDWATER.

subterranean stream A stream flowing along a subterranean passage, crevice, fissure or large interstice. *See* DISAPPEARING STREAMS.

suction-cutter dredger A SUCTION DREDGER fitted with a CLAY CUTTER at the lower end of its suction pipe. The cutter breaks stiff clayey beds and gravel which are then removed by the dredger.

suction dredger A DREDGER in which the underwater mud and clayey material are removed by powerful suction pumps. *See* BUCKET-LADDER DREDGER, FLOATING PIPELINE, SUCTION-CUTTER DREDGER.

suction head The head or height to which a pump can lift water, on the suction side, by atmospheric pressure, measured from the free water level in the sump. Normal atmospheric pressure sustains in a vacuum a column of water about 10.4 m (34 ft) high. Because of air leakage, etc., a pump is not placed at a greater height than about 6 m (20 ft) above the lowest water level in the sump. *See* STATIC SUCTION HEAD, TOTAL SUCTION HEAD.

suction pipe The pipe extending from the sump or SNORE to the pump chamber.

suction valve A CLACK VALVE or non-return valve placed in the suction pipe of a pump.

suction well A well from which a pump extracts water. The well is located in the river and may be part of a pumping station or may be a separate unit. The use of SUBMERSIBLE PUMPS may be considered for river pumping units. *See* INTAKE WORKS.

sudd A floating mass of decaying trees and vegetation, etc., impeding navigation on the Upper Nile.

sudden drawdown The sudden lowering of tidal water levels; a rapid drop in the water level in a reservoir which could be dangerous in the case of earth dams.

sudden injection method A method of measuring discharge whereby a known weight of solute is injected into the stream at a cross-section. At a second cross-section downstream successive samples are taken over a period including the time taken for all the solute to pass the cross-section. The samples taken should be sufficient to enable the estimation of the average concentration of solute over the sampling period. The discharge is calculated by dividing the weight of solute injected by the product of the total sampling period and the average concentration expressed as weight of solute per unit volume of water. *See* CONSTANT RATE OF INJECTION METHOD.

sullage *See* GREY WATER.

sulphate-bearing soil A soil with a relatively high SO_3 content and which chemically attacks any concrete work in the ground. If the groundwater contains more than 1% SO_3 a high-alumina cement is recommended for all concrete work at the site. In groundwater which contains less than 0.02% SO_3, concrete is not affected.

sulphate test A test to determine whether a soil or its water content will have any chemical effect on concrete foundations. The sulphate in the soil is precipitated as barium sulphate and measured to ascertain the conditions.

summit canal A canal which cuts across a summit and must have water pumped into it.

sump (1) The pit, lodge or catch basin which supplies water to a pump suction or inlet. A sump also serves as a settling basin to remove debris and dirt before the water enters the pump. Where the sump receives gravity drainage, screens may be interposed between the sump and drainage channels to remove wood, straw and trash. It may also contain baffles to still the water and assist the settlement of sediment. (2) A receptacle sunk below the normal invert level of a trench or drain to collect and remove water. (3) A small tank adjacent to a FARM RESERVOIR or near the pump of a farm pondage. *See* FLOATING STRAINER, SNORE.

super-critical flow *See* HYPER-CRITICAL FLOW.

superficial compaction The compaction of soil in layers. The layer thickness is between 100 mm and 200 mm, depending on the type of ROLLER employed. To achieve maximum DRY DENSITY the soil must be compacted at its OPTIMUM MOISTURE CONTENT. *See* PROCTOR TEST.

superficial deposits Mainly, the unconsolidated deposits which generally overlie the rockhead and consist of clay, sand, silt, peat, loam, or gravel or mixtures of same. Their stability and bearing capacity depend largely on the water content. *See* SOIL, UNCONSOLIDATED MATERIAL.

superpermafrost water Water overlying the PERMAFROST.

superposed stream A stream whose course was determined by the slopes of land forms now almost obliterated by erosion; this course is now superposed on the underlying rocks irrespective of their structure. *See* CONSEQUENT STREAMS.

supersaturated air An unstable condition in which the RELATIVE HUMIDITY exceeds 100%. Water vapour in air commonly becomes supersaturated before it condenses into clouds. *See* SATURATED AIR.

supplemental irrigation Irrigation or watering of crops during dry periods to maintain growth, used in regions where normal rainfall is not adequate throughout the year.

supply channel (irrigation) The main channel supplying water to the irrigation area.

supply ditch A ditch from which water is drawn onto the irrigation land; it may form part of the supply channel.

supply pipe That portion of the branch supply pipe lying within the consumer's premises. *See* COMMUNICATION PIPE.

support, removal of *See* REMOVAL OF SUPPORT.

suppressed weir A MEASURING WEIR with sides that are flush with the sides of the channel, thus eliminating end contractions.

surcharge As measured in a sewer manhole, the height of wastewater above the crown when the sewer is flowing full (Figure S.9).

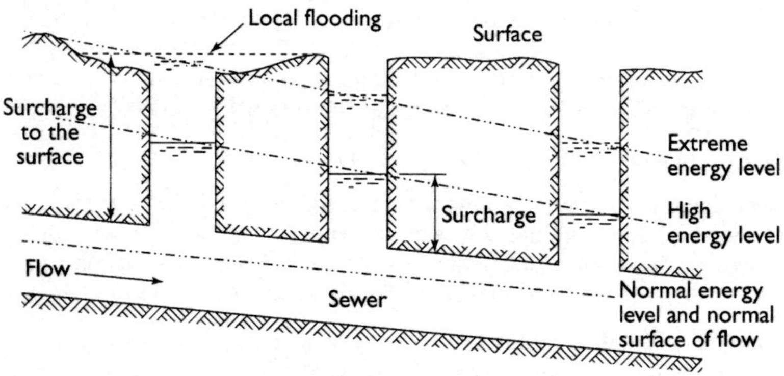

Figure S.9. Surcharge in sewers

surcharge depth The difference between the level of the SURCHARGE STORAGE and the crest level of the spillway or by-wash.

surcharge storage The temporary flood storage built up in a reservoir when water flows over the spillway or by-wash.

surface curve A curve showing elevation of water surface at given horizontal distances along a stream.

surface detention That part of rainfall which forms a thin layer of water over the surface when OVERLAND FLOW occurs. The depth of detention increases until discharge attains equilibrium with rate of supply to surface runoff. The term does not include DEPRESSION STORAGE.

surface drag (or **friction drag** or **skin friction**) The resistance to motion along the contact surface between a fluid and a solid. *See* HYDRAULIC FRICTION, ROUGHNESS COEFFICIENT.

surface drainage The drainage of excess surface water by constructing field ditches connected to larger open ditches, etc. *See* CATCHWATER DRAIN, CONTOUR PLOUGHING, DITCH, DRAIN.

surface float A small wooden or metallic object floating on the water surface and used to determine surface velocity and direction of flow. *See* FLOAT ROD.

surface geology The geology pertaining to the rocks exposed at or near the surface, and also the superficial deposits. The surface and subsurface geology are investigated at new dam and reservoir locations, canals, etc.

surface profile (stream) A longitudinal section or profile of the free surface of a stream or the water surface along an open channel. *See* STREAM GRADIENT.

surface retention *See* INITIAL DETENTION.

surface runoff (or **immediate runoff**) Runoff that has not passed below the surface before it reaches a stream or erodes its own channel.

surface slope *See* GRADE, HYDRAULIC GRADIENT, SLOPE, STREAM GRADIENT, STREAM PROFILE, SURFACE CURVE, WATER-TABLE GRADIENT.

surface storage *See* INITIAL DETENTION.

surface tension The property displayed by the open surface of a liquid as if it were covered by a tight elastic skin: small dry particles will float, largely unwetted, on the surface. *See* BATHOTONIC REAGENT.

surface water All SURFACE RUNOFF, except sewerage or wastewater; broadly, all water which rests on the surface of the earth and may be in liquid or solid state.

surface water drain Any pipe or ditch for the drainage of surface water. *See* DRAIN, SUBSURFACE DRAINAGE CHECK, SURFACE DRAINAGE.

surface water supplies Water stored in reservoirs and lakes, and that obtained from rivers and springs. Towns adjacent to a large river often take their water from that source. The waters from the lower reaches of rivers are more liable to contamination than the upper reaches. River water is filtered and may need both chemical and bacteriological purification before use.

surfactant (surface-active agent) The essential constituent of detergents including soaps and other synthetic formulations, such as coconut oil. The wetting action is improved by lowering the SURFACE TENSION of a liquid; flocculation processes may therefore be inhibited.

surge A large wave or a great rolling swell of water; surging or heaving like a wave; a sudden increase of pressure in a pipeline resulting from closing a valve at its lower end, similar to the process of WATER HAMMER; a sudden change in discharge of water in a channel caused by inflow of water, or by opening or closing of a gate.

surge pipe A standpipe with an open top to release surge pressure.

surge tank An open tank to avoid loss of water during pressure surges. The tank is connected to the top of a surge pipe. It is often used with pressure pipes leading to water turbines.

suspended-frame weir A weir with frames which can be raised and hung above the water from a bridge during floods. *See* FRAMED DAM.

suspended load (1) The fine particles carried by a stream, or other flow of water, in suspension for a considerable period. (2) Material collected in, or measured from, a SUSPENDED LOAD SAMPLER. *See* LOAD.

suspended load sampler A device for taking a sample of water and the particles it carries in suspension without separation of water from the sediment. *See* BOTTLE SILT SAMPLER.

suspended water Water in the ZONE OF AERATION. *See* VADOSE WATER.

suspension Fine particles of matter suspended or held undissolved in water. *See* LOAD.

suspension rod A rod, operated by hand, sometimes used instead of a rack and pinion in shallow water.

sustainability The ultimate goal of development projects ensuring a positive impact on a community by being appropriate, usable, affordable and above all maintainable. Maintenance is the key to sustainability. Evaluation, training and monitoring are essential ingredients.

sustainable development Development that meets the needs of the present without compromising the ability of future generations to meet their own needs.

Sutro weir A type of little-used flat-plate weir (Figure S.10) so shaped that the discharge is directly proportional to the upstream depth above the crest level, i.e. $Q_D = \text{Constant} \times h$.

swabbing The moving of a plunger, fitted with a valve or solid, up and down in a borehole casing. Often used to develop sand screen boreholes, it is also effective in opening choked fissures in chalk and similar rocks. *See* BACKBLOWING.

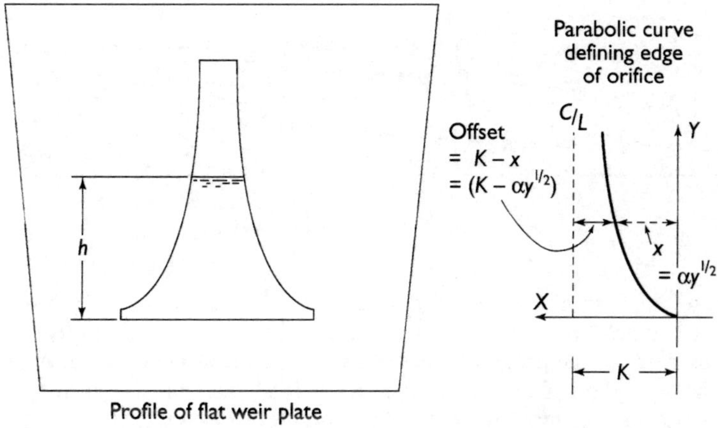

Figure S.10. Sutro weir

swale A vegetated and otherwise unlined open channel that transports and infiltrates runoff waters (Figure S.11). In addition to carrying away drainage flows, swales improve their quality and reduce their volume. They are commonly triangular in cross-section and run along the contours to ensure flow is slow enough to prevent loss of soil and allow sediments to settle out.

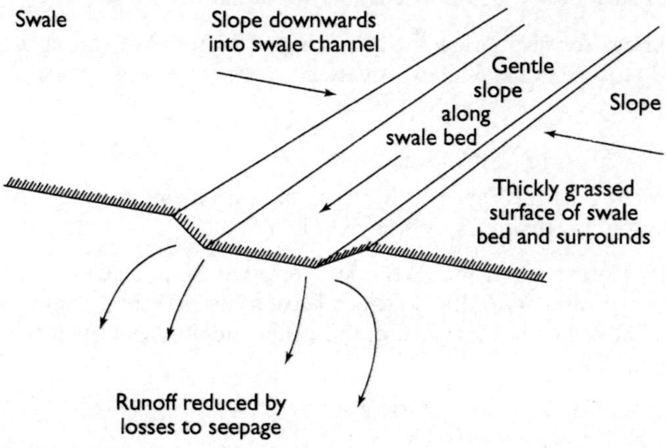

Figure S.11. Swale

swallow holes Roughly circular cavities in limestone country into which streams disappear (Figure S.12). The cavities were dissolved by percolating waters. *See* CHEMICAL WEATHERING, KARST TOPOGRAPHY.

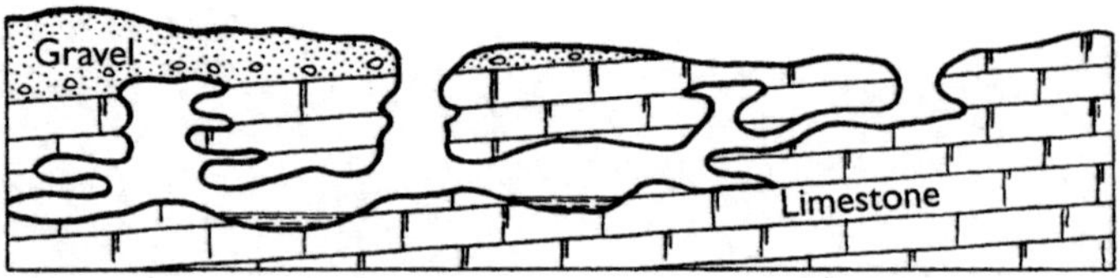

Figure S.12. Swallow holes

swamp A marsh, shallow lake, bog or fen favourable to the growth of aquatic vegetation; soil or ground containing an excess of moisture, due in many cases to the proximity of the water-table. A swamp may be formed in the floodplain of a river, in a depression in glacial drift and in an upland area between streams. *See* CANALIZATION, HILLSIDE SWAMPS.

swamp drainage *See* CANALIZATION.

swash bank The part above the OUTBURST BANK of a sea bank or ridge.

swell (1) *See* FREE WAVES. (2) The increase in volume of a rock or soil when exposed to weathering or to the entry of water. *See* BULKING.

swelling pressure The force exerted by a mass of shale or clay when exposed to water in a confined space. *See* BULKING.

swing bridge (or **pivot bridge** or **turn bridge**) A MOVABLE BRIDGE which is pivoted at its centre and allows a vessel or barge to pass by swinging open horizontally. *See* TRAVERSING BRIDGE.

symbols *See* MAP SYMBOLS.

sympathetic retrogression The lowering of levels in one river as a result of reduction in upstream levels of a river joining it.

synclinal closure (or **canoe fold**) An elongated basin fold with a length much longer than its width, so called because its structural contours on a map are closed; it is of importance in water supply problems if the rocks are favourable.

synoptic meteorology *See* METEOROLOGY (SYNOPTIC).

synoptic weather map A map or chart indicating the weather conditions prevailing over a district at a given time.

synthetic unit hydrograph A UNIT HYDROGRAPH prepared for a drainage basin which is ungauged; it is based on the basin's known physical characteristics.

T

tail bay That part of a canal immediately downstream of the TAILGATE.

tailgate The gate at the downstream end of a LOCK. *See* HEADGATE.

tail race The channel that conducts the spent water from a turbine or water wheel. *See* RACE.

tail water The water or water level immediately downstream of a turbine, water wheel, or structure. *See* HEADWATERS.

Tainter gate *See* RADIAL GATE.

tank An artificial pond or basin for gathering and storing rainfall (especially India); in farms, a small reservoir for liquid manure. *See also* STORMWATER TANKS, WATER TOWER.

tank, balancing *See* EQUALIZING RESERVOIR.

tank, earth (or **livestock reservoir**) An artificial pond or basin formed by an excavation and/or an earthen dam across a drainage channel for gathering and storing water for livestock. *See* FARM POND.

tape gauge *See* CHAIN GAUGE.

TDS Total dissolved solids.

tectonic analysis *See* STRUCTURAL ANALYSIS.

telecontrolled power station A HYDROELECTRIC POWER STATION operated by remote control.

temporary bank protection The protection of river banks from erosion by methods with only a few years life, as opposed to more permanent works such as walls of stone or concrete, or concrete blocks or concrete sheet piling. A temporary protection may suffice along some rivers, to be renewed periodically or until vegetation can be established. *See* ANCHORED TREES.

temporary base level (stream) The water level of a lake or pondage behind a BAR into which a stream flows and erosion proceeds. After a period the lake will be loaded with sediment, or the bar removed by erosion, and a new base level formed, which may be the sea. *See* BASE LEVEL, GENERAL BASE LEVEL OF EROSION.

temporary grade The grade of a stream when it flows into a lake or pondage behind a BAR. In time the lake is filled or the bar removed and the stream will renew erosion to a new grade. *See* GRADED STREAM, LOCAL BASE LEVEL.

temporary hardness (or **carbonate hardness**) *See* HARDNESS OF WATER.

temporary stream An intermittent stream which ceases to flow during part of the year; a stream with high water levels during late winter, followed by a summer drought.

temporary wilting *See* PERMANENT WILTING POINT.

tender (1) A formal offer by a contractor for supply of materials or execution of engineering works as detailed in drawings and specifications for a specified sum of money. (2) A small vessel employed to attend a larger one and supply her with provisions, etc.

tensiometer An instrument which measures the force which plant roots must overcome in order to take in water from the soil. The water required to restore the soil to FIELD CAPACITY may be calculated from the tensiometer readings. Tensiostats, more elaborate forms of this instrument, automatically turn the irrigation water on and off as required.

tensiostats *See* TENSIOMETER.

terminal moraine The ridge of hummocks formed at the end of a glacier as the ice melts. The stages in the recession of the ice are marked by the successive terminal moraines. *See* GROUND MORAINE.

terrace formation (river) The cutting of steps or wide banks in the alluvial deposits of a river valley by the meandering tendency of the current; the cuts tend to deepen with each swing and the channel may become centralized. *See* PLAIN TRACT, TERRACES (1).

terrace height (farm) The vertical distance between the bottom of the terrace channel and the top of the terrace ridge at adjacent points.

terrace interval (farm) The vertical or horizontal distance between corresponding points on two adjacent terraces. In staking terrace lines for construction, the vertical interval, or the difference in elevation between stake lines, is generally used. For the uppermost terrace, the interval is measured from the top of the field to the centre of the terrace.

terrace outlet channel (farm) The channel which takes the discharge from one or more terraces and conveys the water from the area; it is usually covered by dense perennial grass to minimize erosion.

terraces (1) Remnants of gravels deposited by rivers when flowing at higher levels than at present. (2) Ridges of earth or any embankments constructed across a sloping ground to spread runoff and minimize soil erosion. *See* BENCH TERRACE, BROAD-BASE TERRACE, FLOODPLAIN TERRACES, GRADED TERRACE, LEVEL TERRACE, NARROW-BASE TERRACE, NICHOLS TERRACE, RIDGE TERRACE, RIVER TERRACES.

terrace system (farm) All the terraces which occupy a field or slope and are connected to one or more outlet channels.

terrace width The horizontal distance from the high water line on the upper edge of the channel to the foot of the lower slope of the terrace ridge.

terracing *See* CONTOUR PLOUGHING.

tertiary treatment The treatment of wastewater beyond the secondary or biological stage. *See* ADVANCED WATER TREATMENT.

test bore *See* TRIAL BOREHOLES, TRIAL PIT.

test bore pattern *See* BOREHOLE PATTERN.

tested capacity (of well) The maximum rate of yield by pumping from a well without depleting its water supply. The capacity is determined by tests over a specified period.

test pit *See* TRIAL PIT.

tetrahedron A four-sided solid, especially a triangular pyramid. Steel-framed skeleton tetrahedra were used in building the Génissiat dam (1945) in the Rhône Valley to break the force of the water.

tetrapod A pre-cast concrete block consisting of four symmetrically spaced truncated cones (Figure T.1). Tetrapods are used in the construction of marine works, to armour breakwaters and protect them from the force of the waves. The structure of tetrapods facilitates self-interlocking and this desirable attribute increases with the passage of time, because of the wedging action of wave pressure on the breakwater. Reinforced concrete tetrapods can be very large depending on the exposure of the site, and units weighing up to 25 t have been placed. *See* BLOCKWORK.

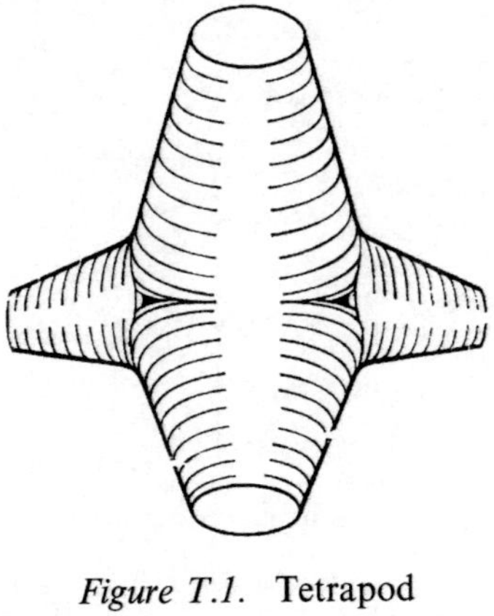

Figure T.1. Tetrapod

thalweg *See* STREAM PROFILE.

thermal spring A SPRING fed by hot groundwater and usually mineralized. In New Zealand, deep boreholes have tapped large reserves of hot water. In one the temperature was 306 °C (582.8 °F). The water is being considered for use in a geothermal power station. *See* HOT SPRINGS.

thermal stratification The formation of separate layers of water with different characteristics based on temperature differences (Figure T.2). In regions where there are pronounced seasonal variations in solar radiation the unusual property of water being at its maximum density at 4 °C (39.2 °F) leads to deep lakes and reservoirs having an annual cycle of first the formation and then the breakdown of distinct water layers at different temperatures. Thermal stratification only occurs in relatively deep bodies of water, as shallow lakes and reservoirs are continuously remixed by wind and wave action. The annual cycle begins in the late autumn when the upper water layers start to cool and so sink towards the bottom. In the spring and summer increased warming of the surface layers causes them to become even less dense and stratification to be more pronounced. The annual cycle ends in the autumn when decreasing solar radiation and falling air temperatures lead to cooling of the upper water layers. They

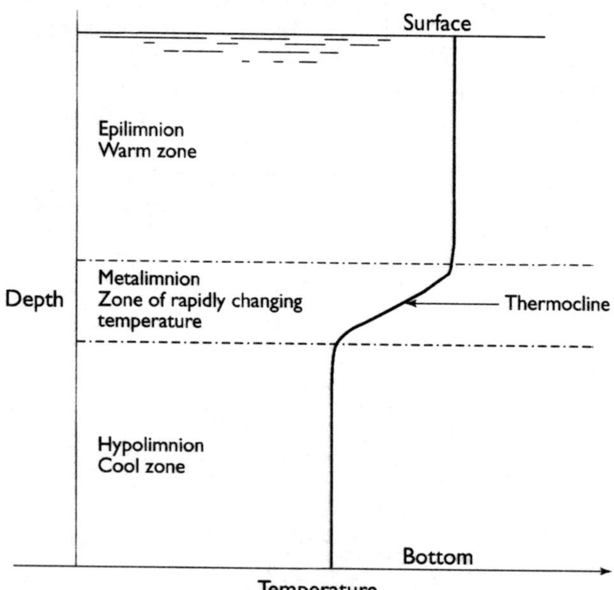

Figure T.2. Thermal stratification in lakes and reservoirs

thus become denser than the bottom layer, and the temperature gradient between the upper EPILIMNION and the lower HYPOLIMNION becomes reversed. The reservoir will then remix, or overturn, often quite suddenly, and the annual cycle begins again. Thermal stratification can have an important effect on the management of water supply reservoirs: the hypolimnion often contains little dissolved oxygen, and at the time of the OVERTURN the remixing of the stored water can cause an abrupt change in the quality of the water being drawn off.

thermocline The layer below the EPILIMNION and above the HYPOLIMNION in a thermally stratified lake. In the thermocline there is rapid decrease in temperature with depth.

thermograph A self-registering thermometer.

thermo-osmosis The natural movement of soil water from a relatively warm part towards a cooler part.

thickening The process of water removal from sludge after normal sedimentation to increase the solid concentration.

Thiessen polygon A polygonal figure formed around rain-gauge stations by perpendicular bisectors of lines which connect the stations on maps.

thin core dams Earth dams with a central impervious core width of about 15 to 20% of the water height.

thin-plate (square-edged) weir A thin-plate vertical weir so shaped that the sheet of water passes clear from the crest. *See* TRIANGULAR NOTCH.

Thomas–Fiering model A type of relatively simple STOCHASTIC MODEL which is suitable for modelling total monthly river flows, and therefore has wide application in applied hydrology and water resource planning. If a monthly flow model is required, the available records are analysed to estimate the average flow and its variation for each month, and also the serial correlation coefficient to one or more preceding months. These data are then used in a computerized Thomas–Fiering model.

Thornthwate index *See* MOISTURE INDEX.

three-throw pump A pump consisting of three single-acting rams, positioned side by side, and operated by cranks set at angles of 120°. The separate suction and delivery pipes join to form common suction and discharge pipes to which air vessels are attached.

threshold velocity The minimum velocity at which wind commences to detach and move particles of soil in a given area and under specified conditions. *See* WIND EROSION.

thunderstorm A local STORM, usually of short duration, with thunder and lightning and often heavy rain or hail. *See* HAILSTORM.

tidal bore *See* TIDAL CURRENT, HYDRAULIC BORE.

tidal current The upstream current generated in narrow channels and shallow estuaries with the rise of the tide. The tidal current often advances with a steep wave at its front, termed a bore. The bore frequently disappears in waters or channels which have been deepened by dredging, etc. Occasionally, a similar wave may advance downstream as a result of a cloudburst or sudden release of water by removal of natural barriers in the valley. In the Mersey estuary (UK) the tidal current advances at about 13 km/h (8 miles per hour).

tidal dock A dock which has no gates, so that the water level inside is the same as that of the outside tides.

tidal flood current The tidal current advancing towards the shore and along shallow estuaries and tidal rivers.

tidal flood interval The time interval between transit of the moon over the meridian of a place and the following strength of flood.

tidal flood strength The tidal flood when current is at peak velocity.

tidal lag The delay between high tide, or low tide, in an estuary, and the corresponding highest or lowest level of the water-table in the area.

tidal marsh A marsh located within the influence of tides and generally flooded during high tides, often traversed by interlacing channels and containing salt-tolerant vegetation such as rushes and grasses.

tidal range Difference in altitude between high and low tidal surface.

tidal range, great diurnal *See* GREAT DIURNAL RANGE OF TIDE.

tidal range, mean *See* MEAN TIDAL RANGE.

tidal river A river within the influence of tides; it periodically rises and falls or flows and ebbs. *See* TIDAL CURRENT.

tidal surge A rapid rise of sea level and water movement, rather like the bore of a tidal current. When superimposed on spring tides the flooding and damage are often extensive. *See* EQUINOCTIAL TIDES.

tidal water level The height or elevation of a tidal surface at a specified stage of the tide. *See* HIGHER HIGH WATER, HIGHER LOW WATER, HIGH WATER, LOWER HIGH WATER, LOW WATER.

tidal wave (1) In popular usage, any abnormal ocean wave of some magnitude; for example, one attributed to an earthquake. (2) In astronomical usage, a periodic variation of sea level as a result of gravitational attraction of sun and moon. (3) A STORM WAVE.

tide The periodic rise and fall of the sea due to attraction of the moon and sun; the swell of water thus generated moves round the earth as the latter rotates. *See* SPRING TIDES.

tide, falling *See* EBB TIDE.

tide gate A swinging gate which excludes water at high tide and allows drainage at low tide, placed below high tide level on the outside of a drainage conduit from a field.

tide gauge (or **tide predictor**) A device with which it is possible to predict the tides in any known set of channels.

tide, rising *See* FLOOD TIDE.

tie bar (or **tombolo**) A deposit or bank of sand connecting the mainland to an island mass. On some coasts, rock reefs off the mainland have aided the formation of tie bars. *See* SPIT.

tile drain Generally, concrete or earthenware pipes laid with open joints to receive and remove excess water from the soil. The internal diameter may be 100, 125 or 150 mm (4, 5, or 6 in), with a wall thickness of about a twelfth of the inside diameter. *See* SUBSOIL DRAINAGE.

tilt dozer A BULLDOZER to which a large degree of tilt can be applied to either side. *See* GRADER, SPREADER (2).

tilting gate A spillway crest gate hinged at the bottom or top, and automatically rising and falling with variations in water level; the gate is counterbalanced by weights.

timbering The operation of setting timber supports in pipe trenches, deep excavations for foundations, large water wells and bank protection works. The timber pieces used include poling boards, walings, struts, runners, wedges, etc.

timber sheet piling A quick method for stopping scour on a river bank. Timber piles are usually driven in a line a little outside the eroding bank, and then the timber sheeting is fixed horizontally from pile to pile or driven vertically against horizontal walings attached to the piles. Shingle or stone is packed in the space between the sheeting and the bank. The bank may be planted with strong-rooting grasses. Timber sheet piling presents some problems such as underscour and the danger of attack by MARINE BORERS. *See* CONCRETE PILING.

time–area method A method of predicting the magnitude of runoff flows which is normally used for urban catchments (Figure T.3). It was developed from the RATIONAL METHOD but gives better predictions as it uses more data and makes more realistic assumptions about rainfall intensities and flows in pipes and culverts. A central feature of applying the time–area method is drawing the time–area diagram, which plots the area contributing to discharge at the outfall against the time since the rainstorm began. This enables changes in rainfall intensity during a storm to be included in the calculations. More computation is thus needed than for the rational method, but in general the time–area method makes more reliable predictions, particularly for long and narrow or irregularly shaped catchments.

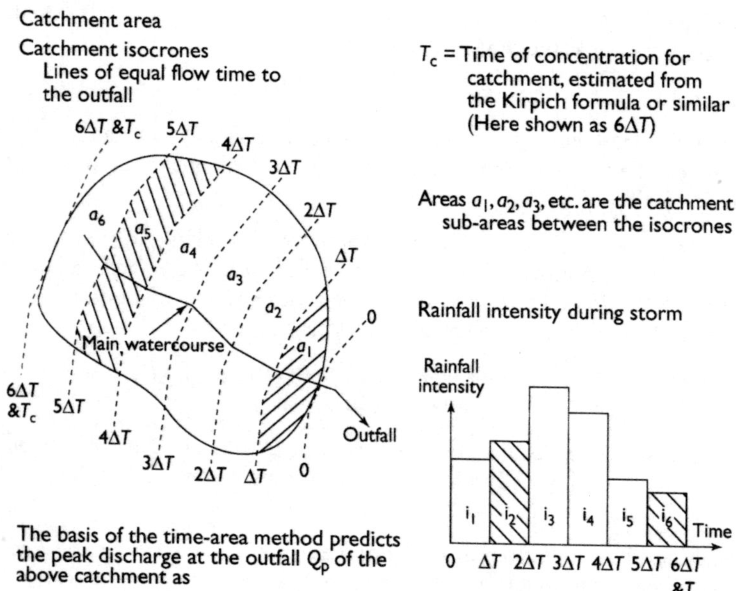

$$Q_p = \sum_{1}^{6} (a_1 \times i_6 + a_2 \times i_5 + a_3 \times i_4 + a_4 \times i_3 + a_5 \times i_2 + a_6 \times i_1)$$

Figure T.3. Time–area method

time lag A term referring to (1) runoff of rainfall, i.e. the time between beginning of rainfall to peak of runoff; (2) discharge or water level, i.e. time elapsing between occurrence of corresponding changes in discharge, or water level at two stations in a river; (3) snow melting, i.e. the period between the start of the snow melt and the beginning of the resulting runoff. *See* LAGTIME.

time of concentration The longest time taken for the rain falling on a drainage basin to travel from the most remote point upstream of a point in that drainage basin at which flows are under consideration. The time of concentration is an essential parameter for use in the RATIONAL METHOD for peak flow prediction. For sewer systems it is usually estimated as the TIME OF ENTRY plus the TIME OF FLOW. The KIRPICH FORMULA or the BRANSBY-WILLIAMS FORMULA can be used to estimate the time of concentration for catchments without artificial drainage.

time of entry In areas which are sewered or otherwise artificially drained, the average time taken for rain falling on the drainage area to flow into the system of drains. It is primarily dependent on slope and surface impermeability. *See* TIME OF CONCENTRATION.

time of flow The total time taken for RUNOFF within artificially drained areas to travel along the maximum length of pipes or channels upstream of the point under consideration, when along each length the flow is travelling at the maximum velocity. It is primarily dependent on slope and surface impermeability. *See* TIME OF CONCENTRATION.

time of travel The time a flood wave takes to travel a known distance downstream; the time taken for the passage of a float, solution or other object between the two end points of a known distance along a stream or other open channel.

time series A set of data in which both the magnitude of each observation and its position within the set are significant. For example, a record of annual peak floods in the historical order of the years in which they were observed would form a time series. The same record of floods, but not in historical sequence, would provide a data set but not a time series.

Titan crane A heavy crane, with a lifting capacity of at least 50 t, for handling and setting stone or concrete blocks when building breakwaters and also for ship construction. Consists essentially of a portal frame carrying a swing jib crane.

toe The lowest upstream or downstream part of a dam, sometimes referred to as the upstream toe or the downstream toe. *See* HEEL.

toe drain (or **toe filter**) A GRADED FILTER drain placed on the downstream toe of an earth dam to improve stability and as a protection against PIPING (Figure T.4). *See* APRON.

toe filter *See* TOE DRAIN.

tombolo *See* TIE BAR.

topographical map A map showing primarily the surface features – natural

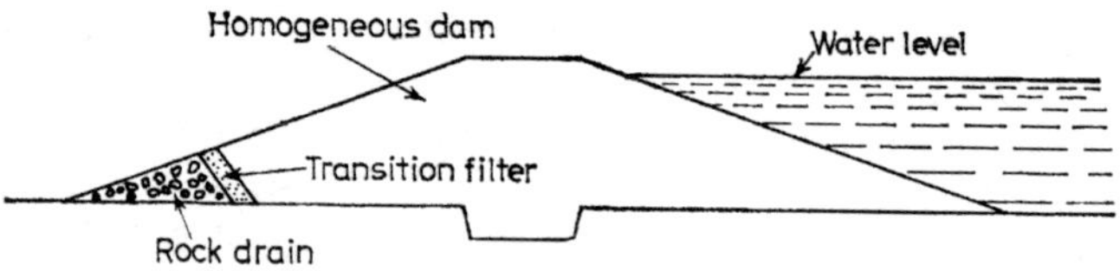

Figure T.4. Toe drain

and artificial – of an area. It shows seas, lakes, ponds, streams, canals, swamps, etc., and is frequently used for plotting or sketching geological data and the position of wells, springs and water-table contours. *See* WATER PROSPECT MAP.

topographic divide *See* DIVIDE.

topset beds *See* DELTA.

topsoil cover A layer of soil over an embankment, by-wash, or other surface to reduce erosion, minimize surface cracking and prevent slaking of the surface material. A cover may be placed over an entire bank to a depth of 230 to 305 mm (9 to 12 in) and grassed (Figure O.1). *See* BANK MATERIALS, SUPERFICIAL COMPACTION.

topsoil replacement An earth dam embankment contract term: the replacement of soil on dam embankment, disposal areas, beautification areas, etc. It is measured in square metres or square yards.

top water level (TWL) *See* FULL SUPPLY LEVEL, NOMINAL RETENTION LEVEL, Figure R.4.

tornado A violent storm of limited extent but of great intensity, characterized by winds of hurricane force, rotary motion, heavy rain or hail, thunder and lightning, and often a funnel-shaped cloud; it is of short duration, not exceeding an hour or two, but causes much damage. Tornadoes occur especially in the USA from April to July and also in west Africa at the beginning and end of the rainy season. *See* WATERSPOUT.

tortuous flow *See* TURBULENT FLOW.

total capacity (of well) The maximum rate of yield by pumping after the water stored in the well has been withdrawn; the yield by pumping when the water level in the well is lowered to the intake point. *See* ECONOMIC YIELD.

total delivery head The sum of the static delivery head and the frictional head loss in the delivery pipe of a pump. Sometimes described as the DYNAMIC DELIVERY HEAD.

total dissolved solids (TDS) The sum of all dissolved solids in a wastewater, expressed as a concentration in mg/L.

total evaporation *See* EVAPO-TRANSPIRATION.

total hardness of water Water hardness which includes the carbonates (temporary hardness) and the sulphates (permanent hardness). *See* HARDNESS OF WATER.

total head (pump) The sum of the delivery pressure head, static pressure head, suction head and friction head.

total head (channel) The HEAD obtained by adding the height of the free surface above the horizontal datum and the velocity head based on the mean velocity at the given point.

total head gradient The total head difference per unit horizontal length along the stream.

total head line The line or curve obtained by plotting the total head against distances measured along the stream.

totalizer A type of RAIN-GAUGE used in isolated or hilly areas where daily measurements would be difficult or inconvenient.

total moisture The moisture content of soil or other material as sampled.

total porosity Sometimes, both capillary porosity (the small voids which hold water by capillarity) and non-capillary porosity (the larger voids which will not hold water by capillarity). A soil or rock with high total porosity is porous or 'open', while a soil with low total porosity is non-porous or 'dense'.

total runoff The sum of the SURFACE RUNOFF and SHALLOW PERCOLATION. *See* DEEP PERCOLATION.

total suction head (or **dynamic suction head**) The sum of the STATIC SUCTION HEAD and the DYNAMIC SUCTION HEAD loss in the suction pipe of a pump.

total surcharge The sum of both the FLOOD SURCHARGE and WAVE SURCHARGE above the FULL SUPPLY LEVEL in a reservoir (Figure R.4).

towed grader *See* GRADER.

toxic water Water containing poisons and tending to give a crop yield which is below normal. *See* ACID WATER, pH VALUE.

trace of rain *See* PRECIPITATION (TRACE OF).

tracer (groundwater) A dye, salt or other substance to trace the movement of water. The tracer is introduced into the groundwater at one point and the time elapsed before it appears at a location down the groundwater gradient is noted. An ideal tracer is not absorbed by the porous material, does not react with the natural water, is not present in the natural water, can be detected in minute concentrations and is available cheaply.

tract (stream) One of the three major stretches or sections of a river from source to mouth. The MOUNTAIN TRACT is along its source in the mountains, followed by the VALLEY TRACT, and finally the PLAIN TRACT where the river approaches its mouth (Figure M.4).

traction (streams) *See* LOAD.

tractive force (water) The force exerted by flowing water on the bed of a stream and in a direction tangential to it.

tractor-driven pump A mobile pumping unit in which the pump is driven through the power take-off of a tractor. Often used in farm irrigation schemes where the water supply enables the pump to be mobile. A trailer is used for carrying the equipment.

training wall A wall built along or near the sides of a river or estuary to confine the waters within the channel. A training wall consisting of a shingle embankment may be satisfactory, provided the current does not undermine it. A more permanent structure is made from piles and sheeting, or a bank of gravel and earth protected on its river side by anchored trees and willows. Another type consists of stonemesh gabions placed between two rows of old rails used as piles. There are many modifications. *See* LOG TRAINING WALL, REVETMENT.

training works All constructions for the regulation and training of rivers, improvement of flow characteristics, and control of scouring and silting. *See* GROYNES, LEVEE.

transmissibility coefficient The product of thickness of saturated portion of aquifer and the FIELD COEFFICIENT OF PERMEABILITY. Also called transmissibility, transmissivity.

transmission coefficient *See* COEFFICIENT OF PERMEABILITY.

transmission constant *See* COEFFICIENT OF PERMEABILITY.

transmissivity *See* TRANSMISSIBILITY COEFFICIENT.

transpiration The emission of watery vapour by living plants into the atmosphere. The energy which converts the water in plants into the vapour emitted by the leaves is obtained mainly from the sun. Hence, practically all transpiration occurs during the hours of daylight. *See* VEGETAL DISCHARGE.

transpiration, potential *See* POTENTIAL TRANSPIRATION.

transpiration ratio (or **water-use ratio**) The relation between weight of water escaping from a living plant (roots usually excluded) during its season of growth, to weight of dry material produced.

transpiration stream The movement of water from the soil into the plant roots, passing up the stems and then out through the leaves as vapour.

transportation (sediment) The movement of fine material and coarse particles by stream action or waves. *See* LITTORAL DRIFT, LOAD, LONGSHORE DRIFT, SEDIMENT TRANSPORT CONCENTRATION.

transport competency The ability of a stream to transport material in suspension in terms of dimensions of particles. *See* CAPACITY OF STREAM, COMPETENCE, LOAD.

transport concentration *See* SEDIMENT TRANSPORT CONCENTRATION.

transposition of storm The use of a storm record obtained in one area at another area within a region with somewhat similar meteorological conditions.

transverse crack A type of crack which runs at roughly right angles to the axis of an earth dam. It is dangerous because it could create a line of concentrated flow through the dam. Transverse cracks are due to DIFFERENTIAL SETTLEMENT.

trapezoidal groyne A four-sided stonemesh groyne with the long sides tapering to a narrow top. The stones are built on a wire mesh mat which is carried up, as the work proceeds, on both sides of the groyne and wired at the joint. *See* STONEMESH GROYNES.

trapezoidal weir A MEASURING WEIR constructed from thin plate with three edges forming a trapezoidal shape in a plane perpendicular to the direction of flow.

trash rack (USA) A screen of mesh or parallel bars placed across the inlet to irrigation siphons, streams, dam outlet pipes and similar hydraulic structures to intercept floating debris and prevent damage.

traveller *See* BONING ROD.

travelling screen (1) A canvas diaphragm in a frame used for the direct measurement of the mean velocity of water. The frame travels with the water and is shaped to fit the channel, which must be of relatively uniform section. (2) A movable screen for extracting debris from flowing water.

travel time *See* TIME OF TRAVEL.

traversing bridge A bridge capable of moving backwards to allow vessels to pass. *See* MOVABLE BRIDGE.

traversing slipway A slipway for hauling vessels, up to about 500 t weight, out of the water for repairs, etc. On the slipway, the vessel can be traversed or moved sideways to another berth, leaving the slipway free for another vessel. *See* SLIP.

treadle pump A foot-operated pump, comprising double cylinders operated by the treadles through a pulley and plunger rods (Figure T.5).

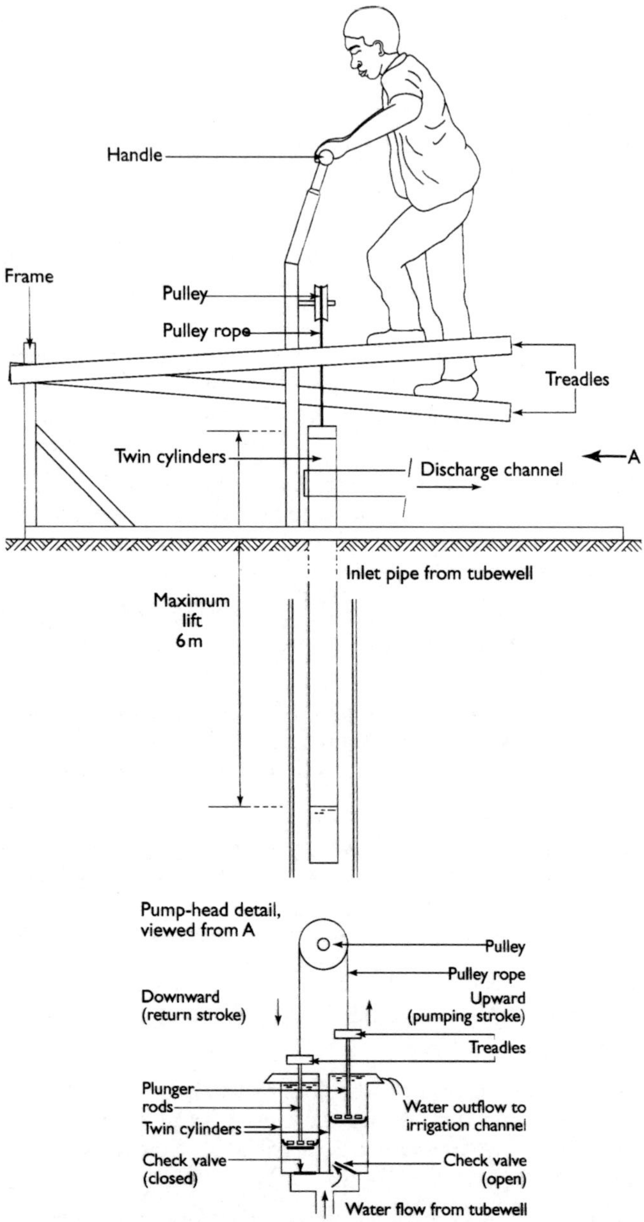

Figure T.5. Treadle pump

Water can be lifted from wells, boreholes, streams and ponds up to 6 m, or at a rate of up to 50 litres per minute.

tree planting Planting shrubs, trees and cuttings of same for controlling gully erosion and protecting river banks, sand dunes and windbreaks. It may be used to prevent excessive runoff and soil erosion, and to stabilize valley slopes liable to landslides. For bank protection, strong-rooting and quick-growing plants are often used. In Australia and New Zealand poplars and willows grow rapidly under a wide range of conditions, and are often used along banks and to protect groynes. The bank is often sloped, and shallow trenches, about 0.9 m (3 ft) apart, are made down the bank. A willow pole is laid in each trench with its foot in wet ground and its head extending up the bank above flood level. The willows usually root where they touch the ground along their length. *See* ANCHORED TREES, POPLARS, TRIPOD PLANTING, WILLOWS.

tree retards A type of RETARD used in some New Zealand rivers consisting of trees with their butts attached to wire cables anchored to concrete blocks or stonemesh gabions, or held by tar drums filled with stone. The retards are constructed out from the eroding bank and direct the current into the desired course. *See* SINKER GROYNE.

trellis drainage (or **grapevine drainage**) A drainage pattern in which streams course parallel to the inclined strata outcrops for long distances, and then cut through the intervening ridges and continue along another course. *See* DRAINAGE PATTERN.

trémie A box or frame used for depositing concrete under water, incorporating a hopper above water level.

trench, cutoff *See* CUTOFF TRENCH.

trench digger *See* TRENCH EXCAVATOR.

trench drain *See* FRENCH DRAIN.

trencher *See* TRENCH EXCAVATOR.

trench excavator (or **ditcher** or **trencher** or **trenching machine**) A self-propelled excavator, usually mounted on crawler tracks, for cutting ditches or trenches for pipelines, drainage, or water supplies; excavation may be by a chain of buckets mounted around a wheel or by a bucket ladder.

trench, groundwater *See* GROUNDWATER TRENCH.

trenching machine *See* TRENCH EXCAVATOR.

trend A long-term and sustained change, usually of a hydrological or meteorological parameter such as mean annual temperature or rainfall. Trends proceed in one direction, but the rate at which they proceed can be very irregular. GLOBAL WARMING is currently a significant trend in the temperature of the earth's surface.

trial boreholes Holes put down as a preliminary to an important well pumping scheme. They may be up to 0.6 m (24 in) in diameter for water, and smaller, not exceeding about 150 mm (6 in) diameter, if only geological information and water levels are required. Chemical and bacteriological analyses of the water are usually made.

trial pit (or **test pit**, USA) Usually, a small shaft excavated to test the superficial deposits and position of the ROCKHEAD and water-table. It may be circular, about 0.9 m (3 ft) in diameter and put down to depths of about 15.2 m (50 ft) or more in firm clayey ground, where water is not troublesome. If rectangular, the pit is made larger, particularly where the ground must be supported. *See* OPEN PIT.

triangular crib groyne A CRIB GROYNE, triangular in shape and filled with stone (Figure T.6). It does not usually exceed about 6 m (20 ft) in length. One side of the triangle is parallel with the river bank, and the angle between the log walls at the nose is about 45°. The walls consist of small logs placed horizontally one over the other. The log ends at the nose are cut or notched to give a locking effect. The entire structure is secured by wire, rods or spikes. The bank behind the groyne is often sloped and grassed. *See* CRIBWORK, RECTANGULAR CRIB GROYNE.

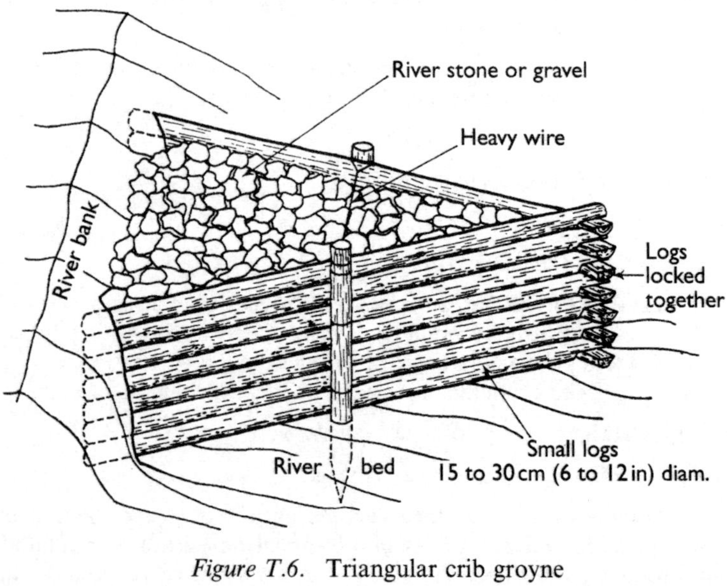

Figure T.6. Triangular crib groyne

triangular notch (or **triangular weir** or **vee notch** or **V notch**) A V-shaped measuring weir or notch with sides that form an angle with its apex downwards (Figure T.7). It measures small flows more accurately than the rectangular-notch weir.

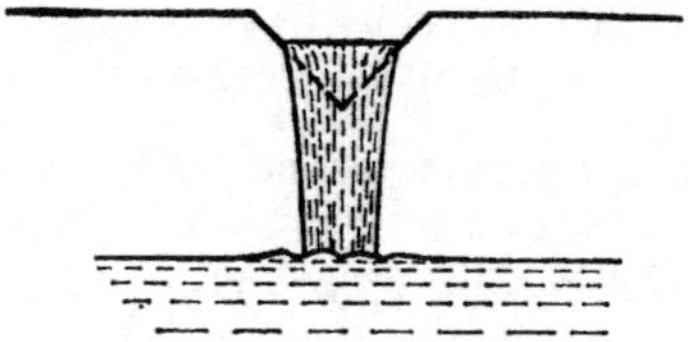

Figure T.7. Triangular notch

triangular weir *See* TRIANGULAR NOTCH.

tribar A reinforced concrete block used in breakwater construction. It consists of three cylinders so spaced that their centres form an equilateral triangle; these cylinders are joined at their midpoints by a three-pronged web (Figure T.8). The tribar was invented by an American, Robert Q. Palmer. *See* BLOCKWORK.

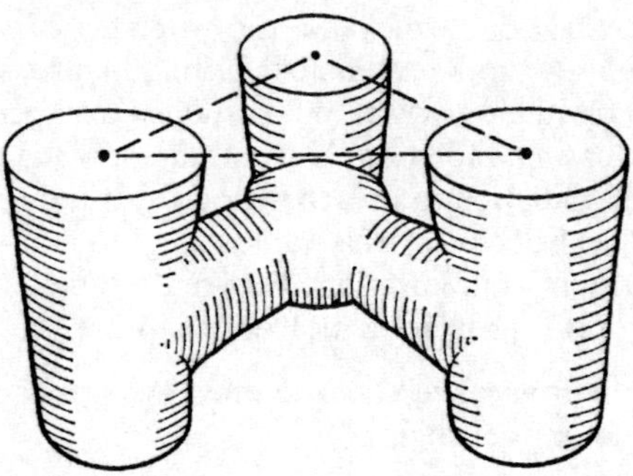

Figure T.8. Tribar

tributary A stream that flows into a larger stream or river; it may be on the surface or underground, and flow continuously or intermittently. *See* CONFLUENCE.

trickle irrigation Irrigation in which water is supplied only to areas close to individual plants. Small-bore flexible pipes are used to convey water to special nozzles or 'emitters', from which it drips onto the soil at a low rate and at a low pressure head. Water may be supplied direct from an overhead tank or direct from the mains. Hard water is unsuitable, as the nozzles become encrusted and blocked. The water from each nozzle spreads downward and outward to form a cone of moist soil. *See* LOW-LEVEL SPRINKLER IRRIGATION.

trickle tube A pipe spillway, usually between 150 and 305 mm (6 and 12 in) in diameter, placed just below the by-wash level of a small earth dam. It prevents erosion in the by-wash by small continuous flows.

Trief process A process to provide a concrete with low heat properties for dam construction. Trief blast-furnace slag is wet-ground, and the resulting slurry used in the concrete mixer. About 70% of the cement content can be replaced by the slag: the rise in temperature of this cement is much lower than that of Portland cement. The Trief process was devised in Belgium by M.V. Trief. *See* LOW-HEAT CEMENT.

trihalomethane Derivatives of methane, CH_4, in which three halogen atoms are substituted for three hydrogen atoms.

tripod planting A recent development in New Zealand, used there to induce siltation or to check the flow of floodwater. It consists of three willow poles planted in the form of an equilateral triangle with the poles inclined inwards. They are tied together just above the ground.

tritium A radioactive isotope of water, the properties of which can enable the age of groundwater to be estimated. Tritium is present in very small quantities in ATMOSPHERIC WATER, and sustained at a generally constant concentration by cosmic radiation. If it falls as rain and seeps below the surface, no further radiation can reach it, and the tritium slowly reverts to normal water. The half-life of this radioactive decay is known, and a comparison of tritium concentrations between a sample of groundwater and atmospheric water provides a measure of its age.

tropical cyclone (or **tropical revolving storm**) A term sometimes used to describe a CYCLONE in the tropics.

troughs Water vessels placed conveniently about a farm to enable stock to be adequately watered; a wooden or other channel for conveying water; any depression of a long and narrow shape for carrying water.

true velocity (groundwater) *See* EFFECTIVE VELOCITY.

truncated soil profile A soil profile which has lost part or all of the A- and sometimes the B-horizon by ACCELERATED EROSION and by mechanical means; the remaining material or C-horizon consists of undeveloped parent material. The degree of truncation is indicated by comparing the eroded soil profile with the virgin profile of the same area.

trunk main Usually a large-diameter pipe for conveying water supplies in bulk from one point to another within an area of supply. *See* AQUEDUCT.

tube well A SHALLOW WELL driven into gravels down to a depth of about 8.0 m (25 ft). It consists of a perforated pipe, usually of less than 100 mm (4 in) diameter, with a pointed shoe, which is forced down into the ground. The water is withdrawn by means of a handpump at the surface. *See* COMPOUND WELL.

tubular spring An outflow of water from tubular channels in drift or soluble rocks such as limestones (Figure T.9). The water may come from a subterranean stream and the flow may be considerable. The water from tubular springs is often muddy after heavy rainfall or floods, because the water has not been filtered by percolation through porous deposits. *See* SWALLOW HOLES.

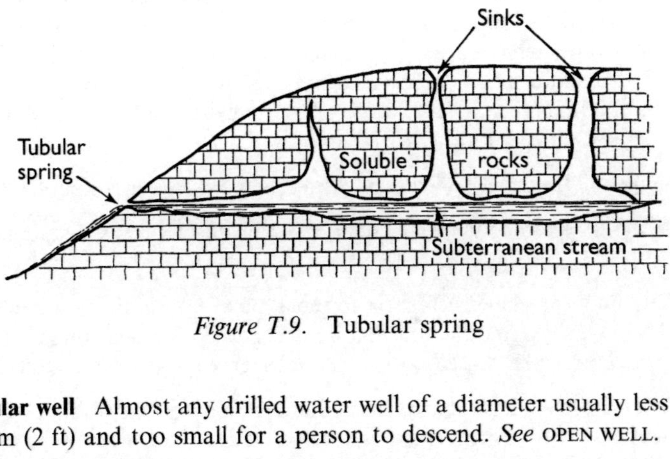

Figure T.9. Tubular spring

tubular well Almost any drilled water well of a diameter usually less than 0.6 m (2 ft) and too small for a person to descend. *See* OPEN WELL.

tunnel A horizontal or inclined passage for drainage or water supplies; it may be provided with a continuous reinforced concrete or steel lining. A subaqueous tunnel may be constructed as an alternative to a bridge over a river to accommodate traffic and other facilities. *See* DRAINAGE TUNNEL, GALLERY, PRESSURE TUNNEL.

tunnel, infiltration *See* INFILTRATION TUNNEL.

tunnelling (1) The development of cracks in earthen dams and embankments under dry conditions. When rapidly brought into contact with water the cracks collapse internally to form a tunnel which could affect the safety of the construction. In dams and banks tunnelling starts at the wet face or upstream side and proceeds downstream. It thus differs from PIPING, which starts downstream and proceeds up to the wet face. Tunnelling is usually a greater danger than piping because its detection is more difficult. (2) The excavation of subways or tunnels to carry trunk mains or other works beneath rivers, through elevated ground, or where surface construction is impossible or too costly. The work is usually entrusted to a specialist contractor.

turbidity The quality of a liquid containing suspended matter. The approximate concentration of suspended matter is ascertained optically from interference to passage of light rays through a sample of the water, and compared with recorded turbidity of standard samples. *See* SUSPENSION.

turbine *See* IMPULSE TURBINE, KAPLAN TURBINE, REACTION TURBINE, TURBINE PUMP, WATER TURBINE.

turbine pump A rotary-type pump with a shrouded IMPELLER with the water entering at its centre. The impeller is surrounded by a diffusion ring containing vanes which direct the impeller discharge into a circular casing, which delivers into the eye of the next impeller in series. The diffusion ring converts the high-velocity discharge of the impeller into PRESSURE HEAD. *See* DIFFUSER CHAMBER, ROTARY PUMP.

turbine pump, deep well *See* PUMP, SHAFT-SINKING PUMP, TURBINE PUMP.

turbulence A state of unsteady flow in which the liquid is disturbed by eddies.

turbulent flow (or **sinuous flow** or **tortuous flow**) A fluid flow in which particles may move in different directions to each other and not in a fixed path. It occurs when the REYNOLDS NUMBER exceeds a certain value.

turbulent velocity In a particular conduit, that velocity above which turbulent flow will always exist, and below which the flow may be either turbulent or streamline, depending on flow conditions.

turfing Covering the banks of a river, irrigation channels and river improvement slopes with grass turfs cut from other areas; it is used as a protection against erosion and to stabilize the slopes. *See* REVETMENT.

turkey's nest tank A TANK formed within a circular continuous embankment built from material excavated from outside the storage basin. The water is entirely above ground level and used for short-term storage. *See* RING DAM.

turn bridge *See* SWING BRIDGE.

twin laterals (streams) A stream which has split and continues its course along each edge of a lava flow which has partially filled a valley. If it continues along only one margin, it is called a LATERAL STREAM.

TWL Top water level.

two-stage dam A dam, wall or barrier containing two distinct drops.

typhoid fever A water-borne bacterial infection which causes a high fever. Without treatment it is often fatal. In the past there was a very high risk of contracting typhoid fever from drinking water but this is now almost unknown wherever effective water treatment has taken place. However, typhoid can also be contracted from food and drink that have been contaminated with human faeces or urine, and outbreaks from such sources are still fairly common.

typhoon A violent hurricane or whirlwind, occurring especially from July to October in China seas. *See* CYCLONE.

U

udometer *See* RAIN-GAUGE.

ultrasonic gauging A technique of stream or pipe flow gauging which uses the difference between the times taken for an ultrasonic pulse to travel across a stream in different directions. Sending and receiving stations are fixed on both sides so the distance for the pulses to travel is the same in both directions. However, in one direction the pulse is made to travel in a diagonal sloping downstream, and upstream in the other. The time differences are very small, so this method requires sensitive electronic equipment. A more sophisticated method of ultrasonic gauging uses the DOPPLER EFFECT.

ultraviolet radiation A type of powerful radiation used to disinfect water both up to drinking water standards and also for the final treatment of wastewater effluent after tertiary treatment (Figure U.1). Ultraviolet (UV) radiation is toxic to most forms of life and can thus destroy micro-organisms suspended in water. Normally the water to be treated is passed slowly through transparent tubes or down shallow channels under an intense source of UV light. There is enough UV radiation in strong direct sunlight to produce partial disinfection, but more intense artificial sources, such as mercury vapour lamps, are more effective. The equipment needed for this method of disinfection can be fairly compact, but the UV lamps require a supply of electricity, and the water to be treated needs to be relatively clear and free of suspended sediments, which often requires filtration.

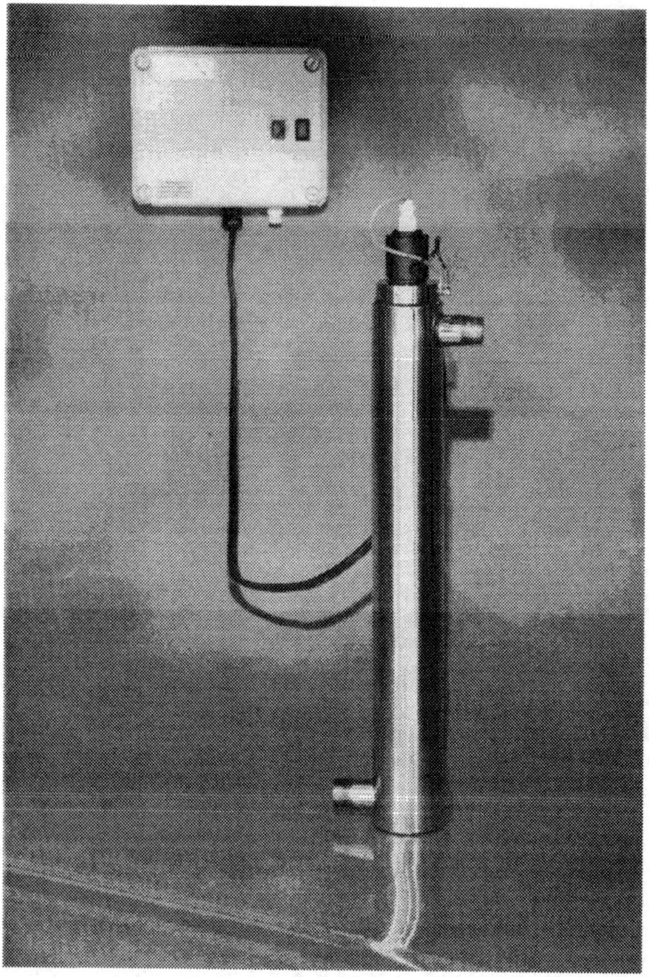

Figure U.1. A small wall-mounted ultraviolet water treatment plant. This unit is just over 500 mm high and can disinfect up to 1000 L/hour

unconfined aquifer An aquifer bounded at the top by the phreatic surface, or water-table (Figure U.2). Flow within an unconfined aquifer towards a riverbank or well can only be sustained by a slope in the phreatic surface as, unlike flow within a confined aquifer, a hydraulic gradient cannot be provided by excess within the water in the pore spaces. See ARTESIAN AQUIFER.

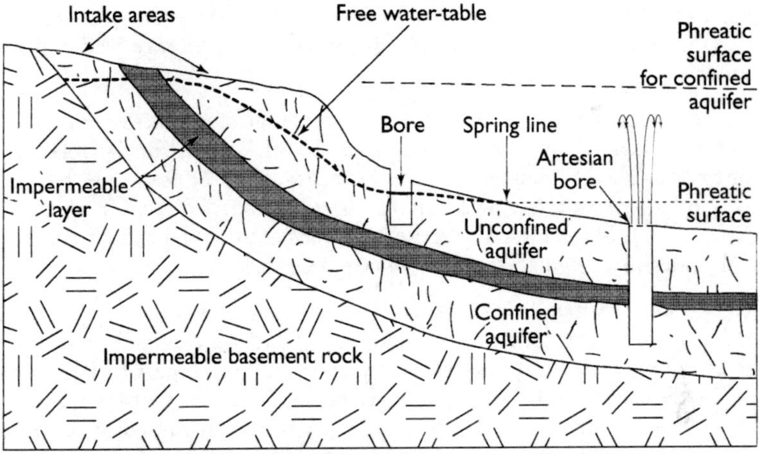

Figure U.2. Unconfined and confined aquifers

unconfined floodplain A broad FLOODPLAIN in which the maximum meanders of the river no longer reach the sides of the valley. Usually bedrock occurs along the scouring depth during floods, and the river is regarded as old. *See* OLD RIVER.

unconfined groundwater Groundwater not restrained in its movement by an impervious or confining bed above or below. *See* ARTESIAN WATER.

unconformity A geological plane or surface along which there is a break in the stratigraphic sequence, and the overlying younger rocks rest upon older rocks which do not immediately precede them in the geological succession.

unconsolidated material The more or less incoherent deposits which overlie the solid rocks in most areas and consist of river alluvium, gravels, sands, clays, raised beaches, peatbogs, etc. The material may be interbedded or intermixed. *See* SUPERFICIAL DEPOSITS.

uncovered open channels *See* OPEN CHANNEL.

undercut slopes Steeply inclined river banks on the outer edge of meanders, formed by erosion along the base and caving of ground. *See* CAVING.

undercutting (or **undermining**) The erosion or wearing away of rocks or soil along the base of cliffs or banks by stream or wave action, and the progressive collapse of the undermined material.

underflow A flow of water under a foundation, a structure or a layer of ice; the movement of water in an UNDERFLOW CONDUIT; the rate of underflow discharge. *See* PIPING.

underflow conduit A deposit of porous material, under a stream, with a flow of water which is separated from the surface stream by relatively impervious clay. In some cases, the surface water is of poor quality while the underflow is clean and can be drawn upon for local supplies. It depends largely upon the thickness of the beds separating the surface flow and underflow; regular quality tests are desirable.

underground drainage (1) Drainage in limestone and other areas where much of the precipitation passes underground along channels or swallow holes, along joints and fissures, or by percolation. In the Chalk districts (UK), the higher lands and valleys are normally dry or perhaps occupied by streams during very wet seasons only. (2) The underground gravity flow of water to sumps or pumping stations remote from the working areas. *See* DRAINAGE, DRY VALLEY.

underground water *See* GROUNDWATER, SUBSURFACE WATER, SUBTERRANEAN STREAM.

undermining *See* UNDERCUTTING.

underseepage *See* UNDERFLOW.

undertow A below-surface current of water running seaward normal to the coastline and of a pulsating character. The undertow helps to scour the submerged shelf along the shore, and it also carries seaward the material eroded by the water. *See* BACKWASH, BEACH RIDGE.

underwater apron A STONE RIPRAP which is carried as a layer or apron well out onto the bed of the river. If bed scour occurs, this extra stone will subside until it roughly beaches the new slope and thus safeguards the main slope above. This is the 'falling apron' principle, used widely in India for protection of river banks. If the river is deep, the stones may be dumped from a barge along the required lines. *See* STONEMESH MATTRESS.

uniform flow A flow which has a constant MEAN VELOCITY.

uniform grade channel *See* VARIABLE-GRADE CHANNEL.

uniformity coefficient The size of aperture through which 60% of the sand by weight will pass, divided by size of aperture through which 10% of the sand by weight will pass. *See* EFFECTIVE SIZE.

unit flow, peak *See* PEAK UNIT FLOW.

unit graph *See* COMPOSITE UNIT GRAPH, DIMENSIONLESS UNIT GRAPH.

unit hydrograph As defined by Sherman, the hydrograph produced by an isolated storm of a given duration and of uniform intensity over a river's whole drainage area with an equivalent runoff of exactly 1 in (25.4 mm) of rain. *See* SYNTHETIC UNIT HYDROGRAPH.

unit of permeability *See* COEFFICIENT OF PERMEABILITY.

unit weight of water The weight per unit volume of water normally equal to 1 g per cm^3 or 62.4 lb per ft^3. *See* WATER.

unsaturated air Air in which the RELATIVE HUMIDITY is less than 100%.

unsaturated zone The layer in a porous medium above the PHREATIC SURFACE within which water drawn up by CAPILLARY ACTION only partially fills the void spaces. This water is available to plants, but DARCY'S LAW does not govern flow in this zone, so calculations of seepage flow within it are complex.

unsteady flow A flow in which the velocity varies in magnitude or direction in relation to time.

unusual floods *See* RETURN PERIOD.

upflow sand filter A filter, either gravity or pressure, in which the water or wastewater flows upward through a lower coarser medium supporting an upper fine medium. *See* SAND FILTER.

uplift (1) The heaving upwards of earth due to the escape of water under high pressure from a dam or other confined area. The entry of such water into sands converts them into the 'quick' condition. The usual remedies are the sealing of leakage fissures in the ground, and drainage. *See* FLOW NET, TUNNELLING. (2) The upward force on the base of a dam or other structure due to water pressure in the foundations.

upstream total head (over a weir) A measurement at a point upstream (depending on the type of weir) of the elevation of the total head relative to the crest. *See* DOWNSTREAM TOTAL HEAD, WEIR HEAD.

USBR method A method for predicting flood flows from ungauged catchments developed by the United States Bureau of Reclamation (USBR). Based on a SYNTHETIC UNIT HYDROGRAPH technique, it enables peak rates of flood flow to be predicted, together with their approximate hydrographs, for any selected return period. Much of the extensive empirical data provided by the USBR is only of relevance within the United States, but the basic method can be successfully applied anywhere in the world, making use of whatever local data may be available on rainfall, rates of runoff, etc.

V

vacuum filtration The passing of wastewater through a rotating cylindrical drum, part of the circumference of which is subjected to a vacuum internally, which draws sludge to the filter medium. The dewatered sludge is released by a scraper.

vacuum tank *See* PNEUMATIC WATER BARREL.

vadose water (USA) Water in the ZONE OF AERATION; sometimes, HELD WATER. Also called suspended water. *See* Figure Z.1.

vadose zone The zone between the top of the water-table or phreatic surface and the surface of the soil. *See* ZONE OF AERATION.

valley An open depression between hills or mountains; it may or may not be traversed by a stream or river. *See* GLACIAL VALLEY.

valley consumptive use The total amount of water evaporated from water and soil surfaces, plus that absorbed and transpired by plants, crops and soils upon which they grow in a valley. *See* CONSUMPTIVE USE.

valley floodplain The flat lowlands of a valley, which are wholly or partly covered with water during floods; the floor is usually covered with waste or debris. *See* UNCONFINED FLOODPLAIN.

valley glaciers Glaciers which occupy valleys and which move slowly down to lower levels. The average rate of movement depends on the slope of the valley, the air temperature and the thickness of ice. The average movement may be about 0.6 m (2 ft) per day, although rates up to 18.3 m (60 ft) per day have been recorded in Greenland. The size of a valley glacier diminishes at lower levels until it ends in a SNOUT.

valley, hanging *See* HANGING VALLEY.

valley-in-valley structure A structure formed when a rejuvenated stream erodes a youthful valley in the floor of the valley in which it formerly flowed. *See* INCISED MEANDERS.

valley spring A spring which occurs in a valley at a point where the surface falls below the WATER-TABLE or a PERCHED WATER-TABLE, or any water-bearing rock. *See* BOURNES.

valley storage (1) The total volume of water stored in a valley between two specified points along a river, including both the channel and the floodplain. (2) The total volume of a stream in flood after it has overflowed its banks. *See* BANKFUL STAGE.

valley tract The channel along which a stream flows in a valley or foothill country, sometimes incised in deposits laid down at an earlier period, where the gradient is low. In this region, some deposition of sediment commences; alluvial flats appear and the river meanders, sometimes from one side of the valley to the other. The valley tract may be viewed as the intermediate stage between the MOUNTAIN TRACT and the PLAIN TRACT (Figure M.4).

valley train A deposit of englacial material beyond the glacier margin in a valley. If the material is spread out over a relatively flat surface, apart from a valley, it is termed an outwash plain or frontal apron.

valve A device, usually in a pipe, to start, stop, or regulate the flow of water. *See* BUTTERFLY VALVE, CLACK VALVE, DISCHARGE VALVE, STOP VALVE, SUCTION VALVE.

valve shaft A shaft connected with a reservoir and from which the outlet pipes run; it also houses most of the valve controls. The upper structure of the shaft may form an exposed platform, or more usually an INTAKE TOWER.

valve, sluice *See* SLUICE VALVE.

valve tower *See* INTAKE TOWER.

vapour concentration The ABSOLUTE HUMIDITY or mass of water present in a cubic metre of air. *See* HUMIDITY.

vapour pressure The pressure exerted by the vapour arising from the surface of a liquid, which increases with temperature. *See* RELATIVE HUMIDITY, SATURATED VAPOUR PRESSURE, UNSATURATED AIR.

variability of a spring The ratio, expressed quantitatively, of the fluctuation of a spring to its average discharge over a specified period.

variable-grade channel A channel along which the slope is not uniform and usually decreases as the distance from the outlet increases. In comparison, a uniform grade channel has a slope which is constant throughout its entire length. *See* STREAM PROFILE.

varves Seasonal deposits from glacial streams; clayey beds containing thin alternate layers of different particle sizes.

vector-borne disease A disease transmitted by an insect, tick or mite from an animal or person to a human host. Water sources are vital for the life cycle of many vectors, such as the malarial mosquito, or for the transmission of disease such as SCHISTOSOMIASIS. Prevention of these diseases often depends on the proper control of water sources.

vee notch *See* TRIANGULAR NOTCH.

vegetal discharge The emission of groundwater due to the physiologic functioning of living plants. *See* TRANSPIRATION.

vegetated channel *See* GRASSED WATERWAY.

vegetative practices All water and soil conservation methods based on the use of vegetation. *See* AFFORESTATION, GULLY CONTROL PLANTINGS, TREE PLANTING.

velocity curve, depth *See* DEPTH VELOCITY CURVE.

velocity gauging *See* CHEMICAL GAUGING, CLOUD VELOCITY GAUGING, COLOUR VELOCITY GAUGING, SALT VELOCITY METHOD.

velocity head (or **kinetic head**) The energy per unit weight of water due to its velocity, V. It is also the vertical distance the fluid must fall freely in a vacuum under the force of gravity to acquire its velocity, V:

$$h_V = V^2/2g$$

where h_V is velocity head, V is the mean velocity of flowing water, and g is acceleration due to gravity. *See* BERNOULLI'S THEOREM.

velocity in pipes The velocity of water in pipes, which in general should not exceed about 1.5 m/s (5 ft/s). This is desirable to keep frictional losses low and also to prevent damage by the WATER HAMMER effect.

velocity measurement (integration method) *See* INTEGRATION METHOD.

velocity of approach *See* APPROACH VELOCITY.

velocity of groundwater *See* APPARENT VELOCITY, AVERAGE VELOCITY, EFFECTIVE VELOCITY.

velocity of retreat The mean velocity on the downstream side of a MEASURING WEIR or structure. *See* TAIL WATER.

velocity rod *See* FLOAT ROD.

vena contracta The minimum cross-sectional area of a jet or nappe of water beyond the orifice, notch or weir through which it emerges (Figure V.1). *See* COEFFICIENT OF CONTRACTION.

ventilation vents (service reservoirs) Cast-iron or other ventilators in the roof, or in the walls above the water level, of a covered service reservoir. Measures are taken to prevent or minimize possible pollution of water through the ventilators.

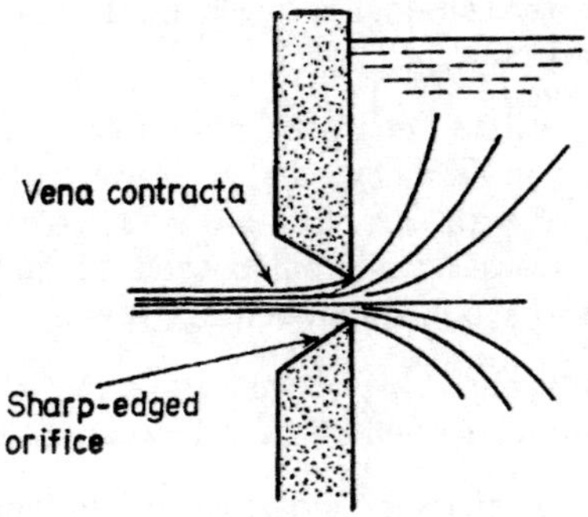

Figure V.1. Vena contracta

Venturi flume A type of control flume used for measuring flow, consisting of a short contraction followed by an expansion to normal width. *See* PARSHALL MEASURING FLUME.

Venturi meter (or **Venturi tube**) A flow meter used for closed pipes, consisting of a constriction or throat followed by an expansion to normal width. The reduced pressure at the throat is measured, as is the pressure upstream where the width is normal; from these points small pipes lead to gauges. The pressure difference between the points is an index of flow or discharge.

Venturi tube *See* VENTURI METER.

vertical alignment *See* REALIGNMENT.

vertical gauge *See* STAFF GAUGE.

vertical sand drain (or **sand drain**) A shallow hole drilled to promote the drainage of water and accelerate the consolidation of a cohesive soil. The hole is put down through a clay or silty soil and filled with sand or gravel. It should reach a permeable bed below the clay, although this is not essential. The drainage of a clay loaded by an earth dam or other heavy structure is accelerated by vertical sand drains if their spacing is less than the thickness of the clay deposit. *See* ARTESIAN WELL DRAINAGE, PILLAR DRAINS.

vertical section (or **columnar section**) A section, drawn to scale, showing the deposits penetrated in a well, borehole, shaft or major excavation. It gives details of the nature and thickness of the soils and rocks as exposed at one spot, and is found useful at all dam sites, cuttings, and proposed water works.

vertical velocity curve *See* DEPTH VELOCITY CURVE.

vibrating roller A ROLLER that is mechanically vibrated. It can be a padfoot or smooth-drum type.

VIP latrine The ventilated improved pit-latrine developed by the Blair Research Laboratory in Zimbabwe and extensively replicated. It is a well-built latrine equipped with a roof, and a screened ventilation pipe that reduces odours and acts as a fly-trap. No water is required for flushing but care must be taken to avoid contaminating groundwater.

virtual slope A graph, curve or slope indicating the loss in pressure due to friction at any point in a system of fluid flow. *See* HYDRAULIC GRADIENT.

virus The smallest particles of biological origin capable of producing infection and diseases in humans or other species. They can only multiply inside a living cell of a suitable host.

viscous flow *See* STREAMLINE FLOW.

VLOM Village level operation and maintenance. *See* HANDPUMP.

V notch *See* TRIANGULAR NOTCH.

voids (or **interstices**) The pores or small cavities between the particles in a rock or soil mass, which may be occupied by water or air, or both, or by other liquid or gaseous substance. The term may also be applied to the spaces between the stones, gravel or shingle in river and shore protection structures. The water velocity is reduced, resulting in the deposition of much of its silt load at and upstream of the structures. *See* INTERSTITIAL WATER, MICROPORES.

voids ratio The ratio of the volume of voids to the volume of solids in a sample of soil or other material. A clay contracts with increase of pressure, and therefore its voids ratio may be defined at a specified pressure.

volatile acids Fatty acids which are soluble in water and which can be steam-distilled at atmospheric pressure. They have pungent odours and are often produced during anaerobic digestion of sludge.

volume curve *See* CAPACITY CURVE.

volumetric efficiency The volume of water entering a pump cylinder for each piston stroke, divided by the volume swept by the piston.

volumetric measurement A method in which the entire flow of water is discharged into a tank, of regular shape, for a measured period. Reliable means are provided for diverting or starting and stopping the flow into the tank. The method is very accurate and is used as a standard to which other measurements are referred. *See* DISCHARGE.

volute The spiral casing of a centrifugal pump, providing an area of passage which gradually reduces the speed of water leaving the IMPELLER and converting it to pressure without shock.

V-shaped valley In general, a valley carved by erosion with comparatively steep sides, such as a ravine, gorge, gully or canyon. The shape of the 'V' depends upon the stage of erosion, etc.

W

wadi (or **wady**) An Arabic term for a valley that is dry for long periods, but becomes a torrent during the rainy season.

wading rod A graduated rod to which is attached a CURRENT METER for measuring the velocity in shallow streams or other channels.

Wallingford procedure A sophisticated method of numerically modelling large urban drainage networks. It predicts the flows at all times during a rainstorm in every pipe or channel that makes up the network. The procedure gives very good results, but requires a large amount of data which must be processed on a computer. It was developed in the UK, where it is normally used with rainfall data first presented in the FLOOD STUDIES REPORT (1975), but if comparable data are available it can be applied worldwide. Many computer programs have been developed to use the Wallingford procedure to model and design urban drainage networks. The development of the Wallingford procedure also led to a modified version of the RATIONAL METHOD which includes several improvements.

wall, training *See* TRAINING WALL.

warnings, flood *See* FLOOD WARNINGS.

warning stage (flood) The stage when a flood begins to cause inconvenience or damage locally near a gauge; the river may be either below or above the BANKFUL STAGE. *See* FLOOD STAGE.

warm front *See* FRONTAL SURFACE.

warp (1) A twist or turn in a river due to an external obstruction or restraint; a MEANDER. (2) A fold in bedded rocks.

wash The dry bed of an intermittent stream; a gully (USA).

wash boring The putting down of a drive pipe or well casing through loose ground to rockhead with the aid of an internal water jet; sometimes penetration is assisted by hammer blows.

wash load The finest part of the SUSPENDED LOAD carried by a river, consisting of very small particles of clay or mud, and too fine to settle out under gravity alone. *See* LOAD.

wash-outs (1) Gutters or gullies cut by swift streams after heavy rainfall, storms or cloudbursts, fairly common in tropical or subtropical countries. *See* GULLY DRAINAGE. (2) Access facilities in main pipelines for scouring, draining or refilling.

washout valve *See* SCOURING SLUICE.

waste The part of the total output of water which is not efficiently used by the consumer, including reservoir and pipe losses, as well as losses at the consumer's end, such as taps not turned off. *See* IRRIGATION EFFICIENCY, LISTENING METHODS.

wastewater The waterborne wastes of a community and industry effluent which contain dissolved and suspended matter.

wasteway A spillway; a channel for conveying water delivered into it from a sluice, escape or spillway.

water H_2O. Chemically pure water consists of hydrogen and oxygen combined in the ratio of about eight parts of oxygen to one part of hydrogen by weight. *See* ACID WATER, FRESH, NATURAL WATER, TOXIC WATER, UNIT WEIGHT OF WATER.

water authority A statutory or municipal body administering a system of water supply to a specified area or district.

water-bailiff An officer who controls and regulates water within a section of an irrigation area.

water balance The accounting for all water inputs (rainfall, snow, etc.) and all water outputs (runoff, evaporation, groundwater seepage, etc.) within a system (e.g. a catchment area). *See* HYDROLOGIC CYCLE.

water barrel A large barrel-shaped bucket, formerly used to hoist water from the bottom of a sinking shaft or deep excavations. *See* PNEUMATIC WATER BARREL.

water-based disease Diseases such as GUINEA WORM DISEASE and SCHISTOSOMIASIS, the pathogens of which spend part of their life cycle in an animal which is water-based.

water-bearing ground The rocks in the ZONE OF SATURATION, or below a PERCHED WATER-TABLE; any porous deposit containing large quantities of water seeping downwards to the zone of saturation. *See* AQUIFER.

waterborne disease Diseases caused by organisms or toxic substances carried by water, including CHOLERA, DYSENTERY, giardiasis, hepatitis A, polio, tetanus, TYPHOID fever and many intestinal disturbances.

water capacity The volume of water a soil or rock can retain; it varies with the texture and composition of the rock. *See* MICROPORES, WATER-HOLDING CAPACITY.

water celerity *See* CELERITY OF WAVE.

water conservation The management, protection, physical control and prudent use of the water resources of a district for the maximum long-term benefits to agriculture, industry, commerce and the national economy in general. *See* RIVER ENGINEERING, SURFACE WATER SUPPLIES.

water consumption The amount of water consumed in a specified area. Consumption varies considerably and depends on many factors, including climate, availability, the type of supply (through standpipes or piped to each house), internal plumbing, anticipated types of water use, including gardening, whether supplies are metered, and the effect of water charges. Provision should be made for commercial and industrial use, for progressive increase in rate of consumption and for projected increase in population. *See* GROUNDWATER BUDGET.

water control The physical control of water by the erection of structures for water retardation and sediment detention; improvement of channels; and land practices to reduce irrigation waste and prevent depletion by seepage or evaporation. *See* RIVER ENGINEERING.

watercourse The path of least elevation in a valley, whether or not the path is occupied by a stream; any artificial or natural channel for the conveyance of water, particularly drainage. It may be a canal, stream, drainage tunnel, ditch, flume or adit.

water creep The movement of water around the base of a structure erected on permeable deposits. *See* PIPING, TUNNELLING, UNDERFLOW.

water cushion *See* STILLING POOL.

water cycle *See* HYDROLOGIC CYCLE.

water demand A schedule of the total quantity of water required for a specific purpose such as municipal supply, power, irrigation or storage. *See* WATER REQUIREMENT.

water disposal Structures and practices for the removal of excess water from land and keeping soil erosion at a minimum. For flat ground these may include surface drains or both surface and subsoil drains. For sloping ground they may include earth dams, grassed waterways, terraces, and terrace outlet channels. *See* SUBSOIL DRAINAGE, SURFACE DRAINAGE, TERRACES, VERTICAL SAND DRAIN.

water divining *See* DIVINING.

water equivalent of snow The depth of water formed when snow melts. *See* SNOW DENSITY, SNOW MELT.

water erosion *See* EROSION.

water face (or **water slopes**) The upstream face, normally in contact with the water, of a dam or embankment. The face is only uncovered after prolonged dry periods or heavy draw-off (Figure A.1). *See* BEACHING, STONE PITCHING.

waterfall A flow of water over a precipice (Figure W.1); a cataract; a cascade. Caused by irregularities in the hardness of the rocks in the river or a dike or sill in softer beds, a waterfall erodes the less resistant underlayers and undermines the more resistant ones, which periodically collapse; thus the waterfall moves upstream. Falls and rapids break the navigable continuity of a river and may need blasting or even a series of locks. Again, a waterfall, if of sufficient volume, and a drop in volume, may be utilized for power generation. *See* CASCADE, CATARACT, RAPIDS.

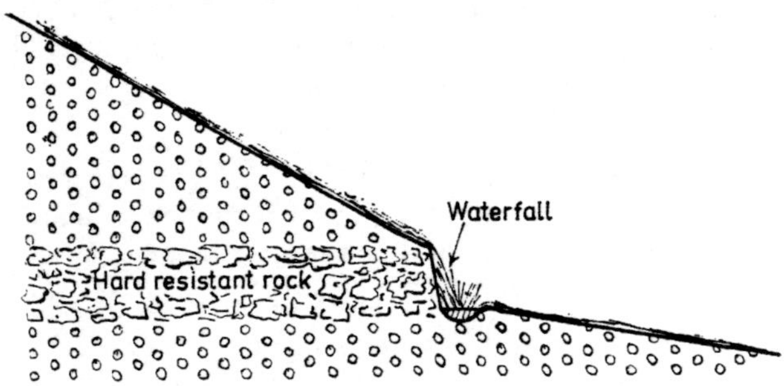

Figure W.1. Waterfall

waterfront The side of a building or structure abutting on a lake, river or the sea; the margin of land or town on the waterside.

water gap A narrow gorge cut by a stream through a hard ridge of rock; often broad lowlands are formed in softer rocks on each side.

water hammer A sudden pressure wave produced in a pipeline when the flow is stopped instantaneously, usually by the closure of valves. The pressure wave results in a hammer-like noise. In large pipelines the water hammer effect may be reduced by standpipes, surge tanks or relief valves.

water hardness *See* HARDNESS OF WATER.

water harvesting The art of collecting and storing water in arid regions by specially treating small catchment areas to increase the runoff. This practice dates back to about 2000 BC, when the inhabitants of the Negev Desert in Israel cleared rocks and gravel from the hillsides and then smoothed the surface to increase runoff. Similar techniques are used today. The simplest is to remove the vegetation by a soil sterilant and then compact the soil surface. More expensive treatments include covering the small catchment with plastic or metal films.

water-holding capacity The minimum possible water content to which a soil sample can be reduced by gravity drainage. *See* CAPILLARY WATER, SPECIFIC RETENTION.

waterhole A shallow depression or cavity in which water collects during a rainy period, or in the bed of a river otherwise dry.

water horsepower A USA and UK term applied to a pump to indicate the actual rate at which useful work is imparted to the water flow, measured in horsepower (1 horsepower = 0.7457 kW).

water infusion Cracking or loosening a coal seam by the injection of water under pressure by means of boreholes and pumps. Some coals are so dense that even high water pressure does not loosen them, while others are so jointed that the water passes through into the roof or floor without loosening the coal.

water level (WL) *See* MEAN WATER, STAGE, STANDING WATER LEVEL.

water lodge *See* LODGE, SUMP.

waterlogged Saturated with water. A depression in the ground is usually waterlogged when it is at, or below, the water-table. The condition may also result from seepage, over-irrigation, or springs.

water lowering *See* GROUNDWATER LOWERING.

water-meadow A meadow or grassland kept fertile by periodic flooding.

water meter An integrating FLOW METER. For measuring irrigation water there are two main types available, namely, the inferential and the rotary-piston. The rotary-piston type can be installed in any position, whereas the inferential type must be fitted in a horizontal pipeline.

water-mill A mill whose machinery is operated by the force of water.

water of capillarity Pore water in rocks or soil above the water-table and held there by capillary forces at less than atmospheric pressure. *See* HELD WATER, INTERSTITIAL WATER.

water of compaction The water given off during the compaction or compression of soils or sediments with a reduction in the volume of voids. *See* CONNATE WATER.

water of crystallization A fixed molecular proportion of water attached to certain minerals when they crystallize. The water can be expelled at a moderate heat as it does not enter into the inner chemical constitution of the mineral.

water of dehydration Chemically combined water which is liberated when the minerals involved undergo chemical changes.

water of infiltration The part of surface water which has entered the voids of a soil or other porous mass by INFILTRATION.

water parting *See* DIVIDE.

water prospect map A map on tracing paper or cloth showing the position of streams and springs, depth to water-table and probable yield of any wells put down in the area. The tracing is placed over the geological map of the same area, and on the same scale, to assist the engineer in selecting sites for wells or in solving drainage or irrigation problems.

water quality The physical and chemical condition of water. To some extent water dissolves all known substances so in its natural state water is never pure. This means it is usually necessary to define its quality with respect to some application. Measures of water quality are used for the assessment of pollution and its modification and control. Water quality is of the greatest importance for drinking water, although other uses of water, such as industrial processes, cooling, agriculture, each have their own requirements. Standards for drinking water quality are commonly based on the World Health Organization (WHO) Guidelines for Drinking Water Quality. These propose acceptable maximum safe concentrations for some 25 key substances. *See* BRACKISH, WATER TREATMENT.

water rate A charge made by a water company, authority or municipality for supply and use of water.

water regime *See* MOISTURE REGIME.

water requirement (agriculture) The total quantity of water required for agriculture, usually assessed at so much per hectare or acre or in relation to the stock population. Typical stock requirements per head per day are: cattle, 45 to 68 L (10 to 15 gal); horses, 36 to 55 L (8 to 12 gal); pigs, 9 to 14 L (2 to 3 gal); sheep, 5 to 9 L (1 to 2 gal).

water requirement (crops) The total quantity of water required by a plant for its growth and maturity under field conditions, including that supplied by rainfall, and any irrigation and groundwater available to the plant.

water requirement (domestic) The total quantity of water required for domestic use, including hot-water systems, baths and waterborne sewerage. The quantity is usually estimated on the basis of population statistics, and consumption is steadily increasing.

water resources The estimated average yield of water from a given catchment area; the availability of surface water or groundwater. *See* GROUNDWATER BUDGET.

water-resource system A term sometimes applied to the entire system which provides water in sufficient quantities, and at reasonable rates, for irrigation, power, domestic, industrial and other uses. Such a system includes the dam, reservoir together with its catchment, and all necessary works, machines, and devices. *See* SIMPLE STORAGE SYSTEM.

water-retaining capacity *See* SPECIFIC RETENTION.

water rights The legal rights possessed or acquired by an individual to the use of water. *See* APPROPRIATED RIGHTS, PRESCRIBED RIGHTS, RIPARIAN RIGHTS.

water–sediment mixture *See* SEDIMENT WATER.

watershed (or **topographic divide**) *See* DIVIDE. (In the USA the term watershed is applied to the catchment area.)

watershed leakage (USA) The flowage of subsurface water from one drainage area to an outlet in a neighbouring drainage area, or directly to the sea.

water slopes *See* WATER FACE.

water-splash Part of a road submerged by a pool or stream.

waterspout A revolving tornado-like cloud descending towards the sea and sucking up a mass of water, the whole forming a whirling column connecting the cloud and sea. *See* TORNADO, WHIRLPOOL.

water spreading The construction of terraces across sloping land to spread and control runoff and to retard soil erosion; water control for irrigation or for subsurface storage and later extracted by pumps for irrigation. *See* WINTER IRRIGATION.

water stage recorder A device which gives a continuous record of the height of water at a point. The level is marked on a clock-operated chart by a float-actuated pen. *See* WIRE WEIGHT GAUGE.

water supply The conserving, pumping and piping of drinkable water, usually by a public authority to consumers. *See* SURFACE WATER SUPPLIES.

water-table The upper groundwater surface of the ZONE OF SATURATION except when under an impermeable layer. It follows in a flatter form the profile of the land surface; it is nearer the surface in valleys and deeper along hills and elevated ground. The water-table will fluctuate as a result of natural or artificial causes. The main natural causes are rainfall, floods, barometric changes and tides. The main artificial causes are well-pumping and dams. *See* CONE OF DEPRESSION, MAIN WATER-TABLE, PERCHED WATER-TABLE.

water-table contour plan A plan showing contour lines on the water-table. The depths are obtained from wells, boreholes and springs, and sometimes by geophysical methods. The plan is useful for estimating the depth of proposed water wells and also when investigating water supplies and drainage schemes.

water-table gradient The slope or inclination of the water-table, from the horizontal plane, at a given place and in a specified direction; it may be obtained from the WATER-TABLE CONTOUR PLAN. *See* PHREATIC WAVE.

water-table levels The levels at which the water-table is encountered in trial pits, boreholes, valleys and deep excavations.

water-table, main *See* MAIN WATER-TABLE.

water-table, perched *See* PERCHED WATER-TABLE.

water-table profile A vertical section of the water-table in a specified direction, constructed from a WATER-TABLE CONTOUR PLAN or from levels in trial pits, wells, boreholes and valley springs.

watertight barrier (earth dam) *See* CORE (2), CUTOFF TRENCH.

watertight seal (irrigation) A synthetic-rubber seal devised to reduce leakage through flow control structures. It ensures complete watertight conditions between the vertical movable gate and the main structure.

water-to-earth ratio *See* STORAGE/EXCAVATION RATIO.

water tower Generally, a cylindrical reinforced concrete tower or tank up to a capacity of 3375 m^3 (750 000 gal) or even more. The tank is elevated above ground level and serves the same purpose as a SERVICE RESERVOIR.

water treatment The purification and treatment of a water supply to ensure that it is drinkable and suitable for domestic and industrial use. It may include (1) neutralizing acid water, (2) softening water of excessive hardness to make it suitable for steam boilers and washing, and (3) removal of iron or other undesirable substance. *See* ACID WATER, HARDNESS OF WATER, PH VALUE, TOXIC WATER, WATER QUALITY.

water turbine An arrangement in which water power rotates a wheel or prime mover, which in turn is coupled to an alternator to generate an alternating current. *See* IMPULSE TURBINE, KAPLAN TURBINE, REACTION TURBINE, TURBINE PUMP.

water-use ratio *See* TRANSPIRATION RATIO.

water-washed disease Diseases of the intestinal tract (such as amoebic dysentery), eye infections (such as conjunctivitis and trachoma) or infections caused by lice or mites (such as scabies) where transmission would be reduced by an increase of water for washing. Soap is a vital factor in combating water-washed disease.

waterway A navigable channel; a watercourse. *See* FARM WATERWAY, RIVER, STREAM.

waterway stabilization structure *See* GULLY STABILIZATION STRUCTURE.

water well (or **water bore**) A well put down (1) to yield a supply of water, (2) to yield groundwater information, or (3) to replenish groundwater. *See* SHALLOW WELL.

water wheel A wheel formerly used for raising water by means of boxes or buckets fitted on its circumference; a wheel to drive machinery with water as the motive power.

waterworks planning Investigations and studies to ensure that the water resources of a region are exploited to the best advantage to avoid wasteful and haphazard schemes. For this purpose the region is divided into areas or units which will be convenient for water supply purposes. Planning may be undertaken for water requirements up to 30 years or even longer. *See* IMPOUNDING SCHEMES.

water year A water record extending over a continuous period of 12 months.

wattle work *See* HURDLE WORK.

wave A moving ridge or swell of water between two depressions, generated by wind action in the sea, lake or other body of water; in relatively shallow water, the ridges or rollers follow each other at regular intervals and successively break on the shore. *See* FORCED WAVES, FREE WAVES.

wave base The lower limit or depth, for any particular series of waves, at which disturbance of the water ceases; erosion of the bed proceeds if the water is shallower than the wave base. *See* BREAKERS.

wave-built terrace Sometimes, the upper coarser portions of a beach along which the waves have thrown the pebbles up into low ridges just above mean high-water level and parallel to the shore line. *See* OFFSHORE TERRACE.

wave-cut terrace The shelf covered by shallow water at the base of a SEA CLIFF. In some areas, the terrace has been elevated to form a bench above sea level, as on the coast of southern California and elsewhere.

wave, dam break The surge wave of water released down a valley by a dam that fails.

wave erosion Erosion by waves, which includes the abrasive action of solid particles in the water and the scouring effect of the water itself. Wave erosion does not necessarily include transportation of the material. Shore currents and wave action are also active on inland seas and lakes. *See* SHORE EROSION.

wave freeboard The vertical height remaining between the CREST of a dam and the maximum level reached by the FLOOD SURCHARGE (Figure W.2).

wave height The height from the trough to the crest; generally ranging up to about 9.0 m (30 ft), it becomes considerably greater in heavy seas. *See* HAWKSLEY'S FORMULA, SIGNIFICANT WAVE HEIGHT, STEVENSON'S FORMULA.

wavelength The distance from crest to crest, varying from 60 m (200 ft) to about 180 m (600 ft) for FORCED WAVES in the open sea; the length of FREE WAVES may be greater.

wave of the water-table *See* PHREATIC WAVE.

wave pressure The pressure exerted by waves along the shore; it is the main cause of coastal erosion. Damage occurs mostly during winter gales. *See* TIDAL RANGE.

wave refraction The swinging around of an originally straight wave on approaching an indented shore line to which it becomes parallel. In the case of FORCED WAVES the effects are modified.

wave run-up (Figure W.2) *See* RUN-UP.

wave surcharge The rise of water against a dam created solely by the run-up of waves of specified probability.

wave surcharge allowance The allowance for FREEBOARD in the design of a dam to ensure that the quantity of water driven over the crest of the dam during extreme events is not large enough to cause significant damage.

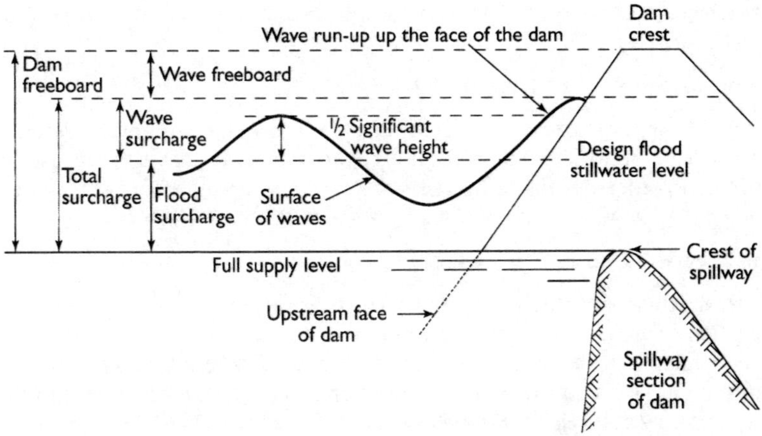

Figure W.2. Wave run-up and significant wave height

wave wall *See* PARAPET WALL.

weather The atmospheric state of a region at any given period, particularly temperature, humidity, cloudiness, precipitation, barometric pressure and wind velocity. *See* CLIMATE.

weathering The changes which occur in soils and rocks as a result of the action of heat and cold, wind, rain, snow, hail, exposure to air and other atmospheric processes. *See* ZONE OF WEATHERING, CHEMICAL WEATHERING.

weep-holes Small apertures or pipes formed during construction of retaining walls, bridge abutments, aprons, or foundations to permit drainage of water and thus reduce the pressure behind the structures.

weighed filter *See* LOADED FILTER.

Weil's disease Leptospiral jaundice, transmitted to humans by rats through contaminated water.

weir A concrete, stone or STONEMESH overflow wall or structure erected across a stream. It is not used to store water but to retard flow, and raise the upstream water level to permit diversions and perhaps to measure the rate of flow. Weirs are sometimes constructed in series, in which case the effect of the lower weir will extend to the next one above it. *See* CIPOLLETTI WEIR, MEASURING WEIR, SHARP-CRESTED WEIR.

weir head The depth of water measured from the bottom of the notch to the water level at a point upstream of the weir. The APPROACH VELOCITY is not included.

weir notch The opening in a weir for the passage of water. *See* MEASURING WEIR, NOTCH.

weir, overflow *See* OVERFLOW WEIR.

well A small pit excavated or drilled from the surface to obtain water; it exceeds 0.9 m (3 ft) in diameter if excavated by hand. A well may also be used for other purposes. *See* DRAINAGE WELL, OBSERVATION WELL, OPEN WELL, SHALLOW WELL, TUBE WELL, WATER WELL.

well, cavity *See* BOULDER WELL.

well development Increasing the yield from a bored well. A method of particular use in water wells is by pumping out sand and silt that may have accumulated in the WELL SCREEN. Gravel may also be introduced into the ground adjacent to the well screen to permit easier flow of water into the well. Successful well development may increase the yield of water by up to 50%.

well drain *See* ABSORBING WELL.

well field A region or locality containing a number of wells which supply water for public, industrial or irrigation purposes. The wells may be pumped individually or in one or more groups or batteries. *See* BATTERY OF WELLS.

well, flowing *See* FLOWING WELL.

well hydrograph A graph showing fluctuations of water level in a well, the levels marked as ordinates against time as the abscissa.

well interference The effect on the CONE OF DEPRESSION of a well caused by that of an adjacent well or wells, resulting in a decrease in its yield of water. The diminution may be expressed as a lowering of water-table in metres or feet, or reduction in yield.

well, inverted *See* INVERTED WELL.

well log A description of the rocks passed through in a well; similar to a borehole log but usually not so detailed. *See* DRILLER'S LOG, WELL RECORD.

wellpoint A shallow well with a pump to drain a water-bearing soil around or along an excavation. A wellpoint strengthens the ground and reduces the pumping required inside the excavation. It consists essentially of a tube of about 50 mm (2 in) diameter which is driven into the ground by JETTING. The tube is fitted with a close-mesh screen at the bottom and connected through a header pipe to a suction pump at the surface. For a major job, a number of wellpoints are sunk and connected to a common header pipe. Fine sands cannot be drained by wellpoints. *See* GROUNDWATER LOWERING.

well pressure The natural pressure of water in a well, which may be sufficient to cause the water to rise to the surface and form a FLOWING WELL. Well pressure is not related to the depth of the well or of the water. *See* ARTESIAN WELL.

well record A brief account of the history of a well and all the available data regarding yield of water, pressure, depth and rocks penetrated. *See* WELL LOG.

well screen A screen or strainer placed in a well when pumping water from a loose, gravelly deposit. The screen excludes solid particles over a certain size but allows fine sand to enter with the water and be removed. The large particles accumulate around the well to form a natural gravel screen. The screen is made from non-ferrous metal; the two main types are tube strainers with milled slots and wire-wound strainers. *See* BONDED GRAVEL SCREEN, GRAVEL FILTER WELL.

well section A vertical section showing the rock deposits passed through in a well. The scale used may be 1:100 (10 ft to 1 in) or more, depending on the depth and the details to be shown. *See* WELL LOG.

well sinker A person skilled in excavating shallow wells, generally with pick and shovel, to yield a local supply of water. Usually a local, the well sinker is familiar with the superficial deposits of the district and also the general lie of the water-table.

well water The water as pumped from a well or obtained from a FLOWING WELL.

wet-bulb thermometer An ordinary thermometer with its bulb covered with muslin which is kept permanently moist with pure water; it is used in conjunction with a dry-bulb thermometer in a HYGROMETER.

wet dock A dock within which the water is maintained at high-tide level by the dock gates, which are opened only at high tide. *See* DRY DOCK.

wet gap A military term used for obstructions involving water, such as streams, ponds, rivers, etc.

wetlands Ecosystems of marshy and/or inundated land.

wet line correction *See* AIRLINE CORRECTION.

wetted perimeter The length of surface in a stream, conduit, or channel in contact with the water, measured at right angles to the flow. *See* HYDRAULIC RADIUS.

wet well *See* DRY WELL (2).

wharf A BERTH or platform of timber, stone or concrete constructed parallel to the waterfront, where ships may be moored for loading or unloading. A solid wharf acts similarly to a retaining wall and holds back all the earth behind it, while an OPEN WHARF does not retain the earth behind it, and is supported on piles of steel, pre-cast concrete, or sometimes timber forced down into the bed.

whirling hygrometer *See* HYGROMETER.

whirlpool A large circular eddy in a stream or the sea, often of great power, which may be caused by an obstruction or spur projecting into the water or channel. *See* MAELSTROM, WATERSPOUT.

whistling buoy A BUOY in which wave action activates an attached whistle to give audible warning.

whole-tide cofferdam *See* FULL-TIDE COFFERDAM.

wicket A small door or gate, especially one forming part of a larger one in any hydraulic structure; a sluice in a lock gate.

wicket dam A movable dam consisting of shutters or wickets which can be revolved about a central axis.

width (or **breadth**) (stream) The linear dimension measured at right angles to the direction of flow. *See* CROSS-SECTION.

willowbox groynes Usually small groynes, similar in construction to FASCINE GROYNES, used for local bank protection. They consist of willow fascines or willow spars, and the interior is filled with shingle. The top is commonly covered over with a layer of fascines.

willow groyne A type of PERMEABLE GROYNE used successfully in New Zealand where the stream bed is favourable for the growth of willows. The groyne is made by wiring green willow stakes to horizontal cables attached to concrete blocks buried in the river shingle. In time, the stakes develop and form a thick barrier of willows. There are structural variations, depending on local conditions and objectives.

willows Trees furnishing pliable shoots and growing in temperate climates near water. They are divided into tree willows, pussy willows and osiers. Tree willows are used for stream stabilization and protection; the shallow roots must be regularly cut back for river protection. Pussy willows are used on drier ground, while osiers may be used at sites where their bitter flavour gives them immunity from grazing stock. *See* TREE PLANTING.

wilting coefficient The moisture, in percentage of dry weight, remaining in the soil when plants reach a state of permanent wilting.

wilting point *See* PERMANENT WILTING POINT.

wind erosion The detachment and removal of soil particles by wind action, a serious problem in some regions during dry periods, particularly where the surface layer is loose and the vegetation is inadequate to bind the soil substance. *See* THRESHOLD VELOCITY.

windlop The generation by wind of short waves in narrow waters, as distinct from ocean waves.

wind-wave flume A special flume used in small-scale models concerned with waves due to wind action. *See* MODEL ANALYSIS.

wing dam *See* SPUR.

wing trenches Long trenches or grout curtains to make the ends of a dam watertight. They are constructed beyond the ends of the dam and continue up the valley sides for distances which depend on the geological conditions at the site.

wing wall (or **abutment wall**) A wall at the side or abutment of a bridge, normally extending beyond the width of the bridge deck to hold back earth; a side wall extending upstream and downstream from the centre line of a dam or barrage structure, particularly downstream for the length of the apron, so as to protect the banks in a zone of possible high turbulence.

winter irrigation The storage of water in the soil and subsoil by irrigation between growing seasons for subsequent use by plants and crops. *See* WATER SPREADING.

wire dam A barrier in which the HEADWALL is constructed from mesh wire.

wire mesh Any wire netting or fabric used to wrap and secure stones or other material in a river or other water control structure. Small-gauge wire with small openings is usually used for light work such as securing brushwork on banks, while heavy wire with large openings is used for heavy structures such as large groynes. *See* STONEMESH CONSTRUCTION, STONEMESH GROYNES, STONEMESH MATTRESS.

wire weight gauge A river gauge consisting of a weight suspended on a wire for measuring the water level under a bridge, often with a datum marked on the bridge structure. *See* CHAIN GAUGE, ELECTRICAL TAPE GAUGE.

WL Water level.

Woolf system (river training) A system often used in Europe for the same purpose as the LOG TRAINING WALL. It employs piles and walings which induce the deposition of shingle and so build up a bank. The system is used sometimes to reclaim an area of land, in which case gaps are left to allow the shingle to be carried through to the area beyond. *See* SLOPING OF BANKS.

working chamber The space in which people work on excavation at the foot of a PNEUMATIC CAISSON.

working head (pump) The sum of the TOTAL SUCTION HEAD and the TOTAL DELIVERY HEAD.

working map A BASE MAP used during GEOLOGICAL MAPPING.

working shaft A small pit to gain access to and to excavate a sewer, drain or other works. After construction, the shaft is usually filled with earth or stone.

Y

yard A British and US unit of length measurement. 1 yard = 0.914 m. *See also* METHOD OF MEASUREMENT.

yard gully A small inlet, covered with a grating, to a drain to receive rainwater and wastewater. *See* GULLY.

yard trap *See* GULLY TRAP.

yield Generally, the ECONOMIC YIELD of a well. The permeability of the rocks, and the probable yield of a new well penetrating such rocks, may be obtained from a short pumping test giving the different values of drawdown for successive increases in rate of pumping. The information enables the probable yield of the well to be estimated. This is a quicker method than carrying out a full-scale pumping test over a period. The term is sometimes applied to CATCHMENT YIELD. *See* OVERPUMPING.

yield of drainage basin The average annual runoff for the basin. *See* DRAINAGE COEFFICIENT.

young stream *See* YOUTHFUL STREAM.

youthful stream A stream which flows swiftly in a narrow steep-sided valley; it cuts vertically rather than laterally, and its course contains many irregularities such as rapids, pools and waterfalls. A youthful stream has a long period of degradation and silting before it attains a PROFILE OF EQUILIBRIUM. *See* EARLY YOUTH.

Z

zeolite process A process to soften hard water (*see* HARDNESS OF WATER) by the base-exchange method in which ZEOLITES are used to remove the salts. Back-flushing with brine regenerates the zeolite layer. The process is not suitable for treating water containing any chloride or sulphate of magnesium.

zeolites (1) Artificial substances used in the base-exchange method of softening hard water. (2) Natural hydrated aluminous silicates of calcium, sodium or potassium, which occur in cracks and cavities in basalts and lavas as white or glassy crystals.

zero moisture index *See* MOISTURE INDEX (ZERO).

zone A layer, belt, or mass of ground or other material bounded by marginal horizons or surfaces which may be plane or curved and flat or inclined, and within which certain physical, chemical or other conditions prevail.

zone, intertropic convergence *See* INTERTROPIC FRONT.

zone of aeration (or **vadose zone**, or **suspended water zone**) The ground above the MAIN WATER-TABLE and extending to the surface in which the soils are not fully saturated (Figure Z.1). In ascending order, the zone may be divided into the CAPILLARY FRINGE, an intermediate belt of ground and the BELT OF SOIL WATER. The zone of aeration may vary in thickness according to local conditions.

zone of capillarity The ground above the water-table containing the WATER OF CAPILLARITY.

zone of saturation The mass of water-bearing ground below the main water-table (Figure Z.1) which may consist of unconsolidated soil or solid rock and within which the water in the soil pores and fissures is at hydrostatic pressure.

zone of weathering (or **belt of weathering**) The layers of superficial deposits which have been altered by the physical and chemical actions of weathering. The depth of this zone depends on the nature of the underlying parent rock and its climatic conditions. Normally it is quite shallow but can be as deep as 30 m. Some or all of it must usually be removed at the start of construction.

zoogloeal film A mass of bacteria covering and embedded in the wetted surfaces of the medium in a mature biological filter.

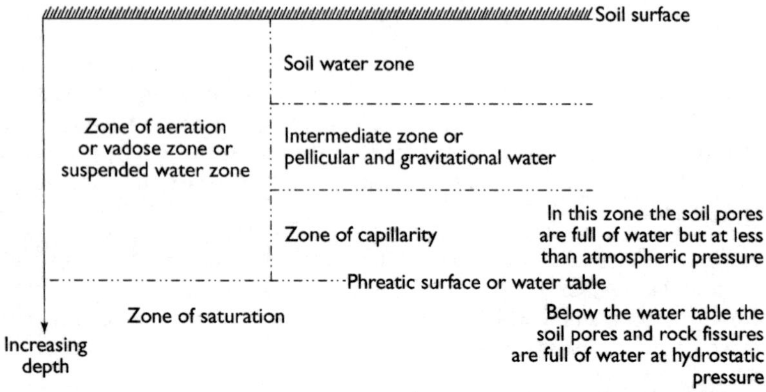

Figure Z.1. Zones of aeration and saturation in the upper layers of the soil

zoogloeal layer (or **schmutzdekke**) A gelatinous-type layer comprising algae, suspended matter, diatoms and micro-organisms, which forms on the surface of the sand of slow sand filters. It is the filtering agent which retains the impurities in the water. *See* FILTER MEDIUM.